Despite an upsurge in national and international debate on environmental issues since the Earth Summit in Rio de Janeiro in 1992, fungi, vital to the functioning of terrestrial and aquatic ecosystems, are rarely mentioned. This volume helps redress this imbalance by considering fungi in the context of the impact of mankind's activity on their habitats. The presentation of experimental evidence is a major feature of the volume. Contributions on the effects of global warming, UV-B radiation, atmospheric and terrestrial pollutants, deforestation in the tropics, loss of biodiversity, genetic engineering and chaos theory ensure a topical and balanced coverage with both ecological and physiological viewpoints being represented. This timely review will be of interest to all mycologists and those ecologists concerned with environmental change.

T0291638

FUNGI AND ENVIRONMENTAL CHANGE

Fungi and environmental change

SYMPOSIUM OF
THE BRITISH MYCOLOGICAL SOCIETY
HELD AT CRANFIELD UNIVERSITY
MARCH 1994

EDITED BY
J. C. FRANKLAND, N. MAGAN &
G. M. GADD

Published for the British Mycological Society

CAMBRIDGE
UNIVERSITY PRESS

CAMBRIDGE UNIVERSITY PRESS
Cambridge, New York, Melbourne, Madrid, Cape Town, Singapore, São Paulo, Delhi

Cambridge University Press
The Edinburgh Building, Cambridge CB2 8RU, UK

Published in the United States of America by Cambridge University Press, New York

www.cambridge.org
Information on this title: www.cambridge.org/9780521106252

First published 1996
This digitally printed version 2009

A catalogue record for this publication is available from the British Library

Library of Congress Cataloguing in Publication data

British Mycological Society. Symposium (1994 : Cranfield University)
    Fungi and environmental change : symposium of the British
Mycological Society, held at Cranfield University, March 1994 /
edited by J.C. Frankland, N. Magan & G.M. Gadd.
        p.   cm.
    Includes index.
    ISBN 0 521 49586 5 (hc)
    1. Fungi – Ecophysiology – Congresses.   I. Frankland, Juliet C.
II. Magan, N. (Naresh)   III. Gadd, Geoffrey M.   IV. Title.
QK604.2.E28B75   1994
589.2′045222–dc20   95-24535 CIP

ISBN 978-0-521-49586-8 hardback
ISBN 978-0-521-10625-2 paperback

Contents

List of contributors

L. Adler
Department of General and Marine Microbiology, Lundberg Laboratory, University of Göteborg, Medicinaregatan 9C, S-413 90 Göteborg, Sweden

P. G. Ayres
Division of Biological Sciences, Institute of Environmental and Biological Sciences, University of Lancaster, Lancaster LA1 4YQ, UK

M. J. Bailey
NERC, Institute of Virology and Environmental Microbiology, Mansfield Road, Oxford OX1 3SR, UK

R. D. Bardgett
Institute of Grassland and Environmental Research, Plas Gogerddan, Aberystwyth, Dyfed SY23 3EB, UK

S. Bermingham
Department of Plant Sciences, University of Oxford, South Parks Road, Oxford OX1 3RB, UK, and School of Biological Sciences, Department of Animal and Plant Sciences, University of Sheffield, Sheffield S10 2TN, UK

L. Boddy
School of Pure and Applied Biology, University of Wales, College of Cardiff, Main Building, Museum Avenue, PO Box 915, Cardiff CF1 3TL, UK

D. H. Brown
School of Biological Sciences, University of Bristol, Woodland Road, Bristol BS8 1UG, UK

J. V. Colpaert
Institute of Botany, Laboratory of Plant Ecology, Katholieke Universiteit Leuven, Kardinaal Mercierlaan 92, B-3001 Leuven, Belgium

F. A. A. M. de Leij
Horticulture Research International, Worthing Road,
Littlehampton, West Sussex BN17 6LP, UK
Present address: School of Biological Sciences, University of
Surrey, Guildford, Surrey GU2 5XH, UK

J. Dighton
Institute of Terrestrial Ecology, Merlewood Research Station,
Grange-over-Sands, Cumbria LA11 6JU, UK.
Present address: Division of Pinelands Research, Institute of
Marine and Coastal Science, Department of Biology, Rutgers
University, Camden, NJ 08102, USA

S. Dursun
School of Pure and Applied Biology, University of Wales, College
of Cardiff, Main Building, Museum Avenue, PO Box 915, Cardiff
CF1 3TL, UK and Institute of Terrestrial Ecology, Merlewood
Research Station, Grange-over-Sands, Cumbria LA11 6JU, UK
Present address: Ondokuzmayis University Department of
Environmental Engineering, Samsun, Turkey

J. C. Frankland
Institute of Terrestrial Ecology, Merlewood Research Station,
Grange-over-Sands, Cumbria LA11 6JU, UK

G. M. Gadd
Department of Biological Sciences, University of Dundee, Dundee
DD1 4HN, UK

M. M. Gharieb
Botany Department, Menoufia University, Shebein El-Koom,
Egypt

J. N. Gibbs
Forestry Authority Research Station, Alice Holt Lodge,
Wrecclesham, Farnham, Surrey GU10 4LH, UK

T. S. Gunasekera
Division of Biological Sciences, University of Lancaster, Lancaster
LA1 4YQ, UK

P. Ineson
Institute of Terrestrial Ecology, Merlewood Research Station,
Grange-over-Sands, Cumbria, LA11 6JU, UK

B. Ing
Chester College of Higher Education, Cheyney Road, Chester
CH1 4BJ, UK
Present address: 24 Avon Court, Mold CH7 1JP, UK

D. P. Janos
Department of Biology, University of Miami, PO Box 249118, Coral Gables, Florida 33124-0421, USA

I. A. Kirkwood
Scottish Agricultural Science Agency, East Craigs, Edinburgh EH12 8NJ, UK

D. Lonsdale
Forestry Authority Research Station, Alice Holt Lodge, Wrecclesham, Farnham, Surrey GU10 4LH, UK

J. M. Lynch
Horticulture Research International, Littlehampton, West Sussex BN17 6LP, UK
Present address: School of Biological Sciences, University of Surrey, Guildford, Surrey GU2 5XH, UK

N. Magan
Biotechnology Centre, Cranfield University, Cranfield, Bedford MK43 0AL, UK

G. F. Morley
Department of Biological Sciences, University of Dundee, Dundee DD1 4HN, UK

K. K. Newsham
Institute of Terrestrial Ecology, Monks Wood Experimental Station, Abbots Ripton, Huntingdon PE17 2LS, UK

D. W. Parry
Crop & Environmental Research Centre, Harper Adams College, Newport, Shropshire TF10 8NB, UK

N. D. Paul
Division of Biological Sciences, University of Lancaster, Lancaster LA1 4YQ, UK

T. R. Pettitt
Crop & Environment Research Centre, Harper Adams College, Newport, Shropshire TF10 8NB, UK
Present address: Horticulture Research International, Efford, Lymington, Hampshire SO41 0LZ, UK

M. S. Rasanayagam
Division of Biological Sciences, University of Lancaster, Lancaster LA1 4YQ, UK

A. D. M. Rayner
School of Biology and Biochemistry, University of Bath, Claverton Down, Bath BA2 7AY, UK

M. Rotheroe
*Cambrian Institute of Mycology, Fern Cottage, Falcondale,
Lampeter, Dyfed SA48 7RX, UK*

J. A. Sayer
*Department of Biological Sciences, University of Dundee, Dundee
DD1 4HN, UK*

P. J. A. Shaw
*Biology Laboratories, Central Electricity Research Laboratories,
Kelvin Avenue, Leatherhead, Surrey KT22 7SE, UK
Present address: Department of Environmental Studies,
Southlands College, Wimbledon Parkside, London SW19 5NN,
UK*

I. Singleton
*Department of Industrial Microbiology, University College
Dublin, Ardmore, Stillorgan Road, Dublin 4, Ireland
Present address Department of Soil Science, Waite Campus,
University of Adelaide, Glen Osmond, SA 5064, Australia*

M. K. Smith
*Institution of Chemical Engineers, Davis Building, 165–171
Railway Terrace, Rugby CV21 3HQ, UK*

G. M. Terry
*Division of Biological Sciences, Institute of Environmental and
Biological Sciences, University of Lancaster, Lancaster LA1 4YQ,
UK*

J. M. Tobin
*School of Biological Sciences, Dublin City University, Dublin 9,
Ireland*

K. K. Van Tichelen
*Institute of Botany, Laboratory of Plant Ecology, Katholieke
Universiteit Leuven, Kardinaal Mercierlaan 92, B-3001 Leuven,
Belgium*

J. M. Whipps
*Horticulture Research International, Worthing Road,
Littlehampton, West Sussex BN17 6LP, UK
Present address: Horticulture Research International,
Wellesbourne, Warwick CV35 9EF, UK*

S. C. Wilkinson
*Department of Biological Sciences, University of Dundee
Dundee DD1 4HN, UK*

Preface

The 'Environment' is now on political agendas, and it is time the mycological voice was heard in the upsurge of national and international
debates that have followed in the wake of the 1992 Earth Summit at
Rio de Janeiro. Despite worldwide concern over environmental changes,
fungi vital to the functioning of ecosystems are rarely mentioned.

This is the first symposium volume to focus on fungi in relation to
man-made changes in the natural environment. It comprises papers presented at a British Mycological Society Symposium held at Cranfield
University, UK, in 1994. The authors, all actively engaged in mycological
research, cover widely diverse but highly topical subjects such as global
warming, rising sea levels and destruction of rainforests. Speculation is
bound to be found, but experimental evidence has been included wher-
ever possible. Selection will also be apparent. The number of mycologists
in this field is not great and many environmental problems remain
untouched. Our aim is to stimulate thought on some of the issues of
the day, and to point to the need for more research at every level, from
field recording to cell physiology.

In Chapter 1 Lonsdale and Gibbs discuss predicted changes in global
climate in relation to associations between fungal pathogens and perennial, woody hosts, and the extent to which the geographic range and
pathogenic activity of the fungus (they do not always coincide) might
alter. This contrasts with Pettitt and Parry's account in Chapter 2 of
the potential effects of long-term climatic change, particularly temperature, on a disease of an annual, herbaceous plant, *Fusarium* foot rot of
winter wheat in the UK, for which predictive modelling is more
feasible.

Depletion of ozone in the stratosphere by the use of chlorofluorocarbons and other chemicals, leading to an increase in UV-B radiation,
is a major threat to all living organisms. Experiments on the sensitivity of
both pathogenic and saprotrophic foliar fungi to UV-B are described by
Ayres and his co-authors (Chapter 3). They also draw attention to

indirect effects that this radiation can have on both living and dead plant materials on which these fungi are growing and competing.

A rise in sea levels in certain regions is forecast to be one of the most dramatic consequences of global warming, and Rotheroe (Chapter 4) describes the particular vulnerability of the mycoflora of UK sand dunes to such events. Ing in Chapter 5 also warns of threats to fungal diversity, quoting evidence from long-term recording of European macrofungi. A decline in the numbers of fruit bodies is ascribed to both loss of specific habitats and pollution. To what extent fruiting reflects the abundance and vigour of vegetative mycelia in the natural environment is still largely unknown.

Pollution is by no means a new area of research for mycologists, but changes in the relative importance of different types of pollutant and the increasing efficiency of analytical techniques justify the inclusion of several chapters on this theme. Chapters 6, 7 and 8 are all on atmospheric pollutants, including dry-deposited sulphur dioxide, which has been less well studied than wet deposition, although probably of greater relevance (see Boddy *et al.*). Ozone and nitrogen oxide, both currently of particular concern, are also among the pollutants discussed by Magan *et al.* and by Shaw.

The effects of environmental change on mycorrhizal associations are particularly difficult to interpret. Typically, mycorrhizas are present when mineral nutrients are already limiting to plant growth. Furthermore, stress factors can act directly or indirectly on the symbiotic partners. Colpaert and Van Tichelin (Chapter 9) discuss these complex interactions in relation to both arbuscular-vesicular and ectomycorrhizal fungi, emphasising the importance of the external mycelium, and supporting their observations with case studies on excess nitrogen, elevated carbon dioxide and metal toxicity.

Chapter 10 is also concerned with mycorrhizas but from the practical viewpoint of how to manage the ecological disaster regions of the humid tropics. In this extensive review, Janos discusses rehabilitation of the deforested, low productivity land that has been stripped of surface soil and of its normal reservoirs of mycorrhizal inoculum.

The impact of changes in land use on beneficial fungi in soil continues to be the theme in Chapter 11. Here Bardgett shows how communities and interactions of saprotrophic fungi, bacteria and fauna could be altered if the UK government implements a policy to reduce overgrazing by sheep and resulting loss of biodiversity in the hill grasslands of northern England. As he points out, the changes could have profound effects on nutrient cycling and organic matter decomposition.

The Chernobyl accident of 1986 brought home to many the irrelevance of national boundaries to the spread of aerial pollution, the ramifications of contamination within the various food chains, and the inequalities in pollutant accumulation by different components of the biota. For example, some fungi accumulate exceptionally high amounts of radiocaesium, particularly in the fruit bodies (basidiomes) of certain basidiomycetes. This has been a useful attribute in studies of Chernobyl fallout, reviewed by Dighton and Terry (Chapter 12). These authors have also explored in laboratory experiments the influx and immobilisation of caesium by grassland and forest fungi, comparing mycorrhizal and saprotrophic species, and also uptake by mycorrhizal and non-mycorrhizal plants. They suggest that, on some sites, much of the fallout could be immobilised in fungal mycelium with implications for grazing animals.

Although aquatic fungi are far outnumbered by terrestrial species, they fulfil a vital function as saprotrophs in freshwater and marine ecosystems, and any perturbations to their activities can affect plant and animal communities. Bermingham in Chapter 13 reviews previous studies on the effects of pollutants on freshwater hyphomycetes, and discusses the results of her own investigations on the effects of effluent from abandoned coal mines contaminated with iron and manganese.

Fundamental physiological processes that underlie the reactions and resistance of fungi to some stress factors in the environment are the subject of Chapters 14 and 15. First, Adler reviews the adaptations of fungi to high concentrations of sodium chloride, important with respect to soil salinisation. Secondly, Gadd and his co-workers discuss sequestration, mobilisation and transformation of toxic metals and metalloids, all processes of environmental importance since they influence the mobility and toxicity of these pollutants.

This volume would be unbalanced without some reference to lichens, well known to be sensitive indicators of pollution, although, as shown in Brown's review (Chapter 16), investigators have concentrated almost entirely on the photosynthetic component of the symbiosis. He shows that there is scope for more research on the fungal partner, which acts as the interface with the immediate environment.

The potential role of fungi in environmental 'bioremediation' is still largely untapped. Their particular versatility, simple structure, and the ease with which they can be cultured in bulk quantities fit them well for commercial use. Singleton and Tobin in Chapter 17 discuss the possibility of using them to remove metals from effluents, pointing out that some species are as effective in removing metals, including radionuclides, as commercial ion-exchange resins.

Uncontrolled environmental hazards of anthropogenic origin have been the subject of most of the preceding chapters, but advances in genetic engineering of microorganisms, as yet almost confined to bacteria and unicellular yeasts, are likely to expose filamentous fungi in the near future to more monitored perturbations. Whipps and his co-authors (Chapter 18) discuss the state of the art and consider it is only a matter of time before genetically manipulated fungi are ready for 'risk assessment'. The need to increase our understanding of the ecology of populations and communities of fungi and associated organisms will then be paramount.

Rayner was challenged to answer the question: 'Has chaos theory a place in environmental mycology?' In an intellectually stimulating, final chapter, he argues that, as mycelial fungi are non-linear, indeterminate systems, chaos (non-linear) theory does not have just a place in environmental mycology but is fundamental to it.

The British Mycological Society gratefully acknowledges sponsorship of this Symposium by Glaxo Holdings plc and Cranfield Environment. The first editor (JCF) also thanks the staff of Merlewood Research Station for all their support, and NM is grateful for assistance given by students in the Applied Mycology Group, Biotechnology Centre, Cranfield University during the Symposium.

Juliet C. Frankland
Naresh Magan
Geoffrey M. Gadd

1
Effects of climate change on fungal diseases of trees

D. LONSDALE AND J.N. GIBBS

Introduction

Climate has been of great importance in the development of associations between trees and pathogenic fungi. In particular, the geographic range of each species of tree or fungus is delimited by factors such as temperature, moisture, snowfall and windiness which affect growth, reproduction and dispersal. Such factors affect the incidence of diseases by determining the distribution of a particular pathogen in relation to the geographic range of a potential host. Also, within a region where both host and pathogen are present, the severity of disease can vary with climate. Such variations can result from the direct effects of climatic factors on the pathogen, or from their effects on aspects of host physiology which determine resistance to attack. Other effects may involve other organisms with which either the host or pathogen interact.

In natural ecosystems, associations between particular tree and fungal species are often of great antiquity and have evolved in ways which tend to avoid mutual destruction. Environmental stability may have been a prerequisite for the development of many of these host–pathogen associations and, if that is the case, it follows that they will be perturbed by major climate change. Less stable relationships tend to occur in the simpler ecosystems that initially exist in man-made plantations, often involving new combinations of host and pathogen species that have artificially been transported beyond their natural geographic ranges. In such cases, it can be envisaged that climate change would encourage major changes in disease incidence and severity.

In an attempt to make a profitable evaluation of the effects of climate change in these diverse situations, we have narrowed our scope to consider only some of the most widely predicted changes – namely that winter temperatures in the temperate regions of the world can be expected to be a few degrees higher than at present and that there will be greater climate instability, including in particular more frequent summer droughts in the middle latitudes (Kräuchi, 1993). Secondly, we have concentrated attention predominantly on diseases of the woody tissues, since it is the perennial nature of these tissues that marks the difference between woody and herbaceous plants (see Pettitt & Parry, Chapter 2).

Finally, we have excluded from consideration the whole topic of saprotrophic survival, as this would demand a chapter on its own.

Direct effects on the pathogens

Effects on the geographic range of pathogens

The geographic ranges of fungal pathogens are, to some extent, determined by the temperature ranges over which they can grow, although many species are prevalent only in regions where temperature and other climatic factors are sufficiently close to optimal values to allow rapid growth and reproduction during part of the year. A very wide range of pathogens could be expected to show alteration of their geographic ranges in response to climate change, and the potential for this is best exemplified by those that respond to the year-to-year fluctuations that already occur.

Leaf rust of poplars (*Populus* spp.), caused by *Melampsora allii-populina*, is an example of a disease which, near the edge of its present climatic range, appears only sporadically due to temperature fluctuation. It is a topical example, since poplar growing is now being encouraged in many European countries as an alternative to producing agricultural surpluses. Many of the new fast-growing clones that are favoured for this purpose were bred in Belgium, where they were screened for field resistance to rust in the 1970s (Pinon *et al.*, 1987). It appears that *M. allii-populina* was virtually absent from the trial grounds in central Belgium at this time. Thus, the clones were in effect screened only against another rust species, *M. larici-populina*, which, unlike *M. allii-populina*, is well established throughout Belgium and in much of northern Europe. In 1985, some of the clones were quite heavily infected by rust in Belgium, and the fungus was found to be *M. allii-populina* which, as shown in Fig.1.1, occurs regularly only in regions further south (Somda & Pinon, 1981).

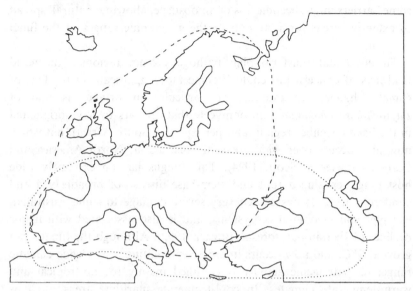

Fig. 1.1. European distribution of the poplar rust fungi *Melampsora larici-populina* (dashed line) and *M. allii-populina* (dotted line). (After Somda & Pinon, 1981.)

There have been similar outbreaks in southern England following the importation of these clones for commercial use.

As shown by Somda & Pinon (1981), *M. allii-populina* is more thermophilic than *M. larici-populina* at some stages of its life cycle – especially urediniospore germination – and, as a wind-dispersed foliar pathogen, it can become prevalent north of its usual range during years with warmer than average temperatures. Other poplar rusts are also quite temperature-sensitive, including one, *M. medusae*, which has been accidentally imported into south-west France and which has so far shown no sign of spreading from this region into other climate zones. However, the possibility of future climate change has important implications for poplar breeding programmes, as well as for plant quarantine controls.

The example of *M. medusae* raises a further issue in relation to certain pathogens which are climatically confined to certain regions but which could find suitable conditions elsewhere, if they were able to 'vault' natural geographical barriers such as mountain ranges. Barriers can, of course, already be circumvented by human interference, as has been the case with the introduction of *M. medusae* into Europe and with the recent appearance of *M. larici-populina* in North America (Newcombe & Chastagner, 1993). However, under conditions of climate amelioration, some barriers might become less of an obstacle, allowing 'natural' spread to extensive areas formerly outside the geographic ranges of the fungi concerned.

The accidental transfer of plant pathogens to new regions of the world is always of concern but could become more significant in the face of climate change. An example of particular importance is that of *Phytophthora cinnamomi*, an oomycete that appears to have originated in the Pacific Celibes region (and perhaps also South Africa), but which now also occurs over wide areas of Australasia, North America and Europe (Brasier & Scott, 1994). This fungus has an extremely wide host range, causing a root and stem-base disease of broadleaved and coniferous trees. It has caused very severe damage to some Australian eucalypt and heathland ecosystems, and is also associated with major declines of Iberian oak forests (Brasier, 1992). Although the fungus can grow at 5 °C, and now occurs in areas representing a very wide climatic range, its pathogenic activity is confined mainly to sub-tropical and warm-temperate climates. In cool-temperate maritime areas, such as Britain, it causes occasional disease and could be expected to become more prevalent with global warming (Fig. 1.2) (Brasier & Scott, 1994). This expectation is based both on the direct response of the fungus to temperature, which has been experimentally modelled using the 'CLIMEX' model for climate matching (Sutherst, Maywald & Bottomley, 1991) and also to changes in soil moisture and to the incidence of drought-induced susceptibility in the host. The question of climate change affecting host susceptibility will be considered in a later section of this chapter.

Effects on the reproduction and dispersal of pathogens

A wide range of pathogens, especially those that infect leaves or green shoots, show large annual fluctuations in their incidence and severity of attack, and these events can often be attributed to weather conditions. Many fungi are favoured by moist conditions during the growing season, due to an enhancement of spore production and, in many cases, dispersal by rain-splash. Others, such as some of the powdery mildews, are favoured by low humidity. Winter conditions are also important in determining the success of the saprotrophic survival of many leaf-infecting fungi.

In view of the effects of existing weather fluctuations on the severity of various foliar diseases, climate change could be expected to affect their

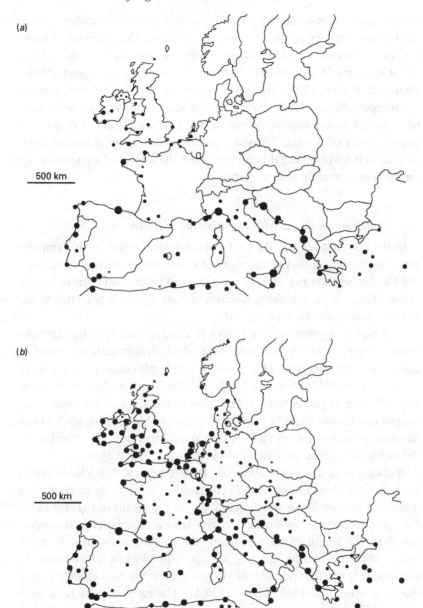

(a)

500 km

(b)

500 km

Fig. 1.2. (a) Current activity of *Phytophthora cinnamomi* in Europe, estimated from its distribution and present-day climate; (b) activity of the fungus predicted after a warming of 3 °C. Dot-size signifies relative suitability of climate for survival and growth of the fungus. Maps pre-date recent national boundary changes. (After Brasier & Scott, 1994.)

relative prevalence in the long-term. A reduction in the number of rain-days in the summer might, for example, decrease the dispersal of many leaf-spot fungi such as *Marssonina* species on poplars (Cellerino, 1979) and *Cristulariella pyramidalis* on black walnut (*Juglans nigra*) (Neely, Phares & Weber, 1976). However, wetter weather in the spring might encourage infection early in the growing season. Any predictions must be regarded as speculative, owing to the complex effects of climate on annual cycles of disease. There are, additionally, many microbial inter-actions that involve fungal pathogens, and the effects of climate change on these are virtually unpredictable.

Effects on the activity of pathogens in winter

In the dormant season, the host's physiological responses to tempera-ture and day-length may to some extent inactivate its defensive reactions, but the temperature can be high enough to allow the pathogen to remain active. Thus, there are many diseases caused by weak parasites which develop mainly at this time of year.

Although many diseases are known to develop mainly in the dormant season, there are few quantitative data which demonstrate the onset or cessation of pathogenesis. One interesting example comes from the work of van Vloten (1952) on the bark-killing pathogen *Phacidium coniferarum* (syn. *Phomopsis pseudotsugae*). In Japanese larch, *Larix kaempferi*, this fungus can invade wounds made during the winter months, such as can be created by pruning operations, then developing until it is checked by the onset of renewed cambial activity in spring (Fig. 1.3).

Working in a stand of 13-year-old Japanese larch, van Vloten (1952) made wound inoculations with *P. coniferarum* at monthly intervals and measured the size of the resulting lesions. As the results in Fig. 1.4 show, the size of the lesions following dormant-season inoculation was propor-tional to the length of time available to the fungus for host invasion before the onset of the growing season. This work was conducted in the relatively maritime climate of Wageningen in the Netherlands during the mild winters of 1949/50 and 1950/51. The results might have been rather different during colder winters when the limiting effects of low temperature on the fungus might have been important.

For pathogens like *P. coniferarum* that have little ability to overcome host resistance during the growing season, winter temperature is likely to be critically important. In climates where temperatures are too low dur-ing most of the dormant season to allow such fungi to grow within host

Fig. 1.3. Canker of Japanese larch caused by *Phacidium coniferarum*, a dormant-season pathogen.

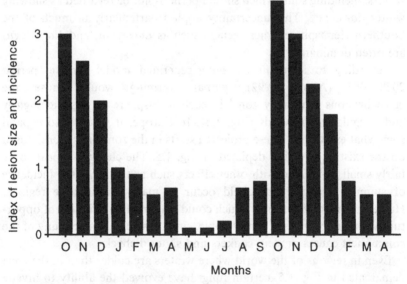

Fig. 1.4. Stem lesions induced by the dormant-season pathogen, *Phacidium coniferarum*, on Japanese larch inoculated at different seasons: bars represent an index of canker incidence and length. (After van Vloten, 1952.)

tissue, there is little opportunity for them to cause disease. However, there are geographic zones, mainly in temperate latitudes, in which winter dormancy of woody plants coincides with periods when temperatures are high enough for fungal activity. Thus, in these zones of 'asynchronous dormancy', disease can be caused by fungi which would otherwise be largely non-pathogenic.

The poleward boundaries of the 'asynchronous dormancy zone' will obviously differ for different host–pathogen combinations. However, the 2 °C isotherm for January in the Northern Hemisphere, and for July in the Southern Hemisphere (adjusted to sea-level) provide a possible demarcation for most diseases, although a more realistic line would need to be based on a detailed analysis of temperature records throughout the winter. Also, there are many cold upland regions within the zone which should be excluded from it. A suggested global 'asynchronous dormancy zone' based on the 2 °C isotherm is shown in Fig. 1.5. Towards the equator, this zone is shown as including all regions with a distinct winter (that is, with the coolest monthly mean below 18 °C). However, it could in reality be much narrower since, as shown here, it includes the subtropical zones, where there are many evergreen tree species, including some which should perhaps not be regarded as showing winter dormancy. This uncertainty applies particularly in much of the Southern Hemisphere, where genera such as *Eucalyptus* and *Nothofagus* are often dominant.

According to a recent computer-generated model for the period 2058–2067 (Anon, 1992), global warming would cause the 'asynchronous dormancy zone' in each hemisphere to migrate slightly and irregularly polewards (Fig. 1.5). In Europe, it would also expand somewhat eastwards. These projected shifts in the zone are superimposed on the existing situation depicted in Fig. 1.5. The changes appear to be fairly small, compared with other effects such as the increased incidence of summer drought that could occur in many mid-latitude regions. However, the range of fungi which could respond to a 'window of opportunity' in mild winters is considerable, since it would include many of the commonest causes of stem cankers and shoot diebacks.

Even in regions of the world where winters are colder than in the zone demarcated in Fig. 1.5, certain fungi have evolved the ability to invade host tissues in the winter. Temperatures rarely fall far below freezing under snow cover, and certain low-temperature pathogens termed 'snow-moulds' have exploited this phenomenon. An example of such a fungus is *Phacidium infestans*, which attacks the needles and shoots of various conifers while they are covered by snow. The snow moulds appear to require the persistently high atmospheric humidity that occurs beneath snow cover, as well as insulation from extreme cold (Björkmann,

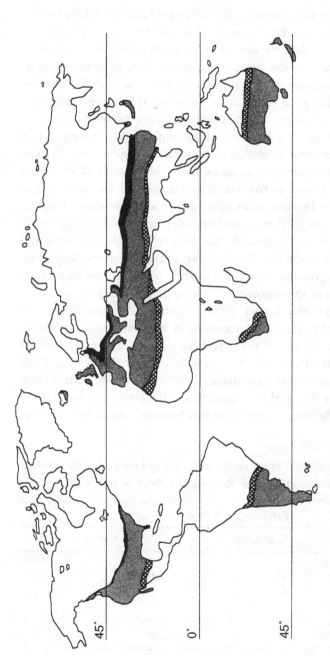

Fig. 1.5. The 'asynchronous dormancy zone' (ADZ) and its possible migration under predicted conditions of global warming for the mid-21st century. *The land areas with the lighter stipple together with their cross-hatched borders represent the 'ADZ', where bark and xylem fungi are often active during winter dormancy of their hosts. The cross-hatched borders represent a possible poleward extension of tropical zones, in which no distinct winter occurs. White land areas together with their darkly stippled borders represent tropical and cold winter zones; in the tropical zones neither hosts nor pathogens haver periods of winter inactivity, while in the cold zones trees and most pathogens are simultaneously inactive. The darkly stippled borders represent a possible poleward extension of the 'ADZ'.*

1948), and so do not usually cause disease in mild climates where snow is absent from the host surface for most of the winter.

The incidence of damage due to snow moulds can be expected to change in response to global warming, since it is likely to involve changes in the amount and persistence of snowfall. This would probably be more important than change in temperature. However, it is interesting to consider the case of an important disease of conifers variously known as Brunchorstia dieback or Scleroderris shoot blight, in which temperature data have been recorded in relation to disease development beneath snow cover. The causal organism is an ascomycete, the teleomorph now most commonly being known as *Gremmeniella abietina* and the anamorph as *Brunchorstia pinea*. In pines, spore infection occurs as the shoot elongates in spring, but the fungus then ceases to develop until the end of the host's growing season, being confined to the dead cells of the epidermis and hypodermis of the shoots and the dead part of the bud scales (Siepmann, 1976). Once the growing season has ended, host invasion can begin, resulting in death of the shoots and buds by the following spring.

Marosy, Patton and Upper (1989) conducted an experiment on Scleroderris shoot blight at two locations in Wisconsin where seedlings of Red pine (*Pinus resinosa*) that had been inoculated in early summer were overwintered either under snow or without snow. The data in Table 1.1 show that there was far more disease in the seedlings kept under snow than in the others. The authors related this to the higher temperatures to be found there: the range below snow was between −6 and 0 °C whereas

Table 1.1. *Effect of snow cover on the incidence of* Gremmeniella abietina *infection of artificially inoculated Red pine seedlings in Wisconsin*

	Percentage infection	
	With snow	Without snow
1984/85 experiment		
Blackhawk	76	0
Copper Falls	48	0
1985/86 experiment		
Blackhawk	83	36*
Copper Falls	62	9

Note: *In this year the seedlings in the no-snow plots spent the winter encased in ice as a result of a fall of wet snow in November.
Source: After Marosy, Patton & Upper, 1989.

the plants in the no-snow plots were exposed to temperatures as low as −36 °C. From this and other work, Marosy *et al.* (1989) developed a concept of 'conducive days', which was based on the idea that a certain number of days between −6 °C and +5 °C is required for disease expression. In winters and locations with insufficient conducive days, the amount of disease would be greatly reduced.

Studies in Japan, where Scleroderris shoot blight occurs on *Abies sachalinensis*, also point in the same direction. Certainly it is known that, once the trees are tall enough for their tops to be above normal snow levels, attacks by *G. abietina* cease to be a problem.

It should be noted that, in the American studies, the upper temperature limit of 5 °C was chosen arbitrarily to exclude days in spring and autumn when the host might be physiologically active and thus perhaps able to resist attack. In the winter, such temperatures are rare in Wisconsin, which has a continental climate, but are common in Britain.

Effects of climate change on the host

Increased climatic stress in the growing season

The evidence from diseases involving native hosts and pathogens suggests that pathogenesis can be exacerbated by stresses induced by extremes of weather, especially drought during the growing season. A classic example of this is the root disease caused by the honey fungus, *Armillaria* spp. Members of this basidiomycete genus are ubiquitous in long-standing woodland, existing in the soil as complicated networks of rhizomorphs sustained by food-bases in the form of colonised stumps and other woody debris. The fungus can establish numerous points of infection on the roots of healthy trees, from which it can extend further into host tissue, sometimes eventually overwhelming the host. Some members of the genus, such as *Armillaria mellea* and *Armillaria ostoyae*, are highly pathogenic (Rishbeth, 1982) but even they may be assisted by the impairment of host resistance through stress from drought and other causes (Wargo, 1984; Rishbeth, 1991). Invasion by less pathogenic species such as *Armillaria gallica* appears to be almost entirely dependent on host stress (Rishbeth, 1982) although it can sometimes be sustained indefinitely after the stress has abated. With an increased frequency of summer drought, therefore, damage caused by *Armillaria* spp may be expected to become more prevalent.

Another disease process that is very much linked to drought stress is the formation of strip-cankers, associated mostly with ascomycetes of the families Xylariaceae and Diatrypaceae. A number of studies (Bassett & Fenn, 1984; Carroll, 1988; Chapela & Boddy, 1988; Hendry, 1993) have indicated that such fungi can exist for many years in the healthy xylem of various broadleaved trees as latent invaders or as endophytes until drought stress, or other damage to the host, allows them to extend by pathogenic growth within the sapwood and into the overlying bark. This pathogenesis is dramatically revealed by the production of very extensive stromatic fruiting structures on the surfaces of the branches or stems that these fungi have helped to kill. An important example is that of *Biscogniauxia mediterranea*, which has caused serious damage to oak species such as *Quercus cerris* and *Q. suber* following droughts in southern Europe (Estanyol & Molinas-de-Ferrer, 1984; Vannini, 1987). In Britain, similar strip cankers on beech (*Fagus sylvatica*) are caused by *Biscogniauxia nummularia* and *Eutypa spinosa* (Lonsdale, 1983; Hendry, 1993), and these were particularly prevalent in southern Britain after the 'double' drought years of 1975/76, 1983/84 and 1989/90.

With climate change in mind, it is interesting to look at the European distributions of *B. mediterranea* and *B. nummularia*. The former causes strip-cankers only in relatively warm climates, such as occur in the Mediterranean countries, while the latter is common as far north as southern Britain but becomes rare in the cooler and damper north and west of Britain, and has its northernmost outposts in southern Scandinavia, where its host is *Prunus* rather than *Fagus* (Granmo *et al.*, 1989). Interestingly, *B. mediterranea* has also been found in southern Britain, though not as the cause of a strip-canker (Spooner, 1986). Increased summer temperatures and droughtiness could be expected to help shift the distributions of these fungi northwards within the range of potential hosts, or at least to increase the geographic range over which they behave as pathogens.

Another interesting stress-related disease akin to the strip cankers is sooty bark disease of sycamore, *Acer pseudoplatanus*, caused by *Cryptostroma corticale* (Fig. 1.6). Here again the fungus is latent or endophytic within the tissues of the healthy tree (Bevercombe & Rayner, 1984; J.N. Gibbs & J. Rose, unpublished observations). In hot dry summers, it can rapidly develop within the xylem and subsequently within the bark (Young, 1978; Dickenson & Wheeler, 1981). Curiously, this disease has been found only in south-east England and northern France (J.N. Gibbs & J. Rose, unpublished observations), even though

Fig. 1.6. Sooty bark disease of sycamore, caused by *Cryptostroma corticale*, a pathogen dependent on host stress.

sycamore occurs as a native over a much wider area of Europe, including the mountains of the south.

In forests where a more frequent incidence of summer drought exacerbates stress-related diseases, there can be a resulting increase in tree mortality which leads to an opening up of the canopy structure. This, in turn, enhances transpirational stress in the remaining trees, due to their increased exposure to insolation and wind.

An exception to the examples described so far appears at first sight to be provided by Dutch elm disease (DED), in which drought conditions suppress the development of foliar symptoms in both the English elm, *Ulmus minor*, and the American elm, *Ulmus americana* (Gibbs & Greig, 1977; Smalley & Kais, 1966; C.M. Brasier, personal communication). However, when in such cases the xylem has been examined, the presence of extensive vascular streaking has indicated that fungal invasion was not prevented, even though foliar symptoms were suppressed (C.M. Brasier, personal communication). This suppression of wilting may be due to stomatal closure during periods of water stress, but there remains some possibility that the fungus could be directly affected by the heat associated with summer droughts. The aggressive DED pathogen,

Ophiostoma novo-ulmi, has a temperature optimum for growth of only *c.* 20–22 °C, as compared with *c.* 30 °C for *Ophiostoma ulmi*, the non-aggressive pathogen (Brasier, Lea & Rawlings, 1981). It is also possible that vascular wilt fungi have a limited ability to colonize drought-stressed trees because of the presence of vessel cavitation (Zimmerman, 1983) which could obstruct their normally rapid invasion of hydraulically functional sapwood.

Effects of altered winter temperatures on the host

During winter dormancy, direct effects of climate on the host are generally less important than those involving the pathogen. However, frost damage has been reported to encourage the development of certain fungal cankers. Examples for which this has been demonstrated include Botryosphaeria canker of rowan, *Sorbus aucuparia*, caused by *Botryosphaeria dothidea* (Wene & Schoeneweiss, 1980) and canker of *Pinus resinosa* caused by *Diplodia pinea* (Palmer, 1991). In areas such as Britain, where climate-change modelling predicts a decrease in the incidence of frost (Anon, 1992), diseases of this type could become less prevalent.

Effects involving interactions between trees or pathogens with other organisms

Interactions with vector organisms

The requirement for an insect or other vector in certain fungal diseases complicates predictions about the possible effects of climate change on the geographic range of disease incidence. Dutch elm disease provides an interesting example, since *Scolytus scolytus* and *Scolytus multistriatus*, the most important of the beetle vectors in western Europe, do not readily fly at a temperature below about 22 °C (Fairhurst & King, 1983) and are therefore less able to act as vectors in cool conditions (Redfern, 1977; Harding & Ravn, 1982). In some parts of the range of elm species, this temperature requirement may have delayed the northward extension of the epidemic until the occurrence of warm summers in particular years and the build-up of beetle breeding habitats in the outbreak areas.

In regions where a pathogen already occurs, weather conditions may favour outbreaks of its vectors in certain years, suggesting that climate change could influence the long term prevalence of the disease. An interesting case is that of the fungus *Ceratocystis laricicola* (Redfern, Stoakley & Steele, 1987), which has been described only in recent years. It infects the bark, cambium and sapwood of larch trees, causing death and dieback, and appears to be transferred between trees by an insect vector, the bark beetle *Ips cembrae*. Redfern *et al.* (1987) found that this disease occurred mainly in areas affected by drought, which favours bark beetle attacks by inducing host stress.

For many fungal diseases involving vectors, the effects of climate and weather on the development of outbreaks and epidemics have not been studied in detail. However, the importance of such effects is to some extent self-evident, since the geographic ranges of insect vectors are determined largely by climate, while their activity and abundance are influenced by the vagaries of weather. Climate change, involving an increase in temperature or in the incidence of drought, could extend the range of any such diseases into areas where the host and fungus can already both exist but in which vectors are not yet operative.

Effects on mycorrhizas

There is evidence that mycorrhizal fungi can protect trees against certain root pathogens (Marx, 1970; Chakravarty & Unestam, 1987; Buscot, Weber & Oberwinkler, 1992), and that some species can be much more effective than others in this respect (Malajczuk, 1988). The incidence of certain root diseases may therefore be influenced by the outcome of interspecific competition between mycorrhizal fungi. Periodic drought tends to provide opportunities for competitive replacement within mycorrhizal communities because it leads to the death and subsequent regeneration of non-woody roots, and thus to the loss and renewal of mycorrhizal associations. Although there appear to have been no studies involving the effects of mycorrhizal replacement on the suppression of disease, replacement *per se* has been observed following environmental change. For example, when seedlings of *Pinus caribaea* were planted in the dry zone of Sri Lanka, their mycorrhizal symbiont, a *Boletus* sp., was totally replaced by the drought-tolerant mycorrhizal fungus *Cenococcum* sp. (Muttiah, 1972). Mycorrhizal replacements might occur not only in seedlings planted out in a new site, but also in established trees exposed to an increased frequency of drought, as has happened during recent years in some countries. It is interesting to ask whether such an effect might partly explain the reported decline in the occurrence of various mycorrhizal fungi in Europe (Arnolds, 1988; Jansen, 1990).

Summary and conclusions

The effects of possible climate change on fungal diseases of trees can, to some extent, be judged by analysing the existing role of climate and of fluctuations in weather. For pathogens whose geographic ranges or pathogenic activity are clearly affected by temperature, the effects of climate warming are probably predictable. These pathogens include both those that are favoured by relatively high summer temperatures and also those that require mild temperatures in the dormant season because it is only then that they can attack the host.

The effects of any increase in the frequency of summer droughts is also reasonably predictable, since the role of host stress in allowing attack by many pathogens, especially root pathogens, is well known. In particular, it can be predicted that such a climate change would alter the stability of associations between tree species and various members of their endophytic mycofloras, some of which would be triggered more frequently into curtailing such associations through pathogenesis.

Prediction is more difficult in the case of pathogens whose reproduction and dispersal is strongly affected by rainfall and atmospheric humidity. This is also true of pathogens that are strongly affected by interactions with other organisms, such as insect vectors or protective mycorrhizal fungi.

Acknowledgements

We are grateful to Dr C.M. Brasier for providing information on *Phytophthora cinnamomi* and to him and Dr D.B. Redfern for comments on the manuscript. We also thank the following individuals and organizations for permission to reproduce data: Dr Robert F. Patton (University of Wisconsin-Madison, USA); J. Pinon (INRA, Nancy, France); Dr C.M. Brasier, Forestry Authority, Great Britain; European Plant Protection Organization, American Phytopathological Society, Blackwell Scientific Publications, Royal Scottish Forestry Society.

References

Anon (1992). *The Hadley Centre Transient Climate Change Experiment.* UK: Hadley Centre, Meteorological Office: Bracknell, 20 pp.
Arnolds, E. (1988). The changing macromycete flora in the Netherlands. *Transactions of the British Mycological Society*, **90**, 391-406.

Bassett, E.N. & Fenn, P. (1984). Latent colonization and pathogenicity of *Hypoxylon atropunctatum* on oaks. *Plant Disease*, **68**, 317-19.

Bevercombe, G.P. & Rayner, A.D.M. (1984). Population structure of *Cryptostroma corticale*, the causal fungus of sooty bark disease of sycamore. *Plant Pathology*, **33**, 211-17.

Björkmann, E. (1948). Studier över snöskyttesvampens (*Phacidium infestans* Karst.) biologi samt metoder för snöskyttets bekämpande. *Meddelanden från Statens Skogsforskningsinstitut, Stockholm*, **37**, 1-136.

Brasier, C.M (1992). Oak tree mortality in Iberia. *Nature*, **360**, 539.

Brasier, C.M., Lea, J. & Rawlings, M.K. (1981). The aggressive and non-aggressive strains of *Ceratocystis ulmi* have different temperature optima for growth. *Transactions of the British Mycological Society*, **76**, 213-18.

Brasier, C.M. & Scott, J.K. (1994). European oak declines and global warming: a theoretical assessment with special reference to the activity of *Phytophthora cinnamomi*. *OEPP/EPPO Bulletin*, **24**, 221-32.

Buscot, F., Weber, G. & Oberwinkler, F. (1992). Interactions between *Cylindrocarpon destructans* and ectomycorrhizas of *Picea abies* with *Laccaria laccata* and *Paxillus involutus*. *Trees: Structure and Function*, **6**, 82-90.

Carroll, G. (1988). Fungal endophytes in stems and leaves: from latent pathogen to mutualistic symbiont. *Ecology*, **69**, 2-9.

Cellerino, G.P. (1979). Le Marssoninae dei pioppi. *Cellulosa e Carta*, **30**, 3-23.

Chakravarty, P. & Unestam, T. (1987). Mycorrhizal fungi prevent disease in stressed pine seedlings. *Journal of Phytopathology*, **118**, 335-40.

Chapela, I.H. & Boddy, L. (1988). Fungal colonisation of attached beech branches, II. Spatial and temporal organisation of communities arising from latent invaders in bark and functional sapwood under different moisture regimes. *New Phytologist*, **110**, 47-57.

Dickenson, S. & Wheeler, B.E.J. (1981). Effects of temperature and water stress in sycamore, on growth of *Cryptostroma corticale*. *Transactions of the British Mycological Society*, **76**, 181-5.

Estanyol, M.O. & Molinas-de-Ferrer, M.L. (1984). Incidencia de *Hypoxylon mediterraneum* en los alcornales gerundenses. *Boletin de la Estacion Central de Ecologia*, **13**, 9-16.

Fairhurst, C.P. & King, C.J. (1983). The effect of climatic factors on the dispersal of elm bark beetles. In *Research on Dutch Elm Disease in Europe*, ed. D. A. Burdekin, Forestry Commission Bulletin, **60**, 40-6. London: Her Majesty's Stationery Office.

Gibbs, J.N. & Greig, B.J.W. (1977). Some consequences of the 1975-1976 drought for Dutch elm disease in southern England. *Forestry*, **50**, 145-54.

Granmo, A., Hammelev, D., Knudsen, H., Læssøe, T., Sasa, M. & Whalley, A.J.S. (1989). The genera *Biscogniauxia* and *Hypoxylon* (Sphaeriales) in the Nordic countries. *Opera Botanica*, **100**, 59-84.

Harding, S. & Ravn, H.P. (1982). Danske fund af de tre elmebarkbillearter i relation til elmesygen. *Tidsskrift for Planteavl*, **86**, 477-95.

Hendry, S.J. (1993). Strip cankering in relation to the ecology of Xylariaceae and Diatrypaceae in beech. PhD Thesis, University of Cardiff, UK.

Jansen, A.E. (1990). Conservation of fungi in Europe. *The Mycologist*, **4**, 83-85.

Kräuchi, N. (1993). Potential impacts of a climate change on forest ecosystems. *European Journal of Forest Pathology*, **23**, 28-50.

Lonsdale, D. (1983). Some aspects of the pathology of environmentally stressed trees. *International Dendrology Society Yearbook*, 1982, 90-7.

Malajczuk, N. (1988). Interaction between *Phytophthora cinnamomi* zoospores and micro-organisms on non-mycorrhizal and ectomycorrhizal roots of *Eucalyptus marginata*. *Transactions of the British Mycological Society*, **90**, 375-82.

Marosy, M. Patton, R.F. & Upper, C.D. (1989). A conducive day concept to explain the effect of low temperature on the development of Scleroderris shoot blight. *Phytopathology*, **79**, 1293-301.

Marx, D.H. (1970). The influence of ectotrophic mycorrhizal fungi on the resistance of pine roots to pathogenic infections. V. Resistance of mycorrhizae to infection by vegetative mycelium of *Phytophthora cinnamomi*. *Phytopathology*, **60**, 1472-3.

Muttiah, S. (1972). Effect of drought on mycorrhizal associations of *Pinus caribaea*. *Commonwealth Forestry Review*, **51**, 116-20.

Neely, D., Phares, R. & Weber, B. (1976). Cristulariella leaf spot associated with defoliation of black walnut plantations in Illinois. *Plant Disease Reporter*, **60**, 587-90.

Newcombe, G. & Chastagner, G.A. (1993). First report of the Eurasian poplar leaf rust fungus, *Melampsora larici-populina*, in North America. *Plant Disease*, **77**, 532-5.

Palmer, M.A. (1991). Isolate types of *Sphaeropsis sapinea* associated with main stem cankers and top-kill of *Pinus resinosa* in Minnesota and Wisconsin. *Plant Disease*, **75**, 507-10.

Pinon, J., van Dam, B.C., Genetet, J. & de Kam, M. (1987). Two pathogenic races of *Melampsora larici-populina* in north-western Europe. *European Journal of Forest Pathology*, **17**, 47-53.

Redfern, D.B. (1977). Dutch elm disease in Scotland. *Scottish Forestry*, **31**, 105-9.

Redfern, D.B., Stoakley, J.T. & Steele, H. (1987). Dieback and death of larch caused by *Ceratocystis laricicola* sp. nov. following attack by *Ips cembrae*. *Plant Pathology*, **36**, 467-80.

Rishbeth, J. (1982). Species of *Armillaria* in southern England. *Plant Pathology*, **31**, 9-17.

Rishbeth, J. (1991). *Armillaria* in an ancient broadleaved woodland. *European Journal of Forest Pathology*, **21**, 239-49.

Siepmann, R. (1976). Ein Beitrag zur Infectionsbiologie des durch *Scleroderris lagerbergii* Gr. verursachten Schwarz Kieferntriebsterbens. *European Journal of Forest Pathology*, **6**, 103-9.

Smalley, E.N., & Kais, A.J. (1966). Seasonal variation in the resistance of various elm species to Dutch elm disease. In *Breeding Pest-Resistant Elm Trees*, pp. 279–287, ed. H.D. Gerhold, E.J. Schreiner, R.E. McDermott & J.A. Winiesky. Oxford, UK: Pergamon Press.

Somda, B. & Pinon, J. (1981). Ecophysiologie du stade urédien de *Melampsora larici-populina* Kleb. et de *M. allii-populina* Kleb. *European Journal of Forest Pathology*, **11**, 243-54.

Spooner, B.M. (1986). New or rare British microfungi from Esher Common, Surrey. *Transactions of the British Mycological Society*, **86**, 401-8.

Sutherst, R.W., Maywald, G.F & Bottomley, W. (1991). From CLIMEX to PESKY, a generic expert system for pest risk assessment. *OEPP/EPPO Bulletin*, **21**, 595-608.

Vannini, A. (1987). Osservazione preliminari sul deperimento del cerro
(*Quercus cerris* L.) nell'Alto Lazio. *Informatore Fitopatologico*, **9/87**, 54-9.

van Vloten, H. (1952). Evidence of host–parasite relations by experiments with
Phomopsis pseudotsugae Wilson. *Scottish Forestry*, **6**, 38-46.

Wargo, P.M. (1984). How stress predisposes trees to attack by *Armillaria
mellea* – a hypothesis. In *Proceedings of the 6th Conference on Root- and
Butt-rots of Forest Trees*, ed. G.A. Kile, pp. 115-122, Melbourne: CSIRO.

Wene, E.G. & Schoeneweiss, D.F. (1980). Localized freezing predisposition to
Botryosphaeria canker in differentially frozen woody stems. *Canadian
Journal of Botany*, **58**, 1455-8.

Young, C.T.W. (1978). *Sooty Bark Disease of Sycamore*. Arboricultural
Leaflet, Forestry Commission, No. 3, Her Majesty's Stationery Office:
London.

Zimmerman, M.H. (1983). *Xylem Structure and the Ascent of Sap*. Springer
series in Wood Science. Berlin: Springer. 139 pp.

2

Effects of climate change on *Fusarium* foot rot of winter wheat in the United Kingdom

T. R. PETTITT AND D. W. PARRY

Much effort has been made in the field of plant disease epidemiology to link disease incidence and severity with short-term weather variables. However, few studies have been published on the effects of longer-term weather patterns or climate change on plant disease, despite an earlier presentation of a good case for this type of approach (Coakley, 1988). The *Fusarium* foot-rot disease complex of wheat provides an interesting opportunity for the study of the impacts of climate on both the severity of a disease of an herbaceous plant and the competition between the various pathogen species capable of causing foot-rot symptoms. This chapter outlines a recent study on the effects of climate, particularly air temperature, on *Fusarium* foot rot in UK cereals, with the aim of developing predictive models.

The disease

Fusarium foot rot of temperate cereals is a disease caused by several *Fusarium* or *Fusarium*-like species, each of which may infect stem bases individually or, in complex infections, involving two or more species. *Fusarium* foot rot in winter wheat is commonly caused in the UK by four species: *Fusarium culmorum*, *Fusarium avenaceum*, *Microdochium nivale* (formerly *Fusarium nivale*) and to a lesser extent *F. graminearum* (Parry *et al.*, 1994). The foot-rot symptoms caused by the four species are indistinguishable (Fig. 2.1) and currently, the individual species present can only be reliably identified by the presence of perithecia in infected material or by isolations from diseased tissue pieces, using selective agar media and identification of emerging fungal colonies.

Foot rot symptoms form part of a seasonal cycle of *Fusarium* disease, governed by the growth stage of the cereal crop (Fig. 2.2). Of the main

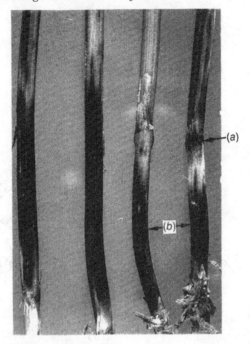

Fig. 2.1. Symptoms of nodal (*a*) and internodal (*b*) *Fusarium* foot rot in winter wheat. (Crown copyright, courtesy of Central Science Laboratory, Harpenden.)

diseases, seedling blight, foot rot and ear blight, foot rot is arguably the least destructive in the UK, although it is undoubtedly the commonest (Polley & Thomas, 1991). However, significant yield losses can result from 'lodging' or collapse of severely diseased cereal stands. Such severe foot-rot symptoms are rare in the UK and are more favoured by the warm, arid conditions prevalent in regions such as eastern Australia (Burgess, Wearing & Toussoun, 1975), and the Pacific Northwest of the USA (Cook, 1980). The importance of the foot rot stage in the disease cycle as a potential inoculum bridge between the seedling blight stage and ear blight should also not be underestimated (Jenkinson & Parry, 1994).

Temperature

The variation in response to temperature of the individual species of *Fusarium* and *M. nivale* is well illustrated by their typical *in vitro* saprophytic growth rates on agar media (Fig. 2.3). The optimum temperature for growth of *F. culmorum* and *F. avenaceum* was between 20 and 25 °C

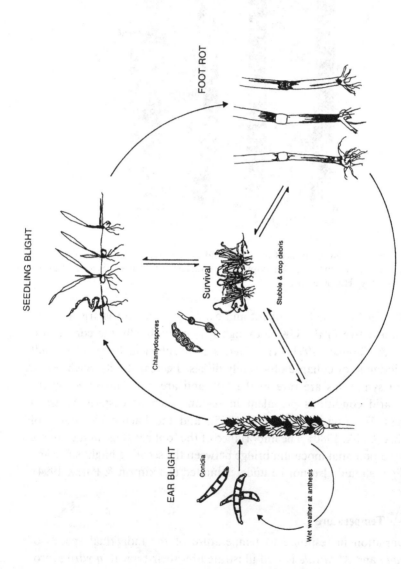

FOOT ROT

SEEDLING BLIGHT

Survival

Chlamydospores

Stubble & crop debris

EAR BLIGHT

Conidia

Wet weather at anthesis

Fig. 2.2 Generalized life-cycle of *Fusarium* diseases in winter wheat (Reproduced with permission of CAB International, from Parry *et al.*, 1994.)

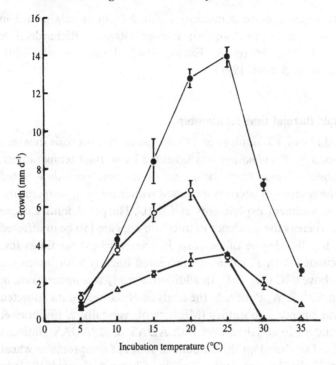

Fig.2.3 The effect of temperature on the rate of mycelial growth of typical isolates of *Fusarium avenaceum* — (-△-), *F. culmorum* (-●-) and *Microdochium nivale* (-○-) on potato dextrose agar. Vertical bars represent 95% confidence limits (df = 5), for points where bars are absent, the confidence limits were too small to accurately represent them   graphically. (After Parry *et al.*, 1994.)

whereas *M. nivale* grew optimally at 20 °C and was significantly (P < 0.05) the fastest growing species at 5 °C (Fig. 2.3). Base temperatures (T_b) for growth *in vitro* of each species can be calculated by extrapolation of the sub-optimal portion of each growth curve. *M. nivale* has the lowest T_b at 1.5 °C with *F. avenaceum* at 3.4 °C and *F. culmorum* has the highest threshold for growth at 5 °C. These *in vitro* observations match results of necrosis measurements on winter wheat seedlings. With *F. culmorum* infections, necrosis was most severe at air temperatures between 16 and 23 °C (Colhoun, Taylor & Tomlinson, 1968) whilst *M. nivale* was favoured by lower temperatures (Millar & Colhoun, 1969). Temperature also seems an important factor in the geographical distribution of species, for example in warmer growing areas such as western USA and eastern Australia, *F. culmorum, F. graminearum* and *F. avenaceum* are found to be predominant (Cook, 1968; Burgess *et al.*, 1975),

24   *T. R. Pettitt and D. W. Parry*

whereas *M. nivale* is more commonly isolated from cereals grown in cooler regions such as northwestern Europe (Rennie, Richardson & Noble, 1983; Locke, Moon & Evans, 1987; Polley *et al.*, 1991, Daamen, Langerak & Stol, 1991).

Simple thermal time relationships

As mentioned above, Colhoun *et al.* (1968) found that necrosis in wheat seedlings caused by *F. culmorum* was favoured by warmer temperatures. They also observed that over the range of temperatures they tested (16–23 °C), the severity of necrosis increased with increasing temperature. Controlled environment experiments at CERC, Harper Adams College showed that, whereas the incidence of infection appeared to be unaffected by temperature, the degree of necrosis in winter wheat seedlings (cv. Mercia) inoculated with *F. culmorum* increased linearly with increasing day degrees above 0 °C (Fig. 2.4). In addition, a study was undertaken, in collaboration with R.W. Polley, of the analysis of national data collected by the Central Science Laboratory (Ministry of Agriculture, Fisheries & Food, Harpenden) in collaboration with ADAS (CSL/ADAS National Cereal Disease Database) on the amount of foot rot symptoms at wheat growth stage 75 ('milky ripe' stage, Zadoks, Chang & Konzak,1974) in

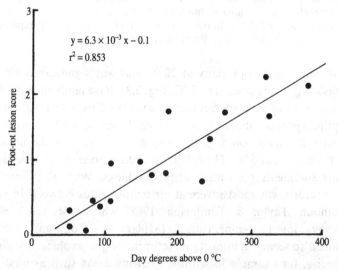

Fig.2.4 Progress of necrosis symptoms of *F. culmorum* infection in winter wheat seedlings (cv. Mercia) with day degrees above 0 °C.

1992 (Pettitt, Parry & Polley, 1995). The incidence of nodal foot rot in each UK county was expressed as a mean and considered in relation to the county mean heat sum calculated in degree days for the period from February to sampling time:

$$S = \sum\nolimits_{m=Feb}^{m=Jul} [\sum\nolimits_{day=1}^{day=j} (T_m - T_b)] \qquad (1)$$

where S is the sum of day degrees, T_b is the base temperature for pathogen development and T_m is the monthly mean air temperature (Pettitt *et al.,* 1995). Assuming a base temperature (T_b) of 0 °C in equation 1, when the mean incidence of foot rot symptoms was considered in relation to heat sum there appeared to be no significant relationship. However, during the 1992 national winter wheat disease survey, stem base samples were sent to CERC, Harper Adams College for isolations and identification of the *Fusarium* and *Microdochium* species present. This allowed the identification of the predominant foot-rot pathogen species present in wheat stem bases in each county assessed for foot rot during 1992 and when this was taken into account in the relationship between symptoms and heat sum (S), an interesting pattern emerged (Fig. 2.5). The relationship between S (degree days) and incidence of symptoms differed between *M. nivale* and *F. culmorum.* Less degree days were required in the case of *M. nivale* to produce the same disease incidence. Hence the T_b value for *M. nivale*-induced symptoms was lower than that for *F. culmorum.* This

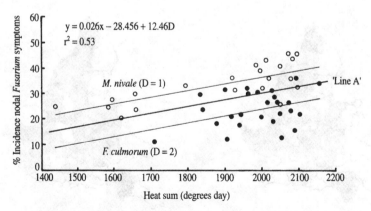

Fig.2.5 Influence of heat sum (S) for the months from February to July (equation 1), in degree days, on the incidence of nodal *Fusarium* foot-rot symptoms in regions where the predominant pathogen in foot-rot infections was *M. nivale* (○) or *F. culmorum* (●). (After Pettitt *et al.,* 1995.)

was in agreement with *in vitro* growth-rate thresholds. Multiple regression analysis, using the dummy variable approach (Gujarati, 1970) to compare regressions of the effect of *F. culmorum* and *M. nivale* on symptoms in relation to heat sum, revealed that although the slopes for symptom incidence with heat sum were the same, the intercepts were significantly (P < 0.05) different.

Assuming the regression lines for the response of foot-rot symptom incidence caused by *F. culmorum* and *M. nivale* to be parallel, a subjective line (Fig. 2.5, line A) can be drawn equidistant between them. This line can be used to discern counties where *M. nivale* was the predominant foot-rot species and therefore the primary cause of symptoms, and similarly those where *F. culmorum* predominated. In Fig. 2.5, points above line A were considered to represent the foot-rot symptoms predominantly caused by *M. nivale* and those below line A, symptoms predominantly caused by *F. culmorum*. Since each point in Fig. 2.5 represents the mean symptom incidence for an individual county, it is possible using this method to make a subjective prediction of the patterns of distribution of *M. nivale* and *F. culmorum* (Fig. 2.6(*a*)). This predicted pattern was compared with 1992 figures for the actual pattern of species distribution obtained by isolations (Fig. 2.6(*b*)). The predominant species in 70% of the counties was correctly predicted using this technique (Fig. 2.6). In 1992 the period from February to July was warmer than average and Zadoks growth stage 75 was reached in UK winter wheat crops several days earlier than usual (Polley & Slough, 1992). In agreement with this

(*a*)   (*b*)

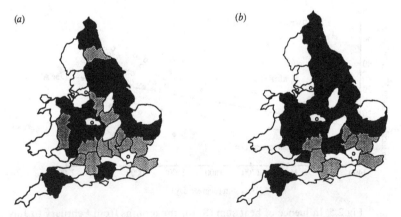

Fig.2.6 Predicted (*a*) and actual (*b*) patterns of the distribution of predominance of *F. culmorum* (■) and *M. nivale* (▨) in winter wheat stem bases in England and Wales during 1992.

observation, the predominant species identified by isolations from 1992 foot-rot infections was the warm-loving species *F. culmorum*, which was present in 55.6% of all stem bases assessed (Pettitt, Parry & Polley, 1993). *F. culmorum* was predominant in 24 out of the 31 counties investigated. Paradoxically, the pattern of distribution of *Fusarium*-like species in stem bases for 1992, as revealed by isolations, showed that the cool-loving species *M. nivale* was most prevalent in the warmer southeastern counties of England (Fig. 2.6(*b*)). Although this pattern could not be explained in terms of the effect of temperature on the distributions of *M. nivale* and *F. culmorum*, a similar pattern was predicted (Fig. 2.6(*a*)) using the procedure described above (Fig. 2.5, 'line A').

A large amount of data on the incidence of foot-rot disease on winter wheat crops from years prior to 1992 (1981, 1982, 1985, 1989, and 1990) was also available from the CSL/ADAS National Cereal Disease Database for analysis. Unfortunately there was insufficient information on the species present during these years for multiple regression analysis by the dummy variable approach mentioned above. However, the annual patterns of distribution of *M. nivale* and *F. culmorum* could be determined using the subjective procedure above with 'line A' (Fig. 2.5), after calculating county means for heat sum and symptom incidence and plotting them against one another. Patterns of distribution obtained by this method (Fig. 2.7) agreed with the findings published in the literature listed below. *M. nivale* was predicted as generally the most predominant species over the years assessed (Fig. 2.7). This matches patterns of *M. nivale* distribution in northwestern Europe, where it is generally considered to be the most prevalent foot rot pathogen (Daamen *et al.*, 1991; Polley *et al.*, 1991). *Fusarium culmorum* was predicted as being more predominant in warmer years such as 1982, 1989 and 1990 (Fig. 2.7). This is supported by observations in the UK of an increase in *F. culmorum* populations in these warmer years (Parry, Bayles & Priestley, 1984; Parry, 1990; Polley *et al.*, 1991).

Conclusions

The relationship between *Fusarium* foot-rot symptoms and temperature cannot be fully understood without determination of the predominant infecting species involved. The use of dummy variables in regression analysis allowed the prediction of disease incidence using monthly mean temperatures calculated for counties in the UK. However, this method requires prior knowledge of the predominant foot-rot pathogen

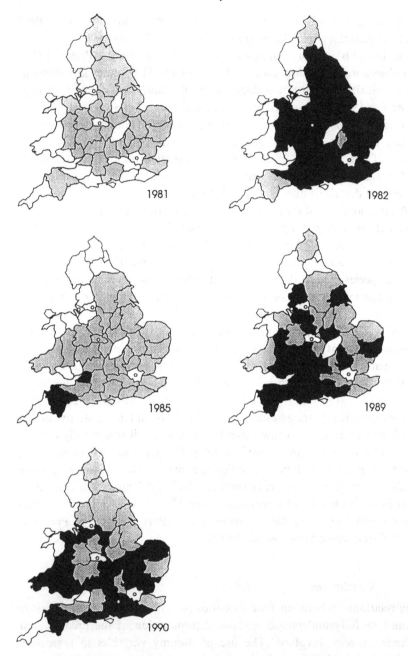

Fig.2.7 Predicted patterns of the distribution of predominance of *F. culmorum* (■) and *M. nivale* (▦) in winter wheat stem bases in England and Wales for the years 1981, 1982, 1985, 1989 and 1990.

species present which unfortunately cannot be determined by using the heat sum alone. The use of a subjective method ('line A' described above) lends some anecdotal support to the idea of taking the predominant species into account in multiple regression analysis of symptoms against heat sum (Fig. 2.5). However, this type of approach, apart from being statistically questionable, would be inappropriate for the prediction of the predominant foot-rot pathogen species present in future wheat crops, as it takes no account of the mechanisms involved in the selection for a particular pathogen, in a particular region or a particular year. The use of 'line A' would also require prior knowledge of the amount of symptoms in each region which cannot be estimated by the multiple regression model without knowledge of the species present.

Use of the CSL/ADAS National Cereal Disease Database has been invaluable to this study, the only shortfall with these data, when used in the type of analysis presented in this chapter, is that they represent only a 'freeze-frame' picture of what has happened at the end of a foot rot epidemic, with no temporal information on how the symptoms developed. Field studies on the sequence of infection support the idea of temperature exerting a strong influence on the distribution of foot-rot pathogen species present in individual seasons (Duben & Fehrmann, 1979; Parry, 1990). However the situation is obviously more complex, and other factors such as the water status of the host plants (Papendick & Cook, 1974; Cook & Christen, 1976; Cook, 1981; Magan & Lacey, 1984) and local fungicide practice (Locke *et al.*, 1987, Parry *et al.*, 1994) are likely to be important species determinants. Controlled environment experiments on the temporal sequence of natural infection and symptom development in collaboration with Reading University (T.R Pettitt *et al.*, unpublished observations) have shown this to be a complex process influenced by seed- and soil-borne inoculum and soil water status in addition to the effect of temperature. Ideally, for the development of accurate predictive models of future foot-rot disease in a warmer UK climate, controlled environment experiments are needed to determine the effects of temperature, moisture, inoculum and stage of plant growth on competition between the various pathogenic species and subsequent cereal foot-rot symptoms.

Acknowledgements

Thanks to R.W. Polley and J.E. Slough of CSL Harpenden for their assistance with the many disease records from the CSL/ADAS National Cereal Disease Database. This study was funded by a MAFF open contract.

30   T. R. Pettitt and D. W. Parry

References

Burgess, L.W., Wearing, A.H. & Toussoun, T.A. (1975). Surveys of fusaria
associated with crown rot of wheat in eastern Australia. *Australian Journal
of Agricultural Research* **26**, 791-9.
Coakley, S.M. (1988). Variation in climate and prediction of disease in plants.
Annual Review of Phytopathology, **26**, 163-81.
Colhoun, J., Taylor, G.S. & Tomlinson, R. (1968). *Fusarium* diseases of cereals
II. Infection of seedlings by *F. culmorum* and *F. avenaceum* in relation to
environmental factors. *Transactions of the British Mycological Society*, **51**,
397-404.
Cook, R.J. (1968). *Fusarium* root and foot rot of cereals in the Pacific
Northwest. *Phytopathology*, **58**, 127-31.
Cook, R.J. (1980). *Fusarium* foot rot of wheat and its control in the Pacific
Northwest. *Plant Disease*, **64**, 1061-6.
Cook, R.J. (1981). Water relations in the biology of *Fusarium*. In *Fusarium:
Diseases, Biology and Taxonomy*, ed. P.E. Nelson, T.A. Toussoun & R.J.
Cook, pp. 236-243. Pennsylvania State University Press: London.
Cook, R.J. & Christen, A.A. (1976). Growth of cereal root rot fungi as
affected by temperature–water potential interactions. *Phytopathology*, **66**,
193-7.
Daamen, R.A., Langerak, C.J. & Stol, W. (1991). Surveys of cereal diseases
and pests in the Netherlands. 3. *Monographella nivalis* and *Fusarium* spp.
in winter wheat fields and seed lots. *Netherlands Journal of Plant
Pathology*, **97**, 105-14.
Duben, J. & Fehrmann, H. (1979). Vorkommen und pathogenität von
Fusarium-Arten an Winterweizen in der Bundesrepublik Deutschland. I.
Artenspektrum und jahreszeitliche Sukzession an der Halmbasis.
Zeitschrift für Pflanzenkrankheiten und Pflanzenschutz, **86**, 638-52.
Gujarati, D.N. (1970). Use of dummy variables in testing for equality between
sets of coefficients in two linear regressions: a note. *American Statistician*,
24, 50-2.
Jenkinson, P. & Parry, D.W. (1994). Splash dispersal of conidia of *Fusarium
culmorum* and *Fusarium avenaceum*. *Mycological Research*, **98**, 506-10.
Locke, T., Moon, L.M. & Evans, J. (1987). Survey of benomyl resistance in
Fusarium species on winter wheat in England and Wales in 1986. *Plant
Pathology*, **36**, 589-93.
Magan, N. & Lacey, J. (1984). Water relations of some *Fusarium* species from
infected wheat ears and grain. *Transactions of the British Mycological
Society*, **83**, 281-5.
Millar, C.S. & Colhoun, J. (1969). *Fusarium* diseases in cereals VI.
Epidemiology of *Fusarium nivale* on wheat. *Transactions of the British
Mycological Society*, **52**, 195-204.
Papendick, R.I. & Cook, R.J. (1974). Plant water stress and development of
Fusarium foot rot in wheat subjected to different cultural practices.
Phytopathology, **64**, 358-63.
Parry, D.W. (1990). The incidence of *Fusarium* spp. in stem bases of selected
crops of winter wheat in the Midlands, UK. *Plant Pathology*, **39**, 619-22.
Parry, D.W., Bayles, R.A. & Priestley, R.H. (1984). Resistance of winter wheat
varieties to ear blight (*Fusarium culmorum*). *Journal of the National
Institute of Agricultural Botany*, **16**, 465-8.

Parry, D.W., Pettitt, T.R., Jenkinson, P. & Lees, A.K. (1994). The cereal *Fusarium* complex. In *Ecology of Plant Pathogens*, ed. J.P. Blakeman & B. Williamson, pp. 301-320. CAB International, UK.

Pettitt, T.R., Parry, D.W. & Polley, R.W. (1993). Improved estimation of the incidence of *Microdochium nivale* in winter wheat stems in England and Wales, during 1992, by use of benomyl agar. *Mycological Research*, **97**, 1172-4.

Pettitt, T.R., Parry, D.W. & Polley, R.W. (1995). Effect of temperature on the incidence of nodal foot rot symptoms in winter wheat crops in England and Wales caused by *Fusarium culmorum* and *Microdochium nivale*. *Agricultural and Forest Meteorology*, In Press.

Polley, R.W. & Slough, J.E. (1992). Survey of winter wheat diseases in England and Wales. Report CSL, MAFF, Harpenden, UK, 44pp.

Polley, R.W. & Thomas, M.R. (1991). Surveys of diseases of winter wheat in England and Wales, 1976-1988. *Annals of Applied Biology*, **119**, 1-20.

Polley, R.W., Turner, J.A., Cockerell, V., Robb, J., Scudamore, K.A., Sanders, M.F. & Magan, N. (1991). Survey of *Fusarium* species infecting winter wheat in England, Wales and Scotland, 1989 & 1990. *Home-Grown Cereals Authority Project Report*, **39**.

Rennie, W.J., Richardson, M.J. & Noble, M. (1983). Seed-borne pathogens and the production of quality seed in Scotland. *Seed Science and Technology*, **11**, 1115-27.

Zadoks, J.C., Chang, T.T. & Konzak, C.F. (1974). A decimal code for the growth stages of cereals. *Weed Research*, **14**, 415-21.

3

Effects of UV-B radiation (280–320 nm) on foliar saprotrophs and pathogens

P. G. AYRES, T. S. GUNASEKERA,
M. S. RASANAYAGAM, AND N. D. PAUL

Ozone (O₃), ultra-violet (UV) radiation and climate change

The earth's atmosphere contains about 3 nl l^{-1} O_3, that fraction being continuously turned over. Approximately 10% is dispersed in the troposphere, which extends 15 km above the ground, and 90% in the stratosphere, which extends up to 50 km. While man's activities are causing some increase in tropospheric O_3, mainly around major urban areas, they are depleting stratospheric O_3 on a global scale (Anon., 1993a,b). Depletion occurs because of the release of chlorine-containing compounds, such as chlorofluorocarbons and carbon tetrachloride, that promote O_3 breakdown.

Solar radiation includes wavelengths as short as 200 nm but that below approximately 290 nm is absorbed in the atmosphere, mainly by O_3. Since energy per quantum of radiation increases as wavelength decreases, O_3 protects organisms from the most energy-rich, and potentially most damaging, wavelengths in solar radiation. The efficiency with which ozone absorbs UV decreases as wavelength increases, so progressive thinning of the O_3 layer will allow shorter wavelengths to reach the earth's surface as well as allowing higher fluxes to be transmitted of the wavelengths already penetrating. It is significant that the cut-off wavelength is not constant but varies with season and time of day. For example, in Reading, England (51.5° N), the shortest detectable wavelength varied from 302 nm in January to 294 nm in July, and from 294 nm at noon to 300 nm at 17.00 hours in July (Anon., 1993b). As these fluctuations indicate, solar elevation affects fluxes. Hence differences in latitude are largely responsible for the range of fluxes that occurs between different parts of the world, while at any location the daily range will depend upon

32

season and time of day and, of course, cloudiness. Over half the UV-B radiation reaching the earth's surface is diffuse (Anon., 1993*b*).

The other factor explaining geographical variation is that the O_3 layer is thinnest in the tropics, in spite of most being formed in that area, and thickest towards the poles, towards which it is carried by stratospheric air currents. Therefore, quite normally, UV fluxes are much greater at the equator than at the poles. Depletion of stratospheric ozone was first noted over Antarctica, but both satellite and ground-based measurements now show that thinning is occurring also over the Arctic and is extending to northern temperate latitudes. From 1979 to 1989, the annual DNA-weighted dose of UV-B (weighting is explained below) increased by about 10% in the northern polar region, 5% at 30° N and 30° S, 15% at 55° S and 45% at 85° S, with no significant trends yet in the tropics (Anon., 1993*b*).

The simple inverse relationship between energy per quantum and wave-length of solar radiation is, unfortunately, not sufficient on its own to explain the injury to biological systems caused by shorter wavelength UV radiation. Responses to UV are strongly wavelength dependent, as are those of photosynthetic organisms to visible light (400–700 nm, photo-synthetically active radiation, PAR), but that dependency is modified by the absorption characteristics of the molecules or structures that are directly affected. The relative effectiveness of different wavelengths in eliciting each particular biological response should be expressed in an action spectrum, but spectra are very difficult to prepare, with the result that few are available to date. The most commonly used are the Erythemal–Diffey (sunburn), the DNA damage, and the plant damage action spectrum (PAS) (Caldwell, 1971) (Fig. 3.1). The practical value of such spectra is that they can be used as weighting functions. Thus, if the relative effectiveness at each wavelength is multiplied by the irradiance (flux) at that wavelength, a measure of the biological effectiveness of each wavelength is obtained. Spectra also make it possible to compare the effects of different lamp systems or to compare lamp systems and natural sunlight. As yet, there is no detailed action spectrum specifically prepared for a fungus, so the general DNA damage spectrum is often used in dose–response studies. The dose is simply the weighted flux multiplied by the period of exposure.

Neither currently, nor in the foreseeable future, does O_3 depletion affect biologically effective fluxes of UV-A wavelengths (320–400 nm). However, because biological phenomena have action spectra with broad, or sometimes several, peaks the processes affected by UV-A, and by

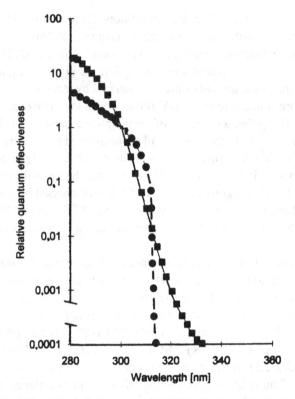

Fig. 3.1. Action spectra for DNA damage (■) (Setlow, 1974) and general plant damage (●) (Caldwell, 1971).

UV-C wavelengths (200–280 nm), can also be affected by UV-B wavelengths and vice versa. Thus, as discussed later, sporulation may be affected by both UV-B and UV-A wavelengths, while some damage to DNA may be caused by UV-B wavelengths as well as UV-C.

Cellular damage and repair mechanisms

Laboratory evidence is overwhelming that killing of fungi by UV-C (usually supplied by a germicidal lamp with an emission peak at 254 nm) results from the formation of thymine dimers within the DNA chain (see Owens & Krizek, 1980 for references). Dimers distort the helix and cause mutation or the failure of transcription. Because the DNA action spectrum is broad, and includes UV-B wavelengths at its upper extremity, UV-B in sunlight can also cause some dimer formation. It is widely believed that small cells are more prone than large cells to damage in this way because the small cell has less volume and a shorter pathlength in which UV-absorbing pigments may be laid down to protect the DNA. Similarly, unicellular organisms may be more at risk of damage than multicellular organisms or structures where outer cells can protect inner cells.

Many organisms, including man, have enzymic repair mechanisms that delete and replace the affected base pairs, indicating that dimer formation can occur continuously at low levels in healthy organisms. Although there is evidence from a few microorganisms, e.g. *Escherichia coli* A (Webb, Brown & Ley, 1982), that large doses of UV radiation at 365 nm (UV-A) can also promote dimer formation, much more commonly repair is promoted by longer wavelength radiation, the photolyase enzymes responding to even very low doses of UV-A or white light. Accordingly, for *Saccharomyces cerevisiae* and the blue-green alga *Gloeocapsa alpicola* (O'Brien & Houghton, 1982), and for organisms such as *E. coli* B (Castellani, Jagger & Setlow, 1964), UV-C followed by UV-A is much less damaging than UV-C alone, and photoreactivation may occur. Studies have failed to detect photoreactivation in other microorganisms, e.g. *Clostridium butyricum* (Carrasco, 1989). The effects that UV radiation has on other cellular structures have been largely ignored but may be at least as important where the effects of UV-B, rather than UV-C, are concerned.

Methods and instrumentation

In field studies of the biological effects of UV, natural sunlight may be either supplemented by lamp sources or depleted by screens (large filters). In the laboratory, natural sunlight has to be replaced by lamps, often used in conjunction with various filters. A problem is that, although lamp types are selected according to the wavelengths at which specified peaks of emission occur, all emit lower fluxes at other wavelengths. For example, fluorescent 'black lights' often used to promote fungal sporulation have an emission peak (typically 360–370 nm) in the UV-A region but emit some radiation as low as 320 nm. There are additional problems associated with the use of filters. Those of the transmission type cannot be assumed to cut off all wavelengths below the value specified for them; the relationship between transmission and wavelength is a sigmoid curve (albeit steep) so they transmit a little radiation beyond the stated cut-off wavelength. Filters of the band-pass type have a transmission vs. wavelength curve that is normal shaped, so although there is an

acknowledged spread of radiation around the wavelength of peak transmission, e.g. ±5 nm at 50% transmission, there is a much greater spread
(wider waveband transmitted) at lower transmissions. Precision can be
achieved using a lamp source and monochromator(s), but this is appropriate only for the most detailed work, e.g. the construction of a full action
spectrum, and is very expensive. A further consideration for laboratory
studies is the emission spectrum of any white light source used to induce
photorepair mechanisms and/or to simulate natural sunlight.

Comparison of UV treatments is facilitated if flux, the rate at which
radiant energy is delivered (W m^{-2} = J m^{-2} s^{-1}), is measured with a
spectroradiometer that contains a double monochromator and expressed
in weighted units, or, less ideally, is measured with a broad band sensor
calibrated for an appropriate action spectrum against a spectroradiometer. Add to this the fact that the lower limit of the UV-B band is
sometimes regarded as 290 nm, and the upper limit as 315 nm, and it
becomes clear that the results of UV research have to be interpreted with
exceptionally careful attention to the methods used.

The effects of a particular dose of UV-B (dose = flux × time) on a
plant may to some extent depend on the accompanying PAR. In field
studies of the phylloplane inhabitant *Sporobolomyces roseus* (pink yeast),
the authors depleted UV-B in sunlight by placing over test plants screens
that held either an acrylic plastic (ICI) that transmitted very little below
320 nm, or Teflon which transmits down to approximately 290 nm. It is
significant that Teflon transmits >90% PAR because the effect of a
particular UV-B flux on a plant is to some extent dependent upon the
accompanying PAR flux. To determine whether the screen had an effect
on the microclimate, the effects of the latter screen were compared with
those obtained under unfiltered sunlight. In support of conclusions from
field experiments, pure cultures of fungi were irradiated with broad band
UV (using either Philips TL UV-B 40/12W or 36W/08 fluorescent tubes
for UV-B and UV-A, respectively) filtered with cellulose diacetate to
remove all radiation below 292 nm. To obtain a more detailed spectral
response for pink yeast, cultures were irradiated with UV from a xenon
arc source after passage through 10 nm band-pass filters.

UV-B and phytopathogenic fungi

Direct effects

There has been a long-held belief, particularly among plant pathologists, that sunlight is detrimental to spore germination, but, apart from the fact that the confounding effects of temperature have often been ignored, the role of UV-B has seldom been distinguished from that of UV-A , or control treatments have been based on the total, and hence unrealistic, exclusion of UV-B, e.g. Rotem, Wooding and Aylor (1985). An early exception was the work of Maddison and Manners (1973) in which, as part of a major investigation of the effects of sunlight and UV-C on rust fungi, they clearly demonstrated by use of cut-off filters to modify both sun and lamp light, that UV-B (0.4 W cm^{-2} at 300 nm and 5.8 W cm^{-2} at 310 nm) inhibited by 90% the germination of spores of *Puccinia striiformis*. Several experiments showed that *P. recondita* and *P. graminis tritici* were rather less sensitive than *P. striiformis* and, whereas the 10 nm interval action spectrum for *P. striiformis* peaked around 265 nm, that for *P. graminis* peaked at 280 nm. In both these species, a mixture of UV-A and PAR caused photoreactivation of UV-C induced damage.

Using Mylar filters of much larger area, held in screens above bean (*Phaseolus vulgaris*) plants in the field, Caesar and Pearson (1983) demonstrated that the UV-B component of sunlight reduced the viability of ascospores of *Sclerotinia sclerotiorum*. The result was confirmed in the laboratory using a xenon arc lamp with cellulose acetate filters to remove all wavelengths below 290 nm. Examining germination of *Cladosporium cucumerinum*, also *in vitro* using filtered radiation from a xenon arc source, Owens and Krizek (1980) found germination was delayed (although the delay was reversed by white light) and final levels of germination were inhibited by 5 min radiation from 265-295 nm (0.57 W m^{-2}), while 5 min radiation from 300–330 nm (520 W m^{-2}) also delayed germination (in a way not reversed by white light) but had little effect on final levels of germination. Construction of a 10 nm interval action spectrum from 260–320 nm indicated that in a natural environment the effects of wavelengths in the region 290–295 nm would be the most effective. Since germination was delayed rather than abolished, they proposed that UV-B affected protein synthesis by post-transcriptional effects. They established, as noted above, the important point that UV-B can have effects other than on DNA. They also established that there was reciprocity between flux and period of exposure in their system (this was true at 325 nm, although not at 265 nm). This implies that a clear day in spring, when solar elevation and flux are low, can be just as damaging as a short period of sunshine on a summer day, when solar elevation and flux are higher, because it delivers the same dose. It is not known yet whether reciprocity extends to other responses in other fungi.

UV-B may have positive as well as negative effects on fungal development. Action spectra prepared at 5 nm intervals show a major peak at 287 nm with a minor shoulder at 300 nm for perithecium production by *Leptosphaerulina trifolii* in culture (Leach & Anderson, 1982) and a major peak at 283 nm with a minor shoulder at 303 nm for conidial production by *Botrytis cinerea* (Honda & Yunoki, 1978). Other peaks occur in the UV-C region, especially for *B. cinerea*, and the unweighted doses for *B. cinerea* appear high, but clearly reproduction of these fungi is likely to be affected in current climate-change scenarios.

It is equally clear that, for some fungi, UV-B has no, or little, effect on development even at relatively high doses. Asthana and Tuveson (1992) found that UV-B (1.1 J m^{-2} s^{-1}), and similarly UV-A, had no effect on colony formation from conidia of the citrus pathogens *Fusarium oxysporum, F. solani, Penicillium italicum* and *P. digitatum*. Exposure of conidia of *Diplocarpon rosae* (black spot) to UV-B (12.2 mW m^{-2}) had no effect on their infectivity on detached rose leaves (Semeniuk & Stewart, 1981), although a high flux given 6–18 h after inoculation did reduce symptom development, an effect attributed to the sensitivity of the germ tubes which would have been exposed on the leaf surface.

Indirect effects

Following their observation of the promotive effects of UV on sporulation of *B. cinerea*, Honda, Toki and Yunoki (1977) achieved control of grey mould of tomato and cucumber by covering an experimental greenhouse in a vinyl film that did not transmit radiation below 290 nm. Orth, Teramura and Sisler (1990) concluded that an unnaturally high dose of UV-B (11.6 kJ m^{-2} d^{-1} weighted to the plant action spectrum) promoted disease if given before inoculation of cucumber leaves with the pathogens *Colletotrichum lagenarium* and *Cladosporium cucumerinum* but reduced infection if given after inoculation, because of its inhibitory effects on fungal development. Their observations were consistent with those of Owens and Krizek (1980) made on the fungus alone, described above, and those of Carns, Graham and Ravitz (1978) on the same host/pathogen combinations. The importance of methodology has been mentioned before, so it is noteworthy that Orth *et al.* (1990) used an unsatisfactory design in which controls had no UV-B, as they were grown in a glasshouse without any supplementary UV-B. Thus, their attribution of the effects of preinoculation UV-B to changes in cuticle or trichomes may have been correct, but it does not bear directly upon natural conditions.

In many other diseases it is more difficult to disentangle the effects of UV on the fungus from the effects UV might have on the plant. The authors have found that, if beans (*Vicia faba*) inoculated with rust (*Uromyces viciae-fabae*) were placed in the field under screens that removed UV-B, sporulating pustules were quicker to develop. Also, if leaf discs taken from these plants were incubated for 24 h under white light in the humid environment of a Petri dish, before spore production was measured, more spores per pustule were produced, in the first days of sporulation, from discs taken from plants screened from UV-B than from those from plants exposed to UV-B (Table 3.1). This is the only demonstration known to the authors that UV-B can affect the sporulation of an obligately biotrophic pathogen. As yet, it is not known whether UV-B influences the host, pathogen or both.

One way in which UV-B can affect the plant's response to infection is by activating particular metabolic pathways concerned with defence, for example, the synthesis of certain isoflavonoid and sesquiterpene phytoalexins requires UV-B (see Downum, 1992 for references). Each photoactivated pathway, or end product, has its own action spectrum, which provides a further reason why the relationship between response and wavelength is not simply linear (see p. 33). The photosensitivity of fungal products may add to the complexity of fungus/plant relationships. Irradiation of sugar beet at a flux equivalent to that expected after 9% ozone depletion in Lund, Sweden, had no damaging effect on plant growth but did exacerbate damage caused by *Cercospora* leaf spot infection (Panagopoulos, Bornman & Bjorn, 1992). The mechanism of that increased injury was not ascertained but, since members of the genus *Cercospora* are known to produce such compounds (Steinkamp *et al.*, 1981), it is possible that UV-B activated a photosensitive toxin.

Phylloplane fungi: species differ in sensitivity to UV

Unlike pathogens or decomposer saprotrophs, which for most of their life can grow within their substratum, phylloplane organisms are typically exposed to long periods of direct and diffuse sunlight. They may be well

Table 3.1. *Effects of screens (C = control, no screen; T = teflon, + UV-B; A = acrylic, − UV-B) over Vicia faba plants in the field on number of pustules of rust Uromyces viciae-fabae, and spores produced per pustule. Mean values ± S.E. (n=4)*

	Days after inoculation								
	23			29			35		
	C	T	A	C	T	A	C	T	A
Pustules cm^{-2}	15 ± 2	24 ± 3	34 ± 3	37 ± 4	35 ± 3	38 ± 4	–	–	–
Spores $\times 10^3$ pustule^{-1}	–	–	–	3.7 ± 1.3	5.0 ± 0.7	15.6 ± 7.1	2.6 ± 0.7	2.5 ± 0.7	5.3 ± 0.8

adapted to the current UV-B environment but could be especially vulnerable to climate change. The authors' studies centred on pink yeast because it is an ubiquitous member of the phylloplane flora, occurring on both dicotyledonous and monocotyledonous plants in temperate and tropical latitudes (Last & Price, 1969), and because it is a known antagonist of several common pathogenic fungi, for example, *Septoria nodorum* (Fokkema & Van der Meulen, 1976).

Whether leaf discs were taken from naturally colonised apple (*Malus domestica*) (Fig. 3.2) and oak (*Quercus robur*) seedlings, or pea (*Pisum sativum*) and broad bean (*Vicia faba*) plants, numbers of ballistospore-derived colonies were consistently most numerous when discs were taken from plants kept under an acrylic screen (-UV-B), and least numerous when discs came from plants kept under a teflon (+ UV-B) screen or that were unscreened (control). There was no consistent difference between colony numbers from control and teflon (+ UV-B) treatments.

In a second experiment, numbers of pink and white yeast (*Bullera alba*) were measured at weekly intervals after *V. faba* plants were sprayed separately with each fungus (monoculture) or with a 1 : 1 mixture.

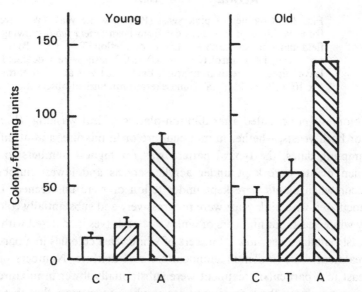

Fig. 3.2. Colonies of pink yeast developing from ballistospores. Discs were taken from young, newly expanded, or old, yellowing leaves of apple seedlings naturally infected in the field before being placed under an acrylic (A = −UV-B) or a teflon (T = + UV-B) screen, or no screen, i.e. control (C = + UV-B). (T.S. Gunasekera, unpublished, observations.)

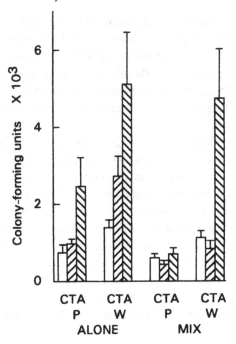

Fig. 3.3. Colonies of pink yeast (P) and white yeast (W) developing from washings of leaf discs cut from bean (*Vicia faba*) growing in the field under an acrylic (A = −UV-B) or a teflon (T = + UV-B) screen, or no screen, i.e. control (C = + UV-B). Yeasts were inoculated on to leaves either alone or in mixture. Each yeast was at a concentration of 5×10^6 cells cm^{-3}. (T.S. Gunasekera, unpublished observations.)

Colonies were counted after dilution plating of leaf washings (Fig. 3.3). For both yeasts, whether in monoculture or in mixture, and at different sampling dates, the typical pattern was for highest numbers to occur when plants were kept under acrylic screens and lowest numbers to occur when plants were kept under teflon or were unscreened. Leaves inoculated with pink yeast were progressively and substantially colonized by white yeast, but numbers of pink yeast on leaves inoculated with white yeast remained very low. Numbers of white yeast colonies in a particular treatment were similar in monoculture and mixture. Numbers of pink yeast in a particular treatment were substantially lower in mixture than in monoculture; the reduction under acrylic was so great that there were no differences between pink yeast numbers under different screening treatments.

Whereas exposure of pink yeast inoculated onto agar to relatively high fluxes of broad band UV-A radiation (5.7–7.7 W m^{-2}) for periods of up to 16 h had no effect on colony numbers, exposure to broad band UV-B at a relatively low flux (0.1 W m^{-2}) delivered across the UV-B range for periods up to 24 h before colonies were returned to the dark showed that survival declined with the period of exposure. Fewer than 5% of colonies survived if the earlier period of exposure was 16 h (dose 4.38 kJ m^{-2} UV-B$_{DNA}$), but more than 50% of colonies survived when the period was only 12 h (3.28 kJ m^{-2}). When exposure was for 8 h (2.19 kJ m^{-2}), as many colonies survived in UV-B as in dark controls. If periods of 4 h white light occurred between four periods of 4 h exposure to UV-B, colony numbers per plate (275 ± 10) were as high as in dark controls (270 ± 16), that is, damage caused by UV-B was repaired.

When pink and white yeast inoculated onto agar were exposed to a range of UV-B fluxes, including current midsummer fluxes at Lancaster (0.076 W m^{-2}), each delivered for 6 h at the beginning of a 48 h period, colony number (Fig. 3.4) and cell number per colony (data not shown) of both yeasts decreased with increasing dose. White yeast was the more sensitive, with no colonies being counted at and above 0.185 W m^{-2} (dose 4.0 kJ m^{-2} d^{-1}) (Fig. 3.4(*c*)) in the dark. White light provided as a background to UV-B had a small but insignificant positive effect on numbers of colonies. However, it significantly increased cell number per colony of pink yeast in the dose range 1.0 to 2.0 kJ m^{-2} d^{-1}, indicating again that in this fungus white light can repair or prevent UV-B induced injury.

When *S. roseus* and *B. alba* were separately inoculated onto agar and exposed to different wavebands of UV radiation at 1.1 mW m^{-2} unweighted for 6 h, before being grown in the dark for a further 18 h, colony survival was, by comparison with controls in the dark, significantly reduced at 310 nm and all shorter wavebands (Fig. 3.5). No colonies were counted at 290 nm or below, although some were detected after longer dark incubation of plates exposed at 290 nm.

UV-B is only one among several environmental factors, for example, temperature and humidity (Bashi & Fokkema, 1977), that naturally regulate pink yeast numbers. Competition with other leaf-surface micro-organisms is another factor determining numbers. However, it cannot be safely concluded that increased UV-B radiation, arising from ozone depletion, will promote those phytopathogens to which pink yeast is antagonistic, for example *Septoria nodorum*, because the effects of UV-B on those other organisms are largely unknown and might be greater than those on pink yeast. It can more safely be inferred that a changing UV-B environment will alter the balance between pink and white yeasts because they occupy the same ecological niche on leaf

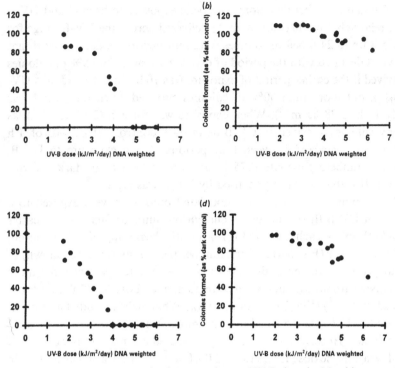

Fig. 3.4 Response of colony number to UV-B dose in (a) white yeast in the light, (b) pink yeast in white light, (c) white yeast in the dark and (d) pink yeast in the dark. Colonies formed are related to those in dark controls. The experiment was repeated with similar results. (T.S. Gunasekera, unpublished observations.)

surfaces. White yeast was dominant in the absence but not in the presence of UV-B. Laboratory studies demonstrated clearly that white yeast was much more sensitive than pink yeast to UV-B, possibly because it lacks critical carotenoid pigments possessed by pink yeast (Simpson, Chichester & Phaff, 1969). It is not known whether UV-B promotes carotenoid synthesis in *S. roseus* (visual observations suggest it does), but in the yeast *Rhodotorula minuta* the action spectrum for photo-induced carotenogenesis has a prominent, broad peak centred on 280 nm (Tada, Watanabe & Tada, 1990). Minor peaks occur at 340, 370 and 400 nm.

Differences in UV sensitivity between fungi have often before been attributed to differences in spore pigmentation (Maddison & Manners, 1973 ; Carns, Graham & Ravitz, 1978; Asthana & Tuveson, 1992), with

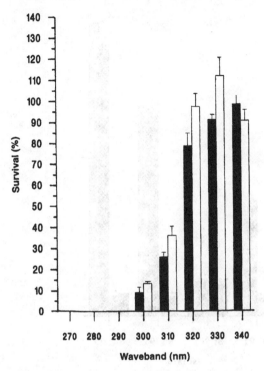

Fig. 3.5. A 10 nm-interval action spectrum for colony development of pink (filled bars) and white (open bars) yeast. Survival is related to that in dark controls. Bars show the S.E. of four determinations. (T.S. Gunasekera, unpublished observations.)

resistance being linked to the presence of carotenoid pigments. The photoreceptors are unknown, although Hsiao and Bjorn (1982) suggested that in *Verticillium agaricinum* the photoreceptor, whose induction spectrum peaked at 290 nm, is a pteridine.

Phytopathogenic fungi; species and isolates differ in sensitivity to UV-B

When pycnidiospores of two species in the same genus, *Septoria tritici* (two isolates from the UK, two from Tunisia) and *S. nodorum* (two isolates from the UK), were exposed to broad band UV-B radiation while germinating on agar, there was no effect on germination after 48 h of either species when a dose of 3 kJ m^{-2} d^{-1} UV-B$_{DNA}$ was given over 4 h (Rasanayagam *et al.*, 1995). The maximum natural dose in Tunisia is

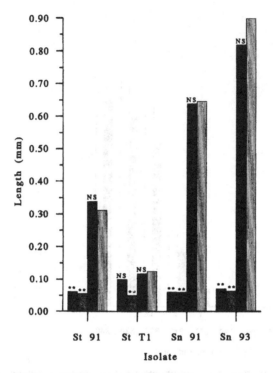

Fig. 3.6. Effect of broad band UV-B radiation on germ-tube growth of *Septoria nodorum* (Sn) and *S.tritici* (St). Treatments were: UV against a white light (■) or dark (▨) background, UV passed through a Mylar filter to remove all wavelengths below 320 nm (▨), and a dark control (▨). Four replicates per treatment.
**, * significantly different at P = 0.01, 0.05, respectively ; NS, not significantly different at P = 0.05. (After Rasanayagam *et al.*, 1995.)

approx. 2.4 kJ m^{-2} d^{-1}. However, a difference arose between species if exposure was continuous (18 kJ m^{-2} d^{-1}), because germination was inhibited in *S. nodorum*. For example, it was 23.5% compared with 79.5% for the dark control in one isolate. The shorter exposure time, 4 h, inhibited extension of germ tubes *in vitro* in all UK isolates of both species but failed to inhibit extension in the Tunisian isolates of *S. tritici* (Fig. 3.6). Thus, differences in sensitivity to UV-B can exist within species. The observation that it was the Tunisian isolates from the higher UV-B environment that had the lower sensitivity raises the intriguing possibility that fungal strains can adapt to local conditions.

Conclusions

The authors in reviewing their own work have concentrated on the direct effects of UV-B on fungi that colonize leaves. It must not be forgotten that UV has additional direct effects on plant material which indirectly affect both saprotrophic and pathogenic fungi.

The effects that UV-B has on some phytoalexins has already been mentioned but of more widespread importance may be the fact that many plants protect themselves from harmful UV by accumulating flavonoids in epidermal cell vacuoles, chemicals which probably also protect against invading fungi. What is unknown is the extent to which UV-induced changes in plant growth rate and form affect the synthesis of materials of nutrient value to pathogens as well as to leaf surface and decomposer saprotrophs. Whether UV-induced changes in the leaf cuticle affect the supply of nutrients from inside the leaf to fungi on the surface is unknown. However, increasing UV-B can alter the structure and composition of leaf waxes. In tobacco, increasing plant-weighted UV-B doses from 4.5 to 5.7 kJ m^{-2} d^{-1} decreased the amount of leaf surface waxes by 45% and substantially increased wettability (static contact angle decreased from 106 °C to 70 °C, Barnes *et al.*, 1994). It is worth noting that comparable changes in leaf surface properties, brought about in oilseed rape by the herbicide dalapon, resulted in a very marked increase in the prevalence and severity of light leaf spot, caused by the splash-dispersed pathogen *Pyrenopeziza brassicae* (Rawlinson, Muthyala & Turner, 1978), Thus, it seems possible that increased UV-B may result in an increase in diseases caused by splash-dispersed pathogens, through increased retention of spores on leaf surfaces. Possibly also altered are those topographical characteristics of the leaf surface that some pathogens use as 'signposts' along the infection pathway.

UV can affect dead as well as living plant material. The contribution of UV-B to the photodegradation of natural materials in the field is poorly characterized but degradation of paper (prepared from *Pinus taeda* pulp) was considerably accelerated by increased UV-B (Andrady *et al.*, 1991), and T.C. Callaghan (pers. comm.) reported that accelerated decomposition in a dry tundra ecosystem was attributable to increased UV-B-induced photodegradation. Thus, UV-B may affect the quality of the resource for decomposer fungi.

In summary, changes in the UV-B component of climate may have both direct and indirect effects on plant–microbe interactions. Indirect effects, mediated through the physiological, chemical and morphological

48   *P. G. Ayres* et al.

responses of the plant to UV-B, may alter the nature of the leaf from birth to decay. Direct effects may involve individual microorganisms, whether specialized pathogens or members of the phyllo- plane community, and may also alter the interaction between species. Direct and indirect effects clearly do not act in isolation, for example, in later stages of the life of the leaf, changes in UV-B could affect the unspecialized pathogens that invade the senescing leaf, hastening its death, but persisting themselves as saprotrophs. Thus, normal patterns of succession of decay-causing organisms might be disturbed both before leaf fall and as long as litter is exposed to sunlight. Investigation of the complex and varied effects of UV-B on plant–microbe interactions is just beginning and poses a notable challenge, not least in terms of the technology to be used. However, without an improved insight into this element of biological responses to UV-B, our overall understanding of the possible consequences of ozone depletion may be significantly compromised.

Acknowledgements

We thank the Tea Research Institute, Sri Lanka (T. S. Gunasekera), Agriculture & Food Research Council (M. S. Rasanayagam) and the Department of the Environment (N. D. Paul) for support.

References

Andrady, A. L., Parthasary, V., Song, Y., Song, Y., Fueki, K. & Torikai, A.. (1991). Photoyellowing of mechanical pulps. Wavelength sensitivity to light-induced yellowing by monochromatic radiation. *Technical Association of the Pulp and Paper Industry, Report.* New York.
Anon (1993*a*). *Environmental UV Radiation. Causes–Effects–Consequences*, ed. J. Acevado & C. Nolan. Report of the European Commission Directorate-General XII for Science, Research and Development. Brussels.
Anon (1993*b*). *Stratospheric Ozone 1993.* Report of the United Kingdom Stratospheric Ozone Review Group. London: Her Majesty's Stationery Office.
Asthana, A. & Tuveson, R. W. (1992). Effects of phototoxins on selected fungal pathogens of citrus. *International Journal of Plant Science*, **153**, 442–52.
Barnes, J., Paul, N. D., Percy, K., Broadbent, P., McLaughlin, C. & Wellburn, A. R. (19944). Effects of UV-B radiation on wax biosynthesis. In *Air Pollutants and the Leaf Surface*, ed. N. Cape, T. Jagels & K. Percy. NATO-ASI Series. Springer-Verlag, Berlin. (in press).
Bashi, E. & Fokkema, N.J. (1977). Environmental factors limiting growth of *Sporobolomyces roseus*, an antagonist of *Cochliobolus sativus* on wheat leaves. *Transactions of the British Mycological Society*, **68**, 17-25.

Caesar, A.J. & Pearson, R.C. (1983). Environmental factors affecting survival of ascospores of *Sclerotinia sclerotiorum*. *Phytopathology*, **73**, 1024-30.
Caldwell, M.M. (1971). Solar UV irradiation and the growth and development of higher plants. In *Photophysiology* vol. 6, ed. A.C. Giese, pp. 131-177. New York: Academic Press.
Carns, H.R., Graham, J.H. & Ravitz, S.J. (1978). Effect of UV-B radiation on selected leaf pathogenic fungi and on disease severity. EPA-IAG-D6-0168 Report. Environmental Protection Agency, Washington, DC.
Castellani, A., Jagger, J. & Setlow, R. B. (1964). Overlap of photo-reactivation and liquid holding recovery in *E. coli* B. *Science*, **143**, 1170–1.
Carrasco, A. (1989). Photoreactivating capacity of *Clostridium butyricum* and *Clostridium acetobutyricum, Letters in Applied Microbiology*, **8**, 131–4.
Downum, K.R. (1992). Light-activated plant defence. *New Phytologist*, **122**, 401-20.
Fokkema, N.J. & Van der Meulen, F. (1976). Antagonism of yeast-like phyllosphere fungi against *Septoria nodorum* on wheat leaves. *Netherlands Journal of Plant Pathology*, **82**, 13-16.
Honda, Y., Toki, T. & Yunoki, T. (1977). Control of gray mould of greenhouse cucumber and tomato by inhibiting sporulation. *Plant Disease Reporter*, **61**, 1041-4.
Honda, Y. & Yunoki, T. (1978). Action spectrum for photosporogenesis in *Botrytis cinerea* Pers. ex Fr. *Plant Physiology*, **61**, 711-13.
Hsiao, K.C. & Bjorn, L.O.(1982). Aspects of photoinduction of carotenogenesis in the fungus *Verticillium agaricinum*. *Physiologia Plantarum*, **54**, 235-8.
Last, F.T. & Price, D. (1969). Yeasts associated with living plants and their environs. In *The Yeasts*, vol.1, ed. A.H. Rose & J.S. Harrison, pp. 183-218. London: Academic Press.
Leach, C.M. & Anderson, A.J. (1982). Radiation quality and plant diseases. In *Biometeorology in Integrated Pest Management*, ed. J.L.Hatfield & I.J. Thomason, pp. 267-306. New York: Academic Press.
Maddison, A.C. & Manners, J.G.(1973). Lethal effects of artificial ultraviolet radiation on cereal rust uredospores. *Transactions of the British Mycological Society*, **60**, 471-94.
O'Brien, P. & Houghton, J.A. (1982). Photoreactivation and excision repair of UV induced pyrimidine dimers in the unicellular cyanobacterium *Gloeocapsa alpicola* (*Synechocystis* PCC 6308). *Photochemistry and Photobiology*, **35**, 359-64.
Orth, A.B., Teramura, A.H. & Sisler, H.D. (1990). Effects of ultraviolet-B radiation on fungal disease development in *Cucumis sativus*. *American Journal of Botany*, **77**, 1188-92.
Owens, O.V.H. & Krizek, D.T. (1980). Multiple effects of UV radiation (265–330 nm) on fungal spore emergence. *Photochemistry and Photobiology*, **32**, 41-9.
Panagopoulos, I., Bornman, J.F. & Bjorn, L.O. (1992). Response of sugar beet plants to ultraviolet B (280–320 nm) radiation and *Cercospora* leaf spot disease. *Physiologia Plantarum*, **84**, 140-5.
Rasanayagam, M.S., Paul, N.D., Royle, D.J. & Ayres, P.G. (1995). Variation in responses of spores of *Septoria tritici* and *S. nodorum* to UV-B irradiation *in vitro*. *Mycological Research* (in press).
Rawlinson, C.J., Muthyala, G. & Turner, R.H. (1978). The effects of herbicides on epicuticular wax of winter oilseed rape (*Brassica napus*) and infection

by *Pyrenopeziza brassicae*. *Transactions of the British Mycological Society*, **71**, 441-51.

Rotem, J., Wooding, B. & Aylor, D.E. (1985). The role of solar radiation, especially ultraviolet, in the mortality of fungal spores. *Phytopathology*, **75**, 510-14.

Semeniuk, P. & Stewart, R.N. (1981). Effect of ultraviolet (UV-B) irradiation on infection of roses by *Diplocarpon rosae* Wolf. *Environmental and Experimental Botany*, **21**, 45-50.

Setlow, R.B. (1974). The wavelengths in sunlight effective in producing skin cancer. A theoretical analysis. *Proceedings of the National Academy of Sciences*, **71**, 3363-6.

Simpson, K.L. Chichester, C.O. & Phaff (1969). Carotenoid pigments of yeasts. In *The Yeasts*, vol. II, ed. A.H. Rose & J.S. Harrison, pp. 493-513. London: Academic Press.

Steinkamp, M.P., Martin, S.S., Hoefert, L.L. & Ruppell, E.G. (1981). Ultrastructure of lesions produced in leaves of *Beta vulgaris* by cercosporin: a toxin from *Cercospora beticola*. *Phytopathology*, **71**, 1272-81.

Tada, M., Watanabe, M. & Tada, Y. (1990). Mechanism of photoregulated carotenogenesis in *Rhodotorula minuta*. VII. Action spectrum for photoinduced carotenogenesis. *Plant Cell Physiology*, **31**, 241-6.

Webb, R.B., Brown, M.S. & Ley, R.D. (1982). Non-reciprocal synergistic lethal interaction between 365 nm and 405 nm radiation in wild type and vvr strains of *E. coli*. *Photochemistry and Photobiology*, **35**, 697-703.

4

Implications of global warming and rising sea-levels for macrofungi in UK dune systems

M. ROTHEROE

Introduction

The effects of environmental change on the sand-dune ecosystem do not have to be considered hypothetically; they can be observed. Sand dunes are, by definition, dynamic systems, being created, shaped, modified, sustained or eroded by continual changes in a wide range of abiotic and biotic environmental factors. The fungi which inhabit this coastal ecosystem are subject to, and are part of, those same changes. Frankland (1981) stated that 'community life for a fungus is dynamic'. This is nowhere more true than in sand-dune habitats where the combination of a dynamic biological organism functioning in a dynamic resource system constitutes a veritable dynamic duo, capable of long-term survival and viability.

Climatic background

Sea levels along the British coastline have been fluctuating for thousands of years. Around 12 000 B.P. they were 50 m lower on the Welsh coast (Savidge, 1983). Some 5000 years ago it would have been possible to cross the Straits of Dover from England to France on dry land (Zuckerman, 1986). Mean global sea-level has risen by about 10–15 cm during the twentieth century (Robin, 1986). Yet sand dunes in various stages of development still occupy 9% of the coastline of mainland Britain (Ranwell & Boor, 1986), in the form of spit dunes, bay dunes, hindshore dunes, prograding dunes or offshore island dunes. They provide all the classic dune habitats for fungi: yellow and grey dune, dune slack, dune grassland, dune scrub and dune heath (Fig. 4.1).

Examples of some of the ecological factors at work can be seen at Ynyslas National Nature Reserve, in Cardiganshire (Nat. Grid Ref.: SN6194). There, at low spring tides, the tree-stump remnants of a submerged forest (Fig. 4.2) emerge as evidence of the fact that the coastline was 11 km farther out to sea 5000 years ago (Anon, 1983).

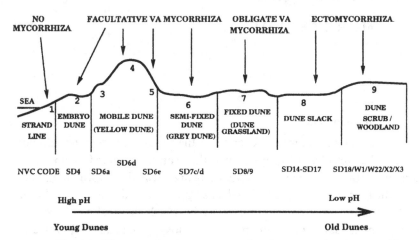

Fig. 4.1. Diagram of a dune transect showing successional zones (after Rotheroe, 1993). NVC code (National Vegetation Classification, Radley, 1988).

Fig. 4.2. Remnants of the submerged forest revealed at low tide at Ynyslas National Nature Reserve. (Photo: J. P. Savidge.)

Ynyslas is a relatively young dune system, having been initiated only between 200 and 400 years ago (Savidge, 1976). Within human memory, the northern end of what is now a continuous spit-dune system was an island. The name Ynyslas means green island in Welsh. In 1985, in what used to be the gap between the green island and the mainland, the first British collection of the agaric species *Melanoleuca cinereifolia* was made on the terrestrial habitat (yellow dune) now established there. Ynyslas is remarkable in that, in spite of its recent formation and its small size (97 ha), it has a particularly rich mycoflora, notable for several rare and uncommon taxa.

Elsewhere around British coasts, dune systems are building (as they are at Ynyslas), remaining at relative equilibrium, or eroding. This latter process may be through a gradual eating away by various physical elements or through individual violent events such as severe storms. Historical records for sand dunes in Britain contain descriptions of long-established dunes (and nearby villages) being devastated and re-shaped in a single gale, as well as examples of slowly encroaching sand inundating and burying villages. Newborough Warren on Anglesey has suffered a series of violent upheavals since the fourteenth century (Anon, 1976). In 1948 the Forestry Commission responded by planting a conifer forest to stabilize the system and protect property. Fierce storms in the winter of 1994 swept away five years' dune restoration work by Northumberland Wildlife Trust at Druridge Bay (Nat. Grid Ref.: NZ2896) on the north-east coast of England (Anon, 1995). In some places, 5 m of dune were lost from the bay. The more gradual process of dune building has been operating for more than 2000 years at Merthyr Mawr (Nat. Grid Ref.: SS8777) in south Wales (Gillham, 1987). There, today, the top cone of a windmill can be seen sticking out of the sand and a mobile yellow dune cliff is advancing on a car park some 1.5 km inland.

Future environmental change

A consequence of the predicted global warming (*c.* 1 °C by the year 2025; Anon, 1991*a*) would be a global rise in mean sea-levels, although opinions vary greatly on the extent of the rise (Boorman, Goss-Custard & McGrorty, 1989). Both thermal expansion of oceans and the melting of glaciers and polar ice sheets are cited as contributory factors. Mean global sea-level could rise by more than 0.5 m in the next 100 years (Anon, 1991*b*). Even greater rises in sea-level have been predicted for British coasts, particularly in the south-east, from the Humber to Poole

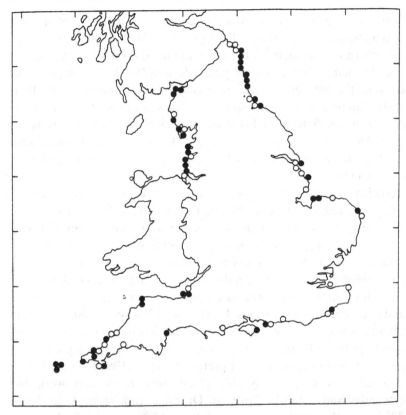

Fig. 4.3. Distribution in England of dunes which appear to show net marine erosion. Symbols refer to 10 km squares. ● Retreating dunes. ○ Others. (From Radley, 1994.)

Harbour, and in other major estuaries, where 'soft coast' as opposed to rocky shore is most prevalent. Dunes in all regions would be subject to some change but those which at present appear to show net marine erosion (Fig. 4.3) could be the earliest and most seriously affected. A rise of 4.5 m has been predicted in the area of the Wash in less than 150 years (Cannell & Hooper, 1990). In Britain, predictive modelling is complicated by the fact that the land is gradually sinking towards the south-east and rising on the Scottish coast to the north-west owing to post-glacial adjustment (Anon, 1991*b*). However, the distribution of sand-dune systems in mainland Britain is such that the south-east corner has only a very small proportion of the total dune coastline, dunes on the west and north-east coasts being the most extensive and important on the mainland (Ranwell & Boor, 1986).

Potential habitat effects

Among the changes to which the mycoflora might be subjected in the event of a rise in mean sea-level are:

1. flooding, resulting in the loss of some existing successional zones. However, in the hygrosere, some species might benefit from a rise in the water table;
2. an increase in rainfall and storms, causing erosion of dunes by water and wind. A consensus of opinion suggests the possibility of an increase in rainfall of some 20% (Boorman *et al.*, 1989). More violent storms have been said to be occurring already in the north-east Atlantic (Carter & Draper, 1988);
3. changes in tidal range (between mean spring high and spring low water marks), accelerating erosion;
4. a rise in mean annual temperature, which would make the environment more hospitable to fungal species at present confined to areas of Europe farther south. A warmer climate would, however, increase pressures on sand dunes, particularly from coastal recreation.

Most low-lying areas in Britain are protected by sea walls, with sand dunes (and also shingle banks) as only a secondary, alternative form of sea defence. Two of the habitats most likely to be affected by rising sea-levels, or by human response to such rises, are yellow dune and dune slack.

Yellow dune

The yellow dunes, where marram grass (*Ammophila arenaria*) is more or less completely dominant and in its most vigorous phase, are characterized by a group of saprotrophic fungi able to thrive at this early stage of dune succession. They appear to be confined to sand dunes and to require continuous input of newly deposited sand, often using marram or buried rabbit dung as a resource. They are rarely or never found farther inland. Species include *Psathyrella ammophila*, the above-mentioned *Melanoleuca*, *Peziza ammophila*, *Phallus hadriani* and the rare taxa *Coprinus ammophilae* and *Hohenbuehelia culmicola*. *Psathyrella ammophila* is the commonest agaric found on intact sand-dune systems, occurring in all months of the year and fruiting in troops several hundred strong when optimal conditions prevail. This community of saprotrophs is clearly involved in nutrient cycling in this zone as well as helping to stabilize the sand particles physically by hypogeal production of mycelia, mycelial cords or pseudostipes. The equally important VA mycorrhizal species which occur in the zone are not included in this chapter.

Dune slack

Dune slacks are undoubtedly the most species-rich habitat in dune systems. Here, grassland saprotrophy is joined by biotrophy, bringing in increasing numbers of species which are ectomycorrhizal with creeping willow (*Salix repens*). The mycobionts have a role in facilitating and sustaining colonisation by the *Salix*. The importance of this successional zone was highlighted by Ranwell (1959), who pointed out that slacks are one of the few remaining lowland damp habitats still relatively little affected by the interference of man.

Prognosis

Left to their own devices, most major British sand-dune systems and their fungal elements would probably be robust and dynamic enough to survive an accelerating rate of sea-level rise by re-forming farther inland, according to the theory of 'sea-level transgression' (Boorman *et al.*, 1989). Predictions vary on the likely fate of individual successional zones. For example, it has been suggested that dune slacks currently threatened by falling water tables would regenerate in the event of a rainfall increase of 20% but might face a new threat in the form of higher inputs of blown sand (Boorman *et al.*, 1989). These authors also pointed out that the processes of natural recovery following a significant rise in sea-level are dependent on there being no artificial barriers to limit the advance of the sea landwards. This is crucial, since man's reaction to such events would be likely to pose a threat several orders of magnitude more damaging than any natural phenomenon.

Coastal sand dunes are already under great pressure from human recreational and other activities (Fig. 4.4). Trampling pressure is a constant threat to vulnerable mobile dunes, sites for golf links are increasingly in demand, while holiday caravan sites and military training areas have sterilized some of the finest 'soft coastline' resources. Stabilizing and protection techniques have included monocultural plantation of trees at many dune locations and increased construction of sea wall defences. While plantations may actually increase the number of fungal species

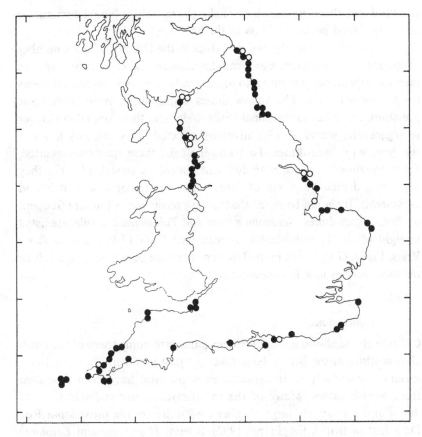

Fig. 4.4. Distribution in England of dunes suffering moderate or severe erosion due to trampling on at least part of the site. Symbols refer to 10 km squares. ● Moderate or severe erosion, ○ Little or no erosion. (From Radley, 1994.)

at a particular location by introducing additional ectomycorrhizal fungi, the previously natural habitat is inevitably destroyed and the newcomers are alien to that habitat. Sea walls cause even more dramatic destruction of the ecosystem. The strandline disappears and, because the gently sloping frontal edge of the foremost dune cliff is removed or distorted, typical mobile-dune formation processes are halted and an important habitat for fungi ceases to exist. *Psathyrella ammophila* and other members of the colonising community mentioned above are among the first casualties of this type of defence fortification. A prime example of the consequences of sea-wall erection can be seen in Jersey at Les Quennevais, one of Britain's largest hindshore-dune systems on St Ouen's Bay. On the decimated frontal dunes there, no members of the characteristic fungal community are to be found because of loss of the habitat.

A warmer climate might well introduce to the UK mycoflora a number of species with a southern European distribution and some extinct species might reappear, but it would also mean greater demands on coastal dunes for leisure activities. The yellow dunes would suffer most from these pressures, but other successional zones, older and therefore taking longer to regenerate, would also be adversely affected. This category includes the grey, semi-fixed dunes. To a non-biologist these sparsely vegetated, lichen-inhabited areas appear dull and almost as wastelands. Yet they support a distinctive group of fungi which is rare or absent at inland ecosystems. It should be noted that two gasteromycetes that are frequent on British grey dunes, *Geastrum nanum* and *Tulostoma brumale*, are listed in eight of the 11 published European Red Data Lists (Arnolds & de Vries, 1993). Grey dunes in the UK can therefore be regarded as a refuge for these species in a European context.

Conclusions

Collectively, sand-dune fungi form an important component of the overall mycoflora of the British Isles. Over the past half century, several new species, particularly in the Agaricales *sensu lato*, have been described from British dunes. Many of the macromycetes are endemic to sand dunes and a relatively large number (>20) are on the provisional Red Data List of British fungi (Ing, 1992). Several *Hygrocybe* and *Entoloma* species which occur in slacks or rabbit-grazed grassland on sand dunes are now rarely recorded outside nature reserves, because of the irreversible loss of traditionally managed pastures.

It takes at least 75 years for a mature sand-dune system to develop, given the appropriate environmental conditions (Ranwell, 1959). However, an entire dune system, together with its fungal and other living elements, can be destroyed 'overnight' by the sea or by the stroke of a bureaucratic pen. Mycologists must join other conservation bodies in arguing the case for protecting this important and unique ecosystem, should global warming accelerate a rise in sea-levels.

Acknowledgements

The author is grateful to the Countryside Council for Wales, the Joint Nature Conservation Committee and the British Ecological Society for grants towards a series of investigations on Welsh and English dune systems; to Dr J. P. Savidge for permission to reproduce the photograph in Fig. 4.2 and to Ms P. M. David for valuable help with fieldwork and editorial advice.

References

Anon (1976). *Newborough Warren National Nature Reserve.* London: Nature Conservancy Council.
Anon (1983). *Ynyslas Dunes, The Submerged Forest.* Aberystwyth: Nature Conservancy Council – Dyfed/Powys Region.
Anon (1991*a*). *Global Climate Change.* 2nd edn. Department of the Environment. London: Her Majesty's Stationery Office.
Anon (1991*b*). *The Potential Effects of Climatic Change in the United Kingdom. First Report UK Climate Change Impacts Review Group.* London: Her Majesty's Stationery Office.
Anon (1995). Dunes damaged. *Natural World*, Spring/Summer, 1995, p.9. Lincoln: The Wildlife Trusts.
Arnolds, E. & de Vries, B. (1993). Conservation of fungi in Europe. In *Fungi of Europe: Investigation, Recording and Conservation*, ed. D. N. Pegler, L. Boddy, B. Ing & P. M. Kirk, pp. 211–230. Kew: Royal Botanic Gardens.
Boorman, L. A., Goss-Custard, J. D. & McGrorty, S. (1989). *Climatic Change, Rising Sea Level and the British Coast.* London: Her Majesty's Stationery Office.
Cannell, M. G. R. & Hooper, M. D. eds. (1990). *The Greenhouse Effect and Terrestrial Ecosystems of the UK.* ITE Research Publication No. 4. London: Her Majesty's Stationery Office.
Carter, D. J. T. & Draper, I. (1988). Has the north-east Atlantic become rougher? *Nature* **332**, 494.
Frankland, J. C. (1981). Mechanisms in fungal succession. In *The Fungal Community. Its Organisation and Role in the Ecosystem*, ed. D. T. Wicklow & G. C. Carroll, pp. 383–401. New York: Marcel Dekker.
Gillham, M. E. (1987). *The Glamorgan Heritage Coast Wildlife Series Volume 1: Sand Dunes.* Heritage Coast Joint Management and Advisory Committee.
Ing, B. (1992). A provisional Red Data List of British fungi. *The Mycologist*, **6**, 124–8.
Radley, G. P. (1988). *National Sand Dune Vegetation Survey, Site Report No. 7. North Walney, 1987.* Peterborough: Nature Conservancy Council.
Radley, G. P. (1994). *Sand Dune Vegetation Survey of Great Britain: A National Inventory. Part 1: England.* Peterborough: English Nature.
Ranwell, D. S. (1959). Newborough Warren, Anglesey 1. The dune system and dune slack habitat. *Journal of Ecology*, **47**, 571–601.
Ranwell, D. S. & Boor, R. (1986). *Coast Dune Management Guide.* Huntingdon: Institute of Terrestrial Ecology.
Robin, G. de Q. (1986). Changing the sea level. In *The Greenhouse Effect, Climatic Change and Ecosystems*, ed. B. Bolin, B. Doos, J. Jager & R. A. Warrick (SCOPE 29). Chichester: Wiley.

Rotheroe, M. (1993). The macrofungi of British sand dunes. In *Fungi of Europe: Investigation, Recording and Conservation*, ed. D. N. Pegler, L. Boddy, B. Ing & P. M. Kirk, pp. 121–137. Kew: Royal Botanic Gardens.

Savidge, J. P. (1976). The sand-dune flora. In *Ynyslas Nature Reserve Handbook*, ed. Anon, pp. 37–64. Aberystwyth: Nature Conservancy Council.

Savidge, J. P. (1983). The effects of climate, past and present, on plant distribution in Wales. In *Flowering Plants of Wales*, ed. R. G. Ellis, pp. 18–32. Cardiff: National Museum of Wales.

Zuckerman of Burnham Thorpe, Lord. (1986). Foreword to British Edition. In *Climatic Change and World Affairs*. Revised Edition. C. Tickell, pp. xi–xiii. Boston & London: University Press of America.

5

Red Data Lists and decline in fruiting of macromycetes in relation to pollution and loss of habitat

B. ING

Introduction

The apparent decline, both in numbers of populations and in the geographical range, of macrofungi over several decades has aroused widespread concern in Europe. One demonstration of this is the publication of lists of species considered to be in danger of near-future extinction as a result of a complex of environmental changes. These Red Data Lists reflect this concern, both in size and composition of the lists of supposedly endangered species. Several European countries have produced such lists of endangered fungi: Austria (Krisai, 1986); the British Isles (Ing, 1992); Denmark (Vesterholt & Knudsen, 1990); Finland (Rassi *et al.*, 1986); Germany (Benkert, 1982, 1993; Kreisel, 1992; Lettau, 1982; Runge, 1987; Schmitt, 1988; Winterhoff, 1984; Winterhoff & Krieglsteiner, 1984; Wöldecke, 1987); the Netherlands (Arnolds, 1989*a*); Norway (Bendiksen & Høiland, 1992); Poland (Wojewoda & Ławrinowicz, 1992) and Sweden (Ingelör, Thor & Hallingbäck, 1991). A provisional list for the whole of Europe has been offered by Ing (1993).

National lists differ in character, some reflecting local criteria based on the cultural importance of fungi rather than ecological priorities, and are necessarily parochial.

Composition of Red Lists

Species included in Red Lists are usually associated with ecosystems that are themselves endangered. In general, the majority are found in ancient woodland – both broad-leaved and coniferous, in unimproved grassland, in lowland bogs and sand dunes. Woodland mycorrhizal species are a group giving a high level of concern, especially boletes, *Cortinarius*,

Hygrophorus, *Tricholoma*, the stipitate hydnoids and *Geastrum*. Vulnerable species in grassland include a large assemblage of *Hygrocybe*, *Entoloma* and the Geoglossaceae. The gastroid basidiomycetes are particularly well represented – their distribution is easier to study than many other groups, with their long-lasting basidiomes and good taxonomic literature (Arnolds & de Vries, 1993).

The high proportion of ectomycorrhizal species in all the lists is shown in Table 5.1 and is further highlighted by Fellner (1993). It is important to include *both* loss of semi-natural, ancient forest *and* changes associated with atmospheric pollutants in this loss. Countries with more woodland are likely to sustain a larger amount of pollution damage, whereas the relatively poorly wooded countries, such as Great Britain, may reflect the greater significance of habitat destruction. The post-war changes in composition of fungal floras in The Netherlands is especially noticeable in grasslands and in the large representation in the lists of ectomycorrhizal species associated with conifers, notably *Cortinarius* spp., and in the stipitate hydnoid fungi (Arnolds, 1988, 1989*b*, 1991).

There has been criticism of both the composition and existence of Red Data Lists (Orton, 1994). It is accepted that our knowledge of fungal taxonomy is imperfect and of distribution and ecology inadequate. However, unless a start is made into drawing up provisional lists of species which satisfy certain stated criteria, that is that they are rare species characteristic of ecosystems which are themselves threatened, then no progress can be made to include fungi in conservation policies at national and international level. Bearing in mind the ecological significance of fungi this would be disastrous. All Red Lists are in essence a

Table 5.1. *Ectomycorrhizal fungi in some European Red Lists*

State	Number of species		
	Macrofungi in Red List	Ectomycorrhizal species	%
Poland	1013	219	21.7
Denmark	994	201	20.2
Netherlands	994	378	40.0
Saarland	981	466	47.5
Norway	676	288	42.6
Sweden	515	112	21.7
United Kingdom	445	199	44.7
Finland	176	47	26.7

statement of concern based on existing knowledge. Where the knowledge is inadequate so will be the list. Such lists are still justified on the grounds that a start must be made sometime and these are the best that can be achieved at present!

Causes of decline

The major cause of loss of fungal communities is loss of the appropriate habitat, as is the case for most organisms. Changes in land-use, such as conversion of forest to agriculture, conversion of broad-leaved woodlands to coniferous plantations, change in forestry management practices, ploughing or fertilising of ancient grasslands (both activities highly destructive of fungal populations), and drainage of wetlands are familiar causes to those involved in the conservation of all life-forms, and fungi are no exception. A summary of the causes of loss in Germany, based on Winterhoff (1986) is shown in Table 5.2. Similar proportions are likely to apply in most of western Europe.

Pollution, whatever the origin or mode of delivery, is now accepted as a major cause of decline. Acid and nitrogen deposition, the increase in soluble aluminium, heavy metals and radionuclides all have been implicated (Høiland, 1993). While it is difficult to evaluate the actual losses caused by these pollutants (and several of them are dealt with elsewhere in this volume), the greatest concerns have centred around the role of soil acidification in the decline of ectomycorrhizal fungi (Kuyper, 1989; Arnolds, 1991; Fellner, 1993). The celebrated case of *Cantharellus cibarius* was first brought to notice by Jansen and van Dobben (1987) who demonstrated a 50% decrease in populations of this popular edible fun-

Table 5.2. *Causes of decline in German fungi*

Cause	Declining species	
	Number	%
Air pollution	114	30
Agricultural conversion	95	25
Forestry	92	24
Building development	37	9
Recreation	25	6
Quarrying	12	3
Road building	10	3
Dam building	2	0.5

gus and linked this directly to quite modest degrees of acidification associated with sulphur dioxide emissions. The increase in numbers of the angiosperms *Vaccinium myrtillus* and *Deschampsia flexuosa*, indicator species for acid soils, was supporting evidence. Such losses are not exhibited by all fungi, and a useful comparison was made of sales of *Cantharellus* and *Armillaria* in Saarbrücken market between 1956 and 1975, in which there was some variation but no reduction in the latter, but the former showed a dramatic decline, even allowing for two dry summers (Derbsch & Schmitt, 1987; Fig. 5.1 and 5.2).

Although the problems experienced by forest fungi are claiming greatest attention, because of the long-term implications for forest health, fungi of unimproved grassland are now seriously affected by the application of artificial, nitrogenous fertilizers and liquid manure (Arnolds, 1989c). Vulnerable species of *Entoloma*, *Geoglossum* and *Hygrocybe* therefore figure prominently in Red Lists as grassland ecosystems lose diversity. Only in the western areas of the British Isles are these fungi apparently maintaining healthy populations, which thus represent a most important European resource.

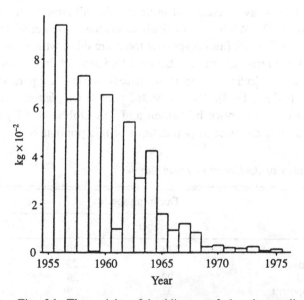

Fig. 5.1. The weight of basidiomes of the chanterelle (*Cantharellus cibarius*) collected in the woods in Saarland, Germany and sold at the market in Saarbrücken from 1956 to 1975. The low yields in 1959 and 1961 were due to very dry summers. (After Demke, in Derbsch & Schmitt, 1987.)

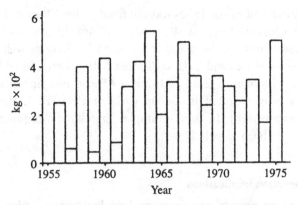

Fig. 5.2. The weight of Honey Fungus (*Armillaria mellea*) sold at the market in Saarbrücken from 1956 to 1975. (After Demke, in Derbsch & Schmitt, 1987.)

An even more recent concern surrounds the impact of climatic change, associated with a postulated global warming, and caused by the increased presence of certain gaseous pollutants. Whether or not there is such a phenomenon as the 'greenhouse effect', and whether or not we are experiencing a shift in the British Isles to a more Mediterranean-type climate, is not yet established. What is clear, however, is that several species of fungi appear to be colonizing suitable areas of these islands, presumably as part of a general move northwards and westwards. They include ectomycorrhizal species such as *Amanita ovoidea*, associated with the evergreen oak *Quercus ilex*. This is a common and conspicuous fungus in the woodlands of the Mediterranean region and France and has recently been found in Surrey and the Isle of Wight, both in the UK (Pope, 1991). *Omphalotus illudens*, normally associated with olive stumps and not previously found north of the native range of the grape-vine, is also becoming established in Sussex and has recently been found near Reading, again in the UK (Marriott, 1994). Two powdery mildews, *Microsphaera platani* on *Platanus* x *hispanica* (London Plane), and *Sawadaea tulasnei* on various *Acer* species including *platanoides*, have arrived recently and are becoming established in eastern and south-eastern England (Strouts, Rose & Refford, 1984; British Mycological Society Autumn Foray Report, Northampton, 1991). The former is characteristic of the eastern Mediterranean and is causing major damage in Italy and the Mediterranean islands whilst the latter is centred on eastern and central Europe. The rust on Mexican and South African species of *Oxalis*, *Puccinia oxalidis*, has been present in the British Isles for thirty years but until the latter half of the 1980s was confined to the Channel Islands and the south coast of England. It is now widespread and no longer restricted to coasts but occurs inland in northern England as well as all round the coasts of Wales and Ireland (Cooper & Ing, 1991). In addition, a species of *Sirobasidium*, a genus of tropical African resupinate hetero-basidiomycetes, was collected in Devon early in 1994 (P. J. Roberts, personal communication). In these examples the climatic requirements have only recently been satisfied.

Conservation implications

The presence in an area of one or more Red List species, whether at national or European level, immediately raises the importance of that area for conservation. Sites can therefore be scored for national or European Red List species and ranked in importance. This may be a useful means of selecting sites for reserve status and should also deter-mine the appropriate management, particularly in relation to forestry practices and the disposal of fallen timber. Where a site has a high pro-portion of species on the European List, it should carry more weight than those with only national Red List species. For example, during the com-pilation of the European list and the British List, it was noticed that the Caledonian pine forests of the Scottish Highlands have a fine representa-tion of the most endangered European species, especially ectomycor-rhiza-formers, and should therefore be graded in terms of European rather than national importance. Similarly, the coniferous forests of northern Fennoscandia have a high proportion of endangered species, notably of old-forest wood-rot species, and should be managed sym-pathetically, with the long-term survival of fungal habitats a major con-sideration. The Atlantic oak woodlands of the British Isles are just as important for fungi on a European scale as they are for those groups of organisms traditionally used as indicators of biodiversity, namely bryo-phytes, lichenized fungi, angiosperms and beetles.

While it is possible to safeguard important fungal habitats from damage, it is not easy to reduce the impact of airborne pollutants or slow the rate of climatic change. This poses important questions for those concerned with the conservation of all aspects of our natural heri-tage. Are we able to justify the expense of conserving areas which we cannot protect from pollution damage? Are we able to predict which vegetation types, with their associated fungi, will be lost as climatic pat-terns change, and, if not, how do we know which areas to designate and protect? Whatever the outcome, however, we can be sure that some fungi will take advantage of the new conditions and replace the less adaptable.

Red Data Lists, therefore, have a role in alerting public bodies to the need to protect certain ecosystems containing vulnerable species, and they reflect a combination of redeemable losses (by the re-creation of habitats) and those which cannot be retrieved, where the chemical environment has become permanently unsuitable or where the total population has been destroyed. Lists themselves should change as the pressures on species change and need regular revision to be of long-term value. The list is only final when all species are extinct!

Acknowledgements

I am grateful to Juliet Frankland for valuable comments on the draft, to Dr J. A. Schmitt for kind permission to use Fig. 5.1 and 5.2, and to numerous field mycologists for their reports and observations which provide the raw material for this discussion.

References

Arnolds, E. (1988). The changing macromycete flora in the Netherlands. *Transactions of the British Mycological Society*, **90**, 391–406.
Arnolds, E. (1989*a*). A preliminary Red Data List of macrofungi in the Netherlands. *Persoonia*, **14**, 77–125.
Arnolds, E. (1989*b*). Former and present distribution of stipitate hydnaceous fungi (Basidiomycetes) in the Netherlands. *Nova Hedwigia*, **48**, 107–42.
Arnolds, E. (1989*c*). The influence of increased fertilisation on the macrofungi of a sheep meadow in Drenthe, the Netherlands. *Opera Botanica*, **100**, 7–21.
Arnolds, E. (1991). Decline of ectomycorrhizal fungi in Europe. *Agriculture, Ecosystems and Environment*, **35**, 209–44.
Arnolds, E. & de Vries, B. (1993). Conservation of fungi in Europe. In *Fungi of Europe: Investigation, Recording and Conservation*, ed. D. N. Pegler, L. Boddy, B. Ing & P. M. Kirk, pp. 211–230. Kew: Royal Botanic Gardens.
Bendiksen, E. & Høiland, K. (1992). Red List of threatened macrofungi in Norway. *Report, Directorate for Nature Management*, **6**, 31–42.
Benkert, D. (1982). Vorläufige Liste der verschollenen und gefährdeten Großpilzarten der DDR. *Boletus*, **6**, 21–32.
Benkert, D. (1993). Großpilze (Makromyzeten). In *Rote Liste der gefährdete Farn- und Blütenpflanzen, Algen und Pilze im Land Brandenburg*, pp. 107–185. Potsdam: Ministerium für Umwelt, Naturschutz und Raumordnung des Landes Brandenburg.
Cooper, J. & Ing, B. (1991). BMS mapping scheme: *Puccinia oxalidis*. *Mycologist*, **5**, 99.

68   *B. Ing*

Derbsch, H. & Schmitt, J. A. (1987). *Atlas der Pilze des Saarlandes*. Teil 2. Saarbrücken, Minister für Umwelt des Saarlandes.

Fellner, R. (1993). Air pollution and mycorrhizal fungi in Central Europe. In *Fungi of Europe: Investigation, Recording and Conservation*, ed. D. N. Pegler, L. Boddy, B. Ing. & P. M. Kirk, pp. 239–250. Kew: Royal Botanic Gardens.

Høiland, K. (1993). Pollution, a great disaster to mycorrhiza? *Agarica*, **12**, 65–88.

Ing, B. (1992). A provisional Red Data List of British fungi. *Mycologist*, **6**, 124–8.

Ing, B. (1993). Towards a Red List of Endangered European Macrofungi. In *Fungi of Europe: Investigation, Recording and Conservation*, ed. D. N. Pegler, L. Boddy, B. Ing & P. M. Kirk, pp. 231–237. Kew: Royal Botanic Gardens.

Ingelör, T., Thor, G. & Hallingbäck, T. (1991). *Hotade växter i Sverige 1990*. Lund: Databanken för hotade arter.

Jansen, E. & van Dobben, H. F. (1987). Is decline of *Cantharellus cibarius* in the Netherlands due to air pollution? *Ambio*, **16**, 27–9.

Kreisel, H. (1992). *Rote Liste der gefährdeten Großpilze Mecklenburg-Vorpommerns*. Schwerin: Die Umweltministerium des Landes Mecklenburg-Vorpommern.

Krisai, I. (1986). Rote Liste gefährdeter Großpilze Österreichs. In *Rote listen gefährdeter Pflanzen Österreichs*. Grüne Reihe Bundesministeriums Gesunheit-Umweltschutz, **5**, 178–92.

Kuyper, T. W. (1989). Auswirkungen der Walddüngung auf die Mykoflora. *Beiträge zur Kenntnis der Pilze Mitteleuropas*, **5**, 5–20.

Lettau, M. (1982). Vorläufige Liste verschollener und gefährdeter Großpilze in Schleswig-Holstein. *Schriftenreihe Landesamtes Naturschutz Landschaftplege Schleswig-Holstein*, **5**, 58–71.

Marriott, J. V. R. (1994). Unusual finds. *Associates Newsletter*, November 1994, 11. Stourbridge: British Mycological Society.

Orton, P. (1994). Some comments on 'A Provisional Red Data List of British Fungi' by B. Ing. *Mycologist*, **8**, 66–67.

Pope, C. R. (1991). Day Foray Report, Isle of Wight. *Mycologist*, **5**, 101.

Rassi, P. Alanen, A., Kemppainen, E., Vickholm, M. & Väisänen, P. (1986). Uhanalaisten eläinten ja kasvien suojelutoimikunnan nietintö. *Komiteamtietintö* **1985**, 62–65 (fungi). Helsinki: Ympäristöministeriö.

Runge, A. (1987). Vorläufige Rote Liste der gefährdeten Großpilze (Makromyzeten) in Nordrhein-Westfalen. In *Landensanstalt für Ökologie, Landschaftsentwicklung und Forstplanung NW, Rote Liste der in Nordrhein-Westfalen gefährdeten Pflanzen und Tiere*, **2** (18), 3–15. Recklinghausen.

Schmitt, J. A. (1988). Die Pilze. In *Rote Liste, Bedrohte Tier-und Pflanzen-arten im Saarland*, pp. 77–116. Saarbrucken: Minister für Umwelt.

Strouts, R. G., Rose, D. R. & Reffold, T. C. (1984). Pathology Report 1984. *Report on Forest Research*, **1984**, 33.

Vesterholt, J. & Knudsen, H. (1990). *Truede storsvampe i Danmark – en rødliste*. København: Foreningen til Svampekundskabens Fremme.

Winterhoff, W. (1984). Vorläufige Rote Liste der Großpilze (Makromyzeten). In *Rote Liste der gefährdeten Tiere und Pflanzen in der Bundesrepublik Deutschland*, ed. J. Blab, E. Nowak, W. Trautman, & H. Sukopp. **4**, 162–184. Greven.

Winterhoff, W. (1986). Auswertung von Roten Listen der verswschollenen und gefährdeten Großpilze. *Sächriftenereich Vegetationskundes*, **18**, 135–46.

Winterhoff, W. & Krieglsteiner, G. J. (1984). Gefährdete Pilze in Baden-Württemberg. *Beiheft Veröffentlichüngen Naturschutz für Landschaftspflege in Baden-Württemberg*, **40**, 1–120.

Wojewoda, W. & Ławrinowicz, M. (1992). Red List of threatened macrofungi in Poland. In *List of Threatened Plants in Poland* (ed. 2) ed. K. Zarzycki & W. Wojewoda, pp. 27–56. Cracow: Polish Academy of Sciences.

Wöldecke, K. (1987). Rote Liste der in Niedesachsen und Bremen gefährdeten Großpilze. In *Informationsdienst Naturschutz Niedersachsen*, **7**, 1–28.

6

Effects of dry-deposited SO_2 and sulphite on saprotrophic fungi and decomposition of tree leaf litter

L. BODDY, J. C. FRANKLAND,
S. DURSUN, K. K. NEWSHAM AND
P. INESON

Sulphur pollution – past trends, current levels

Sulphur (S) is an essential element for the growth and activity of organisms, occuring free in abundant quantities. It is found in valence states ranging from +6 in sulphates to −2 in sulphides, the most stable state being S^{6+}. Globally, natural emissions of S into the atmosphere from biogenic sources have been estimated to range between 78.9 and 142.6 Tg yr^{-1}, of which 5.0 to 63.9 Tg yr^{-1} arises from land (Schlesinger, 1991; Andreae & Jaeschke, 1992; Germida, Wainwright & Gupta, 1992). Estimates of the relative significance of anthropogenic to natural sulphur emissions are variable, but globally the ratio is probably about 4:1 (Möller, 1984), atmospheric S emissions from land being in the order: anthropogenic 93 Tg yr^{-1} > biogenic gases 22 > dust 20 > volcanoes 10 (Brimblecombe et al., 1989). Sulphur emissions to the atmosphere are not constant with time, having increased dramatically since the industrial revolution as a result of combustion of fossil fuels. Global anthropogenic emissions have increased from an estimated 5 Tg in 1860 to 180 Tg in 1985, with considerable increases predicted up to 2000 (Fig. 6.1; Möller, 1984; Schlesinger, 1991), despite electives by European countries to have reduced emissions by 1993 to 30% of 1980 emissions.

Of anthropogenic sulphur emissions to the atmosphere, 95% is in the form of SO_2 (Kellogg et al., 1972). SO_2 emissions and deposition on land vary considerably, both spatially and temporally, as does the form in which it is deposited, that is as dry or wet deposition (see below). For example, former Czechoslovakia, eastern Germany and Poland are exposed to higher atmospheric SO_2 concentrations (up to 9 nl l^{-1} = 9 ppb) than some parts of Europe (< 4 nl l^{-1}; EMEP, 1989) as their major source of fuel is 'brown coal' which is high in sulphur. Throughout

70

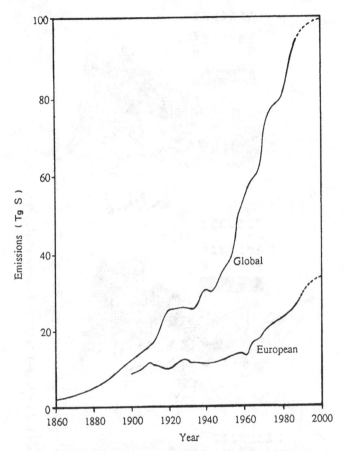

Fig. 6.1. Trends in actual and predicted global and European sulphur emissions (Modified from Möller, 1984.)

Europe SO$_2$ concentrations tend to be higher over urbanized and industrialized areas than other regions (Irwin, 1989). In the UK, the Midlands around Nottingham experience the largest mean annual SO$_2$ concentrations ($> 10\,\mathrm{nl\,l^{-1}}$) and dry-deposited SO$_2$, reflecting the large numbers of coal- and oil-fired power stations, whereas on the west coast and Scottish highlands wet-deposited SO$_2$ predominates, reflecting high rainfall (Fig. 6.2; Williams *et al.*, 1989). On a smaller scale, SO$_2$ concentrations decrease with distance from the source, as indicated, for example, by differences in the lichen flora with distance from the origin of the pollutant (Gilbert, 1968). Temporal variation in SO$_2$ concentrations reflects variation in fuel burning for heating, vehicle emissions and dispersal conditions. Thus peak concentrations of SO$_2$ occur between 0600 h and

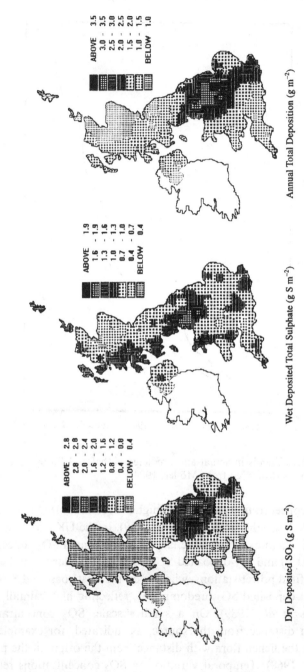

Fig. 6.2. Annual deposition of sulphur in the UK ($g \, m^{-2}$) (After Williams *et al.*, 1989.)

Dry Deposited SO_2 (g S m^{-2})

ABOVE	2.8
2.8 -	2.8
2.0 -	2.4
1.6 -	2.0
1.2 -	1.6
0.8 -	1.2
0.4 -	0.8
BELOW	0.4

Wet Deposited Total Sulphate (g S m^{-2})

ABOVE	1.9
1.6 -	1.9
1.3 -	1.6
1.0 -	1.3
0.7 -	1.0
0.4 -	0.7
BELOW	0.4

Annual Total Deposition ($g \, m^{-2}$)

ABOVE	3.5
3.0 -	3.5
2.5 -	3.0
2.0 -	2.5
1.5 -	2.0
1.0 -	1.5
BELOW	1.0

1200 h and during autumn and winter (Roberts, Darrall & Lane, 1983; EMEP, 1984). Autumn and winter atmospheric concentrations of SO$_2$ can be two-fold greater than the yearly average (EMEP, 1984; Kelly, McLaren & Kadlececek, 1989), and the absence of deciduous tree canopies in winter results in greater exposure of decomposer organisms at this time.

Forest decline processes – significance of litter decomposition

There is considerable evidence in the literature that forest decline in Europe is a pollution-linked phenomenon (Cowling, 1982), and SO$_2$ has been suggested as having a major role (van Breeman, 1985). SO$_2$ may exert a variety of direct and indirect effects on forest soils and vegetation including: (1) direct damage to vegetation, effects usually being exhibited as leaf injury (often as necrotic spots), accelerated crown leaching, and inhibition of metabolic processes (Harvey & Legge, 1979; Anon, 1981); (2) increased incidence of pathogens and putative pathogens (Grzywacz, 1973; Grzywacz & Wazny, 1973; Domanski & Kowalski, 1987), although it is difficult to distinguish between cause and effect; (3) decrease in the presence and activity (Garrett, Carney & Hedrick, 1982) of mycorrhizal fungi can result in loss of plant vigour, and decreased water and nutrient uptake abilities.

Soil acidification, loss of base cations and/or mobilization of aluminium is a well-documented effect of SO$_2$ and its solution products (e.g. Ulrich, Mayer & Khanna, 1980; Abrahamsen, 1984; Haynes & Swift, 1986). As SO$_2$ solubilizes in water on the surface of soil particles, protons are produced (through the formation of sulphate) which displace cations (mainly calcium and magnesium but also potassium and sodium) from the negatively charged soil particles, which are leached along with sulphate from the soil profile. In soils acidified by proton formation (particularly below pH 4.2), aluminium and manganese are released into solution (Ulrich *et al.*, 1980). Toxic effects of these metals to fine roots, some mycorrhizal fungi and other soil microorganisms have been demonstrated and implicated in forest decline (e.g. Ulrich *et al.*, 1980; Thompson & Medve, 1984; Dursun, 1994).

Since soil and litter microorganisms are essential to carbon and mineral nutrient cycling, and hence to continued primary production and ecosystem functioning, deleterious effects on these organisms could have considerable impact on forest ecosystems. SO$_2$ can be highly toxic to microorganisms: it has mutagenic effects; it inactivates mRNA; it reacts with disulphide linkages in proteins, enzyme cofactors, aldehyde and ketone structures of five and six carbon sugars; it deaminates cytosine derivatives to uracil compounds, and it has deleterious effects on membranes (Babich & Stotzky, 1980). The extent to which microorganisms, particularly fungi, are actually affected in litter is the subject of this review.

Trends and limitations of SO_2 pollution research

Several approaches have been adopted for investigating effects of SO_2 on fungal activity and litter decomposition, being either non-manipulative, in which effects have been deduced by comparing sites experiencing different ambient SO_2 concentrations, or experimental, in which SO_2 or solution products have been added at different concentrations to soil/litter in the field, in laboratory microcosms or to agar culture. The former approach suffers from the problem that differences may not solely reflect exposure to different concentrations of SO_2 but may result from differences in, for example soil, microclimate, and other pollutants. With the latter approach, there can be difficulties in relating experimental conditions to those which actually occur in the field.

In any ecological study examining the effects of a particular compound on organisms or processes it is essential that experiments are relevant to natural conditions. In particular, the compound should be in the same form and in *realistic* concentrations, and other abiotic variables which may influence the activity of the compound should be appropriate to the natural situation. Most research into the effects of SO_2 pollution on forest organisms and processes, including the pioneering work of Oden (1968), has simulated the wet deposition (i.e. deposition from the atmosphere in aqueous solution or suspension) of SO_2 by applying acids of sulphur. However, this is largely inappropriate since SO_2 is transferred to the ground, over much of Europe, by dry deposition (direct transfer of gaseous and particulate material from the atmosphere to a surface, not in solution or suspension). In the UK, dry deposition accounts for > 70% of total S deposited in the most SO_2-polluted regions, and between 40 and 70% elsewhere (Fig. 6.2; Williams *et al.*, 1989). Further, 'natural' wet-deposited SO_2 (as sulphate) may not exert such marked effects on microbial activities as equivalent amounts of dry-deposited SO_2, since other cations, for example ammonium, calcium and magnesium, in precipitation may neutralise sulphate to some extent. The omission of counterbalancing cations in simulated wet deposition may lead to further interpretational difficulties. Thus, while several studies have demonstrated decreased decomposition rates and increased fluxes of calcium and magnesium in leachates following application of dilute H_2SO_4 to a variety of soils and litters, the pH values were artificially low, and it is difficult to relate these results to naturally occurring conditions.

The significance of pH lies in the fact that the solubility products of SO_2 in water vary depending on pH (Fig. 6.3; Vas & Ingram, 1949; Babich & Stotzky, 1974), and the toxicity of these products differs, the effects on microorganisms being greatest for undissociated sulphurous acid (H_2SO_3) > bisulphite (HSO_3^-) > sulphite (SO_3^{2-}) (Cruess, Richert & Irish, 1931). Thus above about pH 5 it is the effects of sulphite, the least toxic of the solubility products, that become increasingly important. In northern Britain, rain water samples frequently contain >45 μM sulphite (Cape *et al.*, 1984).

Agar culture must be used with caution because, aside from the usual problems of trying to mimic heterogeneous natural substrata with simple homogeneous media, SO_2 is absorbed readily in water and adsorbs to sugars and other organic compounds (Babich & Stotzky, 1980), both of which are abundant in artificial media, making solution products less available to the decomposer microorganisms.

Clearly then, information on the effects of *dry-deposited* SO_2 and the *less toxic* solubility products is essential, although little emphasis has as

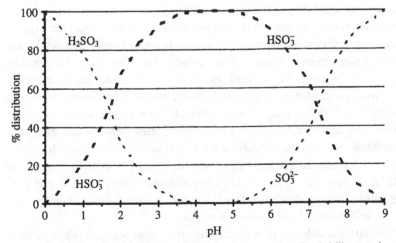

Fig. 6.3. The relationship between solution pH and solubility products of SO_2 (Modified from Babich & Stotzky, 1974, after Vas & Ingram, 1949.)

yet been placed on these components. This review, therefore, largely concerns the effects of dry-deposited SO_2 and sulphite on saprotrophic fungi and the decomposition of tree leaf litter.

Experimental research on SO_2 pollution

Non-manipulative field studies: microbial communities on sites exposed to different concentrations of SO_2

Isolations on agar media and damp chamber incubations have been made from fallen leaf litter (ash, *Fraxinus excelsior*; birch, *Betula* spp.; hazel, *Corylus avellana*; oak, *Quercus robur* and *Q. petraea*; sycamore, *Acer pseudoplatanus*), from three UK woodland sites exposed to different ambient concentrations of SO_2: a relatively unpolluted site, Meathop; a more polluted site, The Knoll; and a highly polluted site, Cefnpennar (Newsham *et al.*, 1992*a*). For these three sites, respectively, the winter SO_2 daily means estimated from the lichen zone scale of Hawksworth & Rose (1970) were 19, 23 and 56 nl l^{-1}, and similar results were obtained with diffusion tubes (Newsham *et al.*, 1992*a*). Species frequently isolated from some of the less polluted litters included *Cladosporium cladosporioides, C. herbarum, Epicoccum nigrum* and *Fusarium avenaceum*, the last being absent from the heavily-polluted Cefnpennar site. In contrast, *Trichoderma* spp., *Penicillium* spp. and *Phoma macrostoma* were most frequently isolated from the more polluted litters. Damp chamber incubations revealed that of the most commonly occurring species *Gliomastix murorum* var. *felina* was most frequent on the less polluted litters (birch, hazel and sycamore), as were *Periconia byssoides* (ash and birch) and *Verticillium lecanii* (birch and sycamore), and the ascomycetes *Gnomonia gnomon* (hazel) and *Venturia fraxini* (ash). In contrast the following were most abundant on polluted litters: *Acremonium persicinum* (all litters), *Colletotrichum dematium* and *Cylindrocarpon orthosporum* (on birch, oak and sycamore), and the ascomycetes *Chaetomium globosum, Gnomonia setacea* and *Venturia ditricha* (on birch). Thus there do appear to be differences in the mycoflora of sites subject to different degrees of SO_2 pollution, but, while all sites experienced similar rainfall and temperatures, soil conditions differed, the less polluted sites being on limestone and the most polluted on sandstone.

The latter problem was reduced when the forest soil microflora of three very similar Canadian sites, situated 2.8, 6.0 and 9.6 km downwind of a sour gas plant emitting SO_2, were compared (Bewley & Parkinson, 1984,

1985). The sites received less SO$_2$ with increasing distance from the pollution source, as indicated by the SO$_4^{2-}$ content of the organic soil which was approximately 1860, 245 and 109 nl l^{-1}, respectively. While there was a significant reduction in both the total numbers of bacteria and starch-utilizing bacteria from the organic soil in the most polluted site, there were no significant differences in the numbers of fungal propagules isolated, and no differences in species composition were reported. Paralleling the differences in microbial numbers, the relative contributions of bacteria and fungi to total respiration of the organic horizons varied slightly between the most polluted and the other two sites, the ratio of bacterial:fungal respiration being 5:95, 14:83 and 18:82 for the most polluted to the least polluted site. Again, the microflora of the organic soil of five Norway spruce (*Picea abies*) stands varied with distance from a pollution source with high SO$_2$ content in western Bohemia (Langkramer & Lettl, 1982). Numbers of aerobic and ammonifying bacteria were considerably reduced while thiobacilli and microfungi, particularly *Mucor* and *Rhizopus* species, increased in the more polluted sites.

Experimental laboratory studies

Effect of SO$_2$ on fungal growth and germination on agar

Mycelial extension and germination of some fungi on agar have been shown to be inhibited by fumigation with SO$_2$ (Gryzywacz, 1973; Magan & McLeod, 1988; Wookey, Ineson & Mansfield, 1990). However, generally, the concentrations of SO$_2$ used were higher than those normally encountered in nature, and the results must therefore be viewed with caution. On tap water agar, *Aureobasidium* sp., *Cladosporium cladosporioides, Mycena galopus* and *Phoma exigua* were inhibited by 40 nl l^{-1} SO$_2$ in the incubation atmosphere, this concencentration being a daily mean commonly encountered in the vicinity of pollution sources (Fig. 6.4; Dursun, 1994). On other media (potato extract agar, potato dextrose and malt extract agar), growth of some of these species was unaffected, probably at least partly due to the adsorption of SO$_2$ on to media components (see p. 75).

Effect of SO$_2$ on activity of fungi inoculated onto leaf litter

In experiments attempting to bridge the gap between completely artificial conditions and the field, activity (indicated by CO$_2$ evolution) of

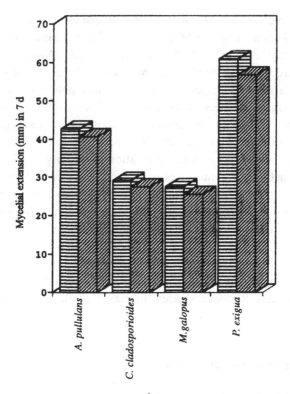

Fig. 6.4. Effect of 40 nl l^{-1} SO$_2$-enriched air on mycelial extension of *Aureobasidium* sp., *Cladosporium cladosporioides*, *Mycena galopus* and *Phoma exigua* growing on tap water agar. SEMS were all within 5% of the mean; extension rate of all species fumigated with SO$_2$ (▤) were significantly different from non-fumigated controls (▨). (From Dursun, 1994.)

Aureobasidium sp., *C. cladosporioides*, *M. galopus* and *P. exigua* colonizing sterilized and inoculated leaf litter was found to be inhibited by exposure to 40 nl l^{-1} SO$_2$ for 4 or 5 wks (Table 6.1; Dursun, 1994). However, whether inhibition occurred depended on the litter species: *P. exigua* was inhibited only on ash and hazel, *C. cladosporioides* only on Scots pine (*Pinus sylvestris*) and ash, *Aureobasidium* sp. only on Sitka spruce (*Picea sitchensis*) but the basidiomycete *M. galopus* on all litters, perhaps reflecting extensive surface growth of mycelium. CO$_2$ evolution from ash and hazel for all test fungi was greater than from spruce and pine litter, probably reflecting the rapidity with which the former two were colonized. In another experiment, CO$_2$ evolution by *Phoma macrostoma* and *P. exigua* colonizing hazel litter was inhibited by only 30 nl l^{-1}

Table 6.1. *Effect of SO₂ fumigation on the mean respiration (CO₂ evolution) rate, measured for 4 or 5 wk, of four litter species inoculated with four decomposer fungal species (3 replicates)*

Litter	Fungi	Mean CO₂ evolution (ml h⁻¹ 100 g⁻¹ litter)			Incubation period (wk)
		Control	Fumigated	Significance	
Pine	*A. pullulans*	57.10 ± 17.89	52.10 ± 18.83	ns	5
Pine	*C. cladosporioides*	45.56 ± 15.39	40.61 ± 15.63	*	5
Pine	*P. exigua*	50.44 ± 14.23	45.34 ± 12.83	ns	5
Pine	*M. galopus*	68.22 ± 9.56	56.67 ± 10.66	***	5
Spruce	*A. pullulans*	62.78 ± 18.31	56.00 ± 21.61	*	5
Spruce	*C. cladosporioides*	51.87 ± 16.34	52.27 ± 17.07	ns	4
Spruce	*P. exigua*	49.11 ± 15.94	47.22 ± 14.50	ns	5
Spruce	*M. galopus*	181.60 ± 74.34	64.40 ± 25.54	***	4
Ash	*A pullulans*	148.93 ± 77.20	145.80 ± 74.55	ns	4
Ash	*C. cladosporioides*	111.00 ± 47.52	96.27 ± 47.11	**	4
Ash	*P. exigua*	153.00 ± 90.35	141.60 ± 88.01	*	4
Ash	*M. galopus*	50.53 ± 25.90	39.80 ± 20.32	***	4
Hazel	*A. pullulans*	57.33 ± 24.05	59.60 ± 20.89	ns	4
Hazel	*C. cladosporioides*	82.40 ± 31.99	81.73 ± 35.70	ns	4
Hazel	*P. exigua*	127.73 ± 49.73	115.73 ± 45.74	*	4
Hazel	*M. galopus*	140.40 ± 51.29	108.07 ± 44.14	***	4

Note: The symbols represent the significance levels of analysis of variance to test overall effect of SO₂ fumigation ($40\,nl\,l^{-1}$) on CO₂ evolution; n.s, not significant; *, $P < 0.05$; **, $P < 0.01$; ***, $P < 0.001$.

SO_2 over 28 d, although *C. cladosporioides* and *Coniothyrium quercinum* var. *glandicola* were not (Newsham *et al.*, 1992*b*).

Effect of SO_2 on microbial activity in naturally colonized leaf litter

The type of experiments just described, involving sterilization followed by inoculation of the substratum, gives no insight into the collective response of a microflora to the pollutant. In an example of a systems approach using naturally colonised material, the decomposition rate (weight loss) of finely ground western wheatgrass (*Agropyron smithii*) was found to be inhibited by exposure to 220 mg m^{-3} (81 nl l^{-1}) SO_2 for 5 wk, initially by 90% and later by 17% reduction. In this experiment significant differences in respiration could not be detected (Leetham, Dodd & Lauenroth, 1983); however, when Austrian pine (*Pinus nigra*) needle litter was constantly exposed to 38 nl l^{-1} SO_2 for 82 d, 33% reduction in respiration was recorded (Ineson, 1983). Manipulation and interpretation of such experiments in the absence of information on the organisms involved are beset with difficulties.

Effect of sulphite on activity of fungi inoculated onto leaf litter

Little work has been done until recently on sulphite – the least toxic of the SO_2 solubility products, despite concentrations in rain water often greater than 45 μM (Cape *et al.*, 1984). The toxicity of sulphite (five concentrations between 0 and 100 μM) to *C. cladosporioides* colonizing Scots pine needles has been investigated at three stages of growth, namely pure cultures of predominantly old spores, mycelium, and fresh spores (Dursun *et al.*, 1993). All three stages were significantly inhibited (as indicated by CO_2 evolution) by 100 μM sulphite and to a lesser extent by lower concentrations, the mycelial stage being less sensitive than spores at lower concentrations (Fig. 6.5). The mycelial stage and freshly added spore stage had recovered from sulphite additions after a fortnight, but the old spore stage had not.

Similar trends were seen when examining the interactive effect of low concentrations of sulphite (0–100 μM) and pH (between 3.0 and 4.5) on CO_2 evolution from Sitka spruce and Scots pine needles, inoculated with pure cultures of *Aureobasidium* sp., *M. galopus* and *P. exigua* as fresh spore suspensions and/or as mycelia (Dursun, 1994). Sulphite inhibited CO_2 evolution for all species, greatest inhibition being with *P. exigua*, then *Aureobasidium* sp., and least with *M. galopus* at all pH values. The effect of sulphite was strongly modified by pH, being greatest at lower pH

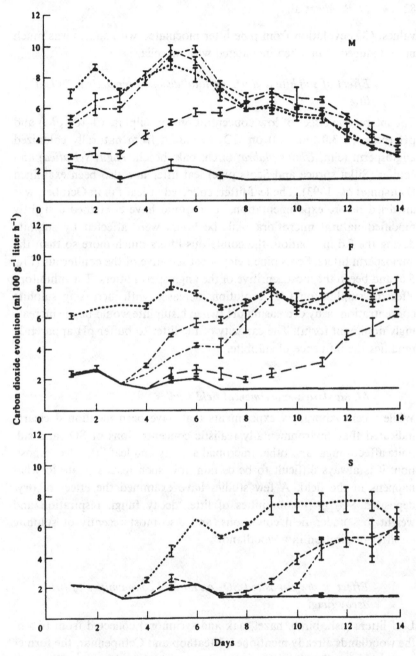

Fig. 6.5. Effect of sulphite on mean (3 replicates; ± SEM) respiration of *Cladosporium cladosporioides* at three growth stages (M, mycelium; F, fresh spores; and S, old spores), inoculated onto Scots pine needle litter at pH 3.0, with time. Sulphite concentration: ·····, 0.0; - - -, 12.5; ----, 25.0; — —, 50.0; ———, 100.0 μM. (From Dursun *et al.*, 1993). (Reprinted by kind permission of Elsevier Science Ltd.)

values. CO_2 evolution from pine litter inoculated with spores was much more reduced than when inoculated with mycelia.

Effect of sulphite and pH on total decomposer activity in leaf litter

The interactive effect of low concentrations of sulphite (0–100 µM) and pH (between 3.0 and 6.0) on CO_2 evolution from naturally colonized angiosperm (elm, *Ulmus glabra*; birch; oak; beech, *Fagus sylvatica*) and conifer (Sitka spruce and Scots pine) leaf litter has also been examined (Dursun *et al.*, 1993). The leaf litter, collected at leaf fall in October, was air dried before experimentation, so it would have contained a slightly modified natural microflora. All the litters were affected by sulphite during the 7 d incubation, the coniferous litters much more so than the angiosperm litters, Scots pine being most sensitive of the coniferous (Fig. 6.6) and beech the most sensitive of the angiosperm litters. The inhibitory effect of sulphite on CO_2 evolution increased with increasing sulphite concentration or by decreasing pH (when bisulphite would be the increasingly dominant form). The capacity of the litter to buffer pH apparently modifies the influence of sulphite.

Manipulative experimental field studies

While recent laboratory experiments that have been mentioned clearly indicated that environmentally realistic concentrations of SO_2 and sulphite affect fungal and other microbial activity and leaf litter decomposition, it is always difficult to be certain how such results relate to what happens in the field. A few studies have examined the effect of dry-deposited SO_2 on communities of litter decay fungi, respiration and weight loss under field conditions, and also most recently of sulphite on litter respiration in a woodland.

Effect of dry-deposited SO_2 on natural communities of litter decay fungi

Leaf litters (ash, birch, hazel, oak and sycamore) collected from two of the woodlands already mentioned, Meathop and Cefnpennar, the former exposed to low (2 nl l^{-1}) SO_2 and the latter to high (> 60 nl l^{-1}) SO_2, were exposed separately at a field fumigation site in trays layered on humus and sand (Newsham *et al.*, 1992*a*). The fumigation system, set up at Liphook, Surrey, by the then Central Electricity Generating Board,

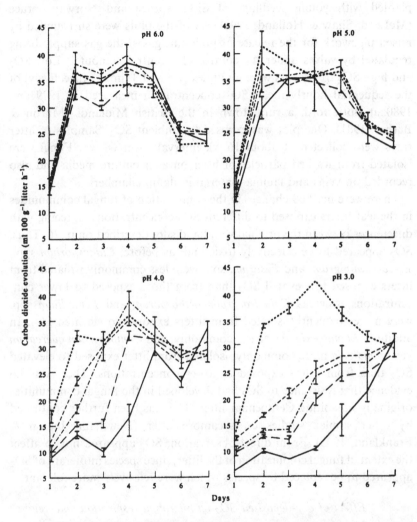

Fig. 6.6. Effect of pH and sulphite on mean (3 replicates; ± SEM) microbial respiration of naturally colonised Scots pine needle litter with time. Sulphite concentration: ·····, 0.0; - - -, 12.5; ----, 25.0; — —, 50.0; ———, 100.0 μM. pH significantly ($P < 0.05$) affected CO_2 evolution at all sulphite concentrations at most times; likewise, sulphite significantly ($P < 0.05$) affected CO_2 evolution at pH 3.0–5.0 but not 6.0, at most times; there was a significant ($P < 0.05$) interaction between pH and sulphite concentration at all times. (From Dursun *et al.*, 1993.) (Reprinted by kind permission of Elsevier Science Ltd.).

comprised seven circular experimental plots, each of 50 m diameter, planted with young seedlings of Sitka spruce and Norway spruce (McLeod, Shaw & Holland, 1992). Six of the plots were surrounded by raised pipework for the release of pollutant gases, the gas supply being regulated by valves under the control of a central computer. Low SO_2 and high SO_2 treatments were multiples (1.5 and 3 times, respectively) of the sequential hourly mean SO_2 concentration measured in 1979 and 1980 at Bottesford, a rural town in the British Midlands (Martin & Barber, 1981). One plot was exposed to ambient SO_2. Samples of litter trays were collected at about 16 wk intervals over 68 wk. Fungi were isolated from washed particles of litter on agar culture media and also recorded on veins and lamina material in damp chambers.

There were marked changes in the composition of fungal communities in the leaf litters exposed to different SO_2 concentrations, agreeing with differences between the original woodland sites described on p. 76. Thus, SO_2 appeared to be selectively toxic and, as before, *Cladosporium* spp., *Epicoccum nigrum*, and *Fusarium* spp. were less commonly isolated from litters exposed to elevated SO_2 than from those exposed to lower concentrations, whereas *Cylindrocarpon orthosporum* and *Penicillium* spp. were more frequently isolated from litters exposed to elevated SO_2. In addition, *Phoma exigua* was less commonly, and *Coniothyrium quercinum* var. *glandicola* more commonly, isolated from litter exposed to elevated SO_2 than from litters exposed to lower concentrations. There was no evidence that resistance to SO_2 had developed in the fungal communities originally inhabiting Cefnpennar litter. This has been further confirmed by other similar studies on sycamore litter (Newsham, Ineson & Frankland, 1995). From direct observation, SO_2 appeared not to affect the extent of fungal colonization of the litter, since species intolerant of SO_2 appeared to be replaced by species which were relatively more tolerant.

Effect of dry-deposited SO_2 on microbial respiration and weight loss of litter

CO_2 evolution and weight loss from the litters exposed in the field experiment just described were determined in the laboratory at each sampling (Newsham *et al.*, 1992*b*). Those litters exposed to elevated SO_2 at Liphook generally respired significantly less CO_2 than those exposed to ambient conditions. There was no evidence to suggest acclimation of microbial communities to SO_2 on the highly polluted site in this study, nor in a similar study of sycamore from fifteen sites subject to a range of SO_2 concentrations (Newsham *et al.*, 1995).

Decomposition rates, estimated from weight loss, in the Liphook experiment were generally lower in the plots with elevated SO$_2$, although significant differences could be detected only infrequently, presumably because of large variation. Similarly, field-fumigated, mixed angiosperm litter from Meathop (29 and 40 nl l^{-1}; Wookey, 1988), western wheat-grass (*Agropyron smithii*; 76 nl l^{-1} SO$_2$ for 153 d; Dodd & Lauenroth, 1981) and Scots pine leaf litter (< 48 nl l^{-1} for 215 d; Wookey, Ineson & Mansfield, 1990) all decomposed more slowly than litter under ambient conditions.

Effect of sulphite on microbial respiration from the leaf litter layer under sycamore, measured using a mobile laboratory

Laboratory studies of sulphite inhibition may not be good predictors of what happens in the field, since sulphite (SO$_3^{2-}$) oxidation to the less toxic sulphate (SO$_4^{2-}$) may occur more rapidly in the field. CO$_2$ flux from the floor of a sycamore woodland close to Merlewood Research Station, Cumbria, was therefore measured using a mobile laboratory. To three replicate plots either 2.5 l deionized water or 50 μM sulphite in 2.5 l water was added at intervals in November/December (see Fig. 6.7; Dursun, 1994). CO$_2$ evolution from the plots was monitored via soil respiration chambers, consisting of closed metal cylinders (18 cm length $\times$ 30 cm diam.) inserted vertically into soil to a depth of 8 cm. Ambient air was

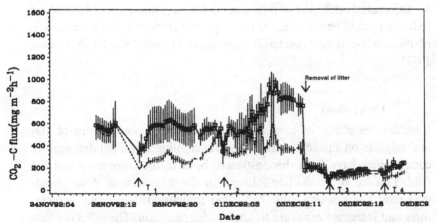

Fig. 6.7. Effect of sulphite on mean (3 replicate plots; ± SEM) CO$_2$ evolution from the floor of a sycamore woodland before and after removal of the litter in November/December 1992. Daily temperature min 3.7 °C, max 9.2 °C; zero precipitation. Sulphite (50 μM in 2.5 l water; - - - -) or 2.5 l deionized water (——) was added at T1 to T4. (From Dursun, 1994).

drawn through the chambers and then routed, via PTFE (polytetrafluoroethylene) pipes, to a gas chromatograph in the mobile laboratory. The air streams were sampled in rotation every 2 h.

Following treatment of the woodland floor, there was initially a slight reduction in CO_2 evolution from the water-treated plots, but after a few hours an increase occurred (Fig. 6.7). The sulphite treatment, however, resulted in a considerable (maximum about 50%) reduction compared with the watered controls. This effect persisted for about 2 d, but CO_2 evolution had returned to the same as that from the controls by 3 d. A subsequent application of sulphite again resulted in a considerable reduction in CO_2 evolution, suggesting that no resistance had developed, but after 3 d CO_2 evolution remained constant and did not return to that of the control plots. The effect of sulphite was suspected to be exerted in the litter layer, so to test this the litter layer was removed. Measurement of CO_2 flux was repeated for 1 d and, not surprisingly, was considerably less than with litter present. The soil was then treated as previously with application of water or sulphite at 3 d intervals. Slightly less CO_2 was evolved from sulphite-treated plots, but there was no significant difference between treatments after the first application, although there was after the second. Thus, environmentally realistic concentrations of sulphite inhibit microbial activity in both the soil and the litter layer. Again, effects on the fungal component were not distinguished from those on other microorganisms, but, assuming (albeit unlikely) there is no differential inhibition of different microbial groups, about 70% of the reduction could be attributable to fungal respiration, based on estimates of the relative fungal : bacterial respiration in soils (Kjøller & Struwe, 1982).

Conclusions

Considerable effects of environmentally-realistic concentrations of SO_2 and sulphite on the composition and functioning of fungal decomposer communities have been demonstrated, both in the laboratory and the field. This has profound implications for the maintenance of our woodland ecosystems, which could be further compounded by rising temperatures and increased exposure to ultra-violet radiation. The effects of SO_2 and sulphite on fungal decomposition may be a significant factor in forest decline, although these effects may be outweighed by the large effects on plant growth of changes in soil and litter cation balances, resulting from acidification.

Acknowledgement

Thanks are offered to the University of Ondokuz Mayis in Turkey for supporting S. Dursun during the course of the work on sulphite.

References

Abrahamsen, G. (1984). Effects of acidic deposition on forest soil and vegetation. *Philosophical Transactions of the Royal Society of London (B)* **305**, 369-82.

Andreae, M.O. & Jaeschke, W.A. (1992). Exchange of sulphur between biosphere and atmosphere over temperate and tropical regions. In *Sulphur Cycling on the Continents. Wetlands, Terrestrial Ecosystems, and Associated Water Bodies*, ed. R.W. Howarth, J.W.B. Stewart & M.V. Ivanov, pp. 27-66. Chichester: John Wiley.

Anon. (1981). *Effects of SO₂ and its Derivatives on Natural Ecosystems, Agriculture, Forestry and Fisheries*. Report of an International Electric Research Exchange (IERE) Working Group. Leatherhead, UK: Central Electricity Research Laboratories.

Babich, H. & Stotzky, G. (1974). Air pollution and microbial ecology. *Critical Reviews in Environmental Control*, **4**, 353-421.

Babich, H. & Stotzky, G. (1980). Environmental factors that influence the toxicity of metal and gaseous pollutants to microorganisms. *CRC Critical Reviews in Microbiology*, **8**, 99-145.

Bewley, R.J. & Parkinson, D. (1984). Effects of sulphur dioxide pollution on forest soil organisms. *Canadian Journal of Microbiology*, **30**, 179-85.

Bewley, R.J. & Parkinson, D. (1985). Bacterial and fungal activity in sulphur dioxide polluted soils. *Canadian Journal of Microbiology*, **31**, 13-15.

van Breeman, N. (1985). Acidification and the decline of central European forests. *Nature*, **315**, 16.

Brimblecombe, P., Hammer, C., Rodhe, H., Ryaboshapko, A. & Boutron, C.F. (1989). Human influence on the sulphur cycle. In *Evolution of the Biogeochemical Sulphur Cycle*, ed. P. Brimblecombe & A.Y. Lein, pp. 77-121. Wiley: New York.

Cape, J.N., Fowler, D., Kinnaird, J.W., Paterson, I.S., Leith, I.D. & Nicholson, I.D. (1984). Chemical composition of rainfall and wet deposition over northern Britain. *Atmospheric Environment*, **18**, 1921-1932.

Cowling, E.B. (1982). Acid precipitation in historical perspective. *Environmental Science and Technology*, **16**, 110-23.

Cruess, W.V., Richert, P.H. & Irish, J.H. (1931). The effect of hydrogen-ion concentration on the toxicity of several preservatives to microorganisms. *Hilgardia*, **6**, 295-314.

Dodd, J.L. & Lauenroth, W.K. (1981). Effects of low-level SO₂ fumigation on decomposition of western wheatgrass litter in a mixed-grass prairie. *Water, Air and Soil Pollution*, **15**, 257-61.

Domanski, S. & Kowalski, T. (1987). Fungi occurring on forests injured by air pollution in the Upper Silesia and Cracow industrial regions. X. Mycoflora of dying young trees of *Alnus incana*. *European Journal of Forest Pathology*, **17**, 337-48.

Dursun, S. (1994). *The effects of sulphur pollution on soil fungi and decomposition of tree leaf litters.* PhD Thesis. University of Wales, Cardiff.

Dursun, S, Ineson, P., Frankland, J.C. & Boddy, L. (1993). Sulphite and pH effects on CO_2 evolution from decomposing angiospermous and coniferous tree leaf litters. *Soil Biology and Biochemistry*, 25, 1513-25.

EMEP. (1984). *Report of the Meteorological Synthesizing Centre West (MSC-W) for 1984.* EMEP/MSC-Report 1/84. Co-operative programme for monitoring and evaluation of the long-range transmission of air pollutants in Europe (EMEP). Oslo: Norwegian Meteorological Institute.

EMEP. (1989). *Airborne Transboundary Transport of Sulphur and Nitrogen Over Europe – Model Descriptions and Calculations.* EMEP/MSC-Report 2/89. Co-operative programme for monitoring and evaluation of the long-range transmission of air pollutants in Europe (EMEP). Oslo: Norwegian Meteorological Institute.

Garrett, H., Carney, J.L. & Hedrick, H.G. (1982). The effects of ozone and sulfur dioxide on respiration of ectomycorrhizal fungi. *Canadian Journal of Forest Research*, 12, 141-5.

Germida, J.J., Wainwright, M. & Gupta, V.V.S.R. (1992). Biochemistry of sulphur cycling in soil. In *Soil Biochemistry*, Vol. 7, ed.G. Stotzky & J-M. Bollag, pp. 1-53. New York: Marcel Dekker, Inc..

Gilbert, O.L. (1968). Bryophytes as indicators of air pollution in the Tyne valley. *New Phytologist*, 67, 15-30.

Grzywacz, A. (1973). Sensitivity of *Fomes annosus* Fr. Cooke and *Schizophyllum commune* Fr. to air pollution with sulphur dioxide. *Acta Societatis Botanicorum Poloniae*, 42, 347-60.

Grzywacz, A. & Wazny, J. (1973). The impact of industrial air pollutants on the occurrence of several important pathogenic fungi of forest treees in Poland. *European Journal of Forest Pathology*, 3, 191-241.

Harvey, G.W. & Legge, A.H. (1979). The effect of sulfur dioxide upon the metabolic level of adenosine triphosphate. *Canadian Journal of Botany*, 57, 759-64.

Hawksworth, D.L. & Rose, F. (1970). Qualitative scale for estimating sulphur dioxide air pollution in England and Wales using epiphytic lichens. *Nature, London*, 227, 145-8.

Haynes, R.J. & Swift, R.S. (1986). Effects of soil acidification and subsequent leaching on levels of extractable nutrients in a soil. *Plant and Soil*, 95, 327-36.

Ineson, P. (1983). *The effect of airborne sulphur pollutants upon decomposition and nutrient release in forest soils.* PhD Thesis, University of Liverpool.

Irwin, J.G. (1989). Acid rain: emissions and deposition. *Archives of Environmental Contamination and Toxicology*, 18, 95-107.

Kelly, T.J., McLaren, S.E. & Kadlececek, J.A. (1989). Seasonal variations in SO_x and NO_y species in the Adirondacks. *Atmospheric Environment*, 23, 1315-32.

Kellogg, W.W., Cadle, R.D., Allen, E.R., Lazrus, A.L. & Martell, E.A. (1972). The sulfur cycle. *Science*, 175, 578-96.

Kjøller, A. & Struwe, S. (1982). Microfungi in ecosystems: fungal occurrence and activity in litter and soil. *Oikos*, 39, 389-422.

Langkramer, O. & Lettl, A. (1982). Influence of industrial atmospheric pollution on soil biotic component of Norway spruce stands. *Zentralblatt für Mikrobiologie*, 137, 180-96.

Leetham, J.W., Dodd, J.L. & Lauenroth, W.K. (1983). Effects of low-level sulphur dioxide exposure on decomposition of *Agropyron smithii* litter under laboratory conditions. *Water, Air and Soil Pollution*, **19**, 247-50.

Magan, N. & McLeod, A.R. (1988). *In vitro* growth and germination of phylloplane fungi in atmospheric sulphur dioxide. *Transactions of the British Mycological Society*, **90**, 571-5.

Martin, A. & Barber, F.R. (1981). Sulphur dioxide, oxides of nitrogen and ozone measured continuously for two years at a rural site. *Atmospheric Environment*, **15**, 567-78.

McLeod, A.R., Shaw, P.J.A. & Holland, M.R. (1992). The Liphook forest fumigation project: studies of sulphur dioxide and ozone effects on coniferous trees. *Forest Ecology and Management*, **51**, 121-7.

Möller, D. (1984). Estimation of the global man-made sulphur emission. *Atmospheric Environment*, **18**, 19-27.

Newsham, K.K., Boddy, L., Frankland, J.C. & Ineson, P. (1992*b*). Effects of dry-deposited sulphur dioxide on fungal decomposition of angiosperm tree leaf litter III. Decomposition rates and fungal respiration. *New Phytologist*, **121**, 127-40.

Newsham, K.K., Frankland, J.C., Boddy, L. & Ineson, P. (1992*a*). Effects of dry-deposited sulphur dioxide on fungal decomposition of angiosperm tree leaf litter I. Changes in communities of fungal saprotrophs. *New Phytologist*, **121**, 97-110.

Newsham, K.K., Ineson, P. & Frankland, J.C. (1995). The effects of open-air fumigation with sulphur dioxide on the decomposition of sycamore (*Acer pseudoplatanus* L.) leaf litters from polluted and unpolluted woodlands. *Plant, Cell and Environment*, **18**, 309–19.

Oden, S. (1968). *The Acidification of Air and Precipitation and its Consequences in the Natural Environment*. Ecology Committee Bulletin No.1, Swedish National Science Research Council: Stockholm.

Roberts, T.M., Darrall, N.M. & Lane, P. (1983). Effects of gaseous air pollutants on agriculture and forestry in the UK. *Advances in Applied Biology*, **9**, 1-142.

Schlesinger, W.H. (1991). *Biogeochemistry. An Analysis of Global Changes*. New York: Academic Press, Inc.

Thompson,G.W and Medve, R.J. (1984). Effects of aluminium and manganese on the growth of ectomycorrhizal fungi. *Applied and Environmental Microbiology*, **48**, 556-60.

Ulrich, B., Mayer, R. & Khanna, P.K. (1980). Chemical changes due to acid precipitation in a loess-derived soil in central Europe. *Soil Science*, **130**, 193-9.

Vas, K. & Ingram, M. (1949). Preservation of fruit juices with less sulphur dioxide. *Food Manufacture*, **24**, 414-16.

Williams, M.L., Atkins, D.H.F., Bower J.S., Campbell, G.W., Irwin, J.G. & Simpson, D. (1989). *A Preliminary Assessment of the Air Pollution Climate of the UK*. Report LR 723(AP) Stevenage, UK: Warren Spring Laboratory, Department of Trade and Industry.

Wookey, P.A. (1988). *Effects of dry deposited sulphur dioxide on the decomposition of forest leaf litter*. PhD Thesis, University of Lancaster.

Wookey, P.A. Ineson, P. & Mansfield, T.A. (1990). Effects of atmospheric sulphur dioxide on microbial activity in decomposing forest litter. *Agriculture, Ecosystems and Environment*, **33**, 263-80.

7

Effects of atmospheric pollutants on phyllosphere and endophytic fungi

N. MAGAN, M. K. SMITH AND
I. A. KIRKWOOD

Introduction

Plant leaf surfaces (here termed the phyllosphere) are colonized by a wide range of bacteria, yeasts and filamentous fungi. They include epiphytic fungi, endophytes and pathogens. The total community and its structure are influenced by a range of abiotic and biotic factors. These include not just atmospheric pollutants but also wind, precipitation, water availability and pH. Microbial community structure may also be influenced by the presence of nutrients from insect honey dew, pollen, leaf exudates and organic debris. Interactions between abiotic and biotic factors can occur. For example, atmospheric pollutants have been demonstrated to markedly increase the weathering of conifer needle surfaces and to decrease the protective wax fibrillar structure (Rinallo et al., 1986), which could increase nutrient exudates in the phyllosphere and thus influence patterns of colonization and perhaps senescence of needles.

The impact of atmospheric pollutants on both epiphytic and endophytic fungi colonizing plant leaves has received increasing attention by being implicated directly or indirectly in 'forest decline' in parts of Europe where significant premature senescence and defoliation of forests occurred in the 1970s and 1980s (Boddy et al., this volume; Schutt & Cowling, 1985; Kandler, 1990). For example, Rehfuess and Rodenkirchen (1985) implicated the endophytes *Lophodermium piceae* and *Rhizosphaera kalkhoffii* in premature senescence and needle reddening disease. However, Butin and Wagner (1985) suggested that they may be only early colonizers of dying or dead needles and therefore not involved in such forest decline syndromes.

The major primary and secondary pollutants implicated in tree decline have been sulphur dioxide (SO_2) (see Chapter 6), nitrogen oxides (NO_x, ozone (O_3), and, in some countries, ammonia (NH_3). Their major anthropogenic sources can be summarized as: (SO_2) electricity generation, fossil fuel combustion and smelter operations; (NO_x) high temperature combustion and fertilizer production; (O_3) atmospheric transformation of NO_2 driven by sunlight in the presence of NO_x and hydrocarbons, and (NH_3) intensive animal husbandry. During long distance transport of SO_2, transformation into SO_4^{2-} with NO_3^- contributes to wet acidic deposition. The effect of dry and wet deposition of SO_2, and O_3 on phyllosphere fungi has been studied in some detail, while that of NO_x and NH_3 has received surprisingly little attention (Magan & McLeod, 1991*a*). It is important to remember that SO_2 is an atmospheric pollutant which dissolves readily in water, and exists in various ionic forms dependent on the pH of the solution, including undissociated sulphurous acid, bisulphite and sulphite ions. These relationships are described in detail elsewhere (see Boddy *et al.*, Chapter 6). The effects of SO_2 on fungi are determined by the pH of the solution, as well as its form. Generally, fungi have been found to be more sensitive to HSO_3^- and SO_3^-, particularly below pH 4.5 (see Babich & Stotzky, 1982; Magan, 1993). The mean levels of exposure to dry atmospheric SO_2 in the UK and parts of Germany have been in the range 5–12 and 6–17 $nl\,l^{-1}$, respectively. While that of ozone, particularly in the summer months can reach peaks of 15–20 $nl\,l^{-1}$.

Because the effects of dry and wet deposition of SO_2 and O_3 on phyllosphere and endophytic fungi have received particular attention, this chapter will consider some aspects of the following: (1) *in vitro* effects of dry and wet SO_2 and O_3 on phyllosphere and endophytic fungi; (2) impact of atmospheric SO_2 on specific phyllosphere fungi and endophytes when pre-inoculated on to seedlings; (3) effect of open-air fumigation with SO_2 and O_3 on phyllosphere and endophytic fungi. A more detailed review of the impact of atmospheric pollutants on phyllosphere fungi and pathogens, and mechanisms involved in sensitivity and tolerance of fungi to SO_2 can be found elsewhere (Magan & McLeod, 1991*a*; Magan, 1993).

In vitro effects of sulphur dioxide and ozone

Work by Magan and McLeod (1988) demonstrated that phyllosphere yeasts from the leaves of cereals, such as *Sporobolomyces roseus* and *Cryptococcus* spp., were particularly sensitive to atmospheric SO_2 up to 200 $nl\,l^{-1}$ while filamentous fungi were much less sensitive. Using 1%

Table 7.1. *The effect of atmospheric sulphur dioxide and ozone on mean percentage (%) spore germination of filamentous phyllosphere fungi isolated from conifer needles and exposed for 24 h on 1% malt agar plates (pH 3.5) at 18–20 °C. Means of 4 × 100 spores examined*

Sulphur dioxide (nl l⁻¹)	0	50	100	200
Fungal species				
Alternaria alternata	100	100	100	98
Arthrinium phaeospermum	95	95	94	89
Botrytis cinerea	100	100	100	100
Cladosporium herbarum	95	94	56	37
Epicoccum nigrum	95	91	83	84
Penicillium sp.	99	89	99	33
Ozone (nl l⁻¹)	0	50	100	200
Fungal species				
A. alternata	100	100	100	100
A. phaeospermum	94	97	91	91
B. cinerea	100	100	100	100
C. herbarum	93	95	97	90-
Epicoccum nigrum	97	97	97	96
Penicillium sp.	100	99	98	99

Note: (From Kirkwood, 1991).

malt extract agar plates and a closed chamber exposure system (Lockyer, Cowling & Jones, 1976), Kirkwood (1991) showed that phyllosphere fungi from conifer needles behave in a very similar manner to concentrations of 200 nl l^{-1} SO$_2$ or O$_3$. Table 7.1 shows that germination of *Alternaria alternata*, *Botrytis cinerea*, *Epicoccum nigrum* and *Arthrinium phaeospermum* were relatively unaffected by either atmopheric pollutant when exposed to < 200 nl l^{-1} for 24 h at 18–20 °C. However, SO$_2$ at 100 and 200 nl l^{-1} concentration did significantly reduce germination of *Cladosporium herbarum* and at 200 nl l^{-1} reduced germination of a *Penicillium* spp. These experiments were all carried out at pH 3.5, so changing the pH or the time of contact might have influenced the outcome of the experiment. Other factors such as water availability (water activity, a_w) may also interact with the adverse effect of the SO$_2$ on phyllosphere fungi including endophytes.

Laboratory studies with conifer needle isolates of epiphytic and endophytic fungi implicated in premature senescence and needle drop such as *L. piceae* (and other *Lophodermium* spp.) and *R. kalkhoffii* have been carried out (Kirkwood, 1991; Smith, 1993) to determine the efficacy of dissolved SO$_2$ (up to 2000 nl l^{-1}) on fungal growth. As before, 1% malt

Table 7.2. *The effect of dissolved sulphur dioxide ($\mu l \, l^{-1}$) on growth (mm day^{-1}) of epiphytic and endophytic fungi grown on 1% malt extract agar at 15 and 25 °C. All data are means of three experiments with three replicates on each occasion*

Fungal species	SO$_2$ optimum	Growth rate	SO$_2$ maximum	Growth rate
15 °C				
Aureobasidium pullulans	0	1.50	150	0.45
Arthrinium phaeospermum	0	4.0	75	0.95
Botrytis cinerea	0–25	6.50	150	0.78
Cladosporium herbarum	0	1.35	75	0.76
Epicoccum nigrum	0	2.95	100	0.45
Sclerophoma phythiophila	0–75	2.10	150	0.95
Lophodermium seditiosum	0	0.50	50	0.23
Lophodermium pinastri	0	0.49	50	0.18
Lophodermium conigenum	0	1.55	50	0.23
Lophodermium piceae	0	0.54	50	0.39
Rhizosphaera kalkhoffii	25–50	1.17	75	1.12
25°C				
A. pullulans	0	1.70	150	0.50
A. phaeospermum	0	2.73	100	0.85
B. cinerea	0	8.20	150	0.70
C. herbarum	0	1.34	200	0.36
E. nigrum	0	3.76	200	0.41
S. pythiophila	0–25	0.62	150	0.21
L. seditiosum	0	1.95	50	0.57
L. pinastri	0	1.96	50	1.15
L. conigenum	0	1.94	10	1.61
L. piceae	0	0.75	50	0.24
R. kalkhoffii	50	0.27	50	0.27

Note: Data on epiphytes from Kirkwood (1991) and endophytes from Smith (1993).

extract agar was used at pH 3.5 and 15 or 25 °C. These showed that a range of filamentous epiphytic fungi and some endophytes were particularly tolerant of elevated concentrations of dissolved SO$_2$ (Table 7.2). The endophyte *R. kalkhoffii* is of particular interest as it was significantly more tolerant of SO$_2$ than was *L. piceae* or other *Lophodermium* spp. Isolates of *R. kalkhoffii* from sites in the UK also behaved in a similar manner to those from forest decline areas in Germany (Smith, 1993).

Further studies suggested that interactions between SO$_2$ and water activity could also influence the ability of such endophytic fungi to become established in a polluted environment. Fig. 7.1 compares the

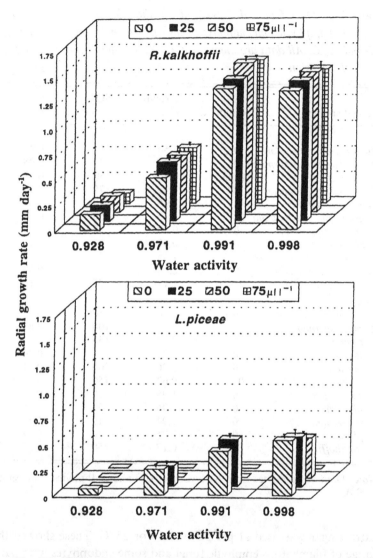

Fig. 7.1. Effect of interaction between dissolved sulphur dioxide concentration (μl l^{-1}) and water activity (a_w) on growth of the needle endophytes *Rhizosphaera kalkhoffii* and *Lophodermium piceae* on 1% malt extract agar (pH 3.5) and 15 °C. (After Smith, 1993). Bars indicate standard errors.

behaviour of a *R. kalkhoffii* and *L. piceae* isolate from the UK at different interacting SO_2 and a_w levels. Again, this showed that *R. kalkhoffii* may have a competitive advantage over *L. piceae* when colonizing conifer needles over a range of interacting environmental conditions. This type of laboratory study has also demonstrated that growth of phyllosphere fungi such as *Aureobasidium pullulans* and *Sclerophoma pythiophila* is unaffected by $250-500\,\text{nl}\,\text{l}^{-1}$ SO_2 in solution at reduced a_w levels (Kirkwood, 1991).

Establishment of phyllosphere and endophytic fungi on conifer needles exposed to atmospheric sulphur dioxide

Very few studies have been carried out to follow the ability of fungi to become established on plant surfaces during controlled exposure to SO_2. Two such experiments have been carried out by inoculating *A. pullulans* or *S. pythiophila* onto seedlings of Scots pine (*Pinus sylvestris*) and *R. kalkhoffii* and *L. piceae* on to those of Sitka spruce (*Picea sitchensis*) and exposing them to 0, 50, 100 and $200\,\text{nl}\,\text{l}^{-1}$ SO_2 in closed fumigation chambers (Lockyer *et al.*, 1976). The phyllosphere fungi were sprayed onto the needles and samples taken immediately before inoculation, and after 1, 3 and 5 days. For endophyte experiments, the seedlings were inoculated and exposed to SO_2 for 2 weeks, then removed from the chambers, and left for a further 2 weeks. Phyllosphere fungi were examined, both by a serial washing technique (Magan & McLeod, 1991*b*) and the fluorescein diacetate (FDA) fungal activity method (Shaw & Johnston, 1993), while, for endophytes, needle segments were surface sterilized and plated onto acidified 1% malt extract agar as described by Magan *et al.* (1995).

Fig. 7.2 shows the effect of SO_2 exposure on changes in populations of *A. pullulans* and on fungal activity (FDA technique) of *S. pythiophila*. There was a more rapid increase in the total colony-forming units (CFUs) of *A. pullulans* in the $50\,\text{nl}\,\text{l}^{-1}$ SO_2 treatment after 3 days' exposure, and the population levels after 5 days were greatest in the same treatment. Very similar results were obtained with *S. pythiophila*, where total fungal activity and CFUs (not shown) were greater in the 50 and $100\,\text{nl}\,\text{l}^{-1}$ SO_2 treatments after 3 days than in the control. Fig. 7.3 shows the frequency of isolation of the two endophytes, *R. kalkhoffii* and *L. piceae*, from the needles of Sitka spruce seedlings. The frequency of isolation of *R. kalkhoffii* remained about 10% at $50-200\,\text{nl}\,\text{l}^{-1}$ SO_2, whereas for *L. piceae* optimum isolation frequency was obtained at $100\,\text{nl}\,\text{l}^{-1}$ SO_2 and a very

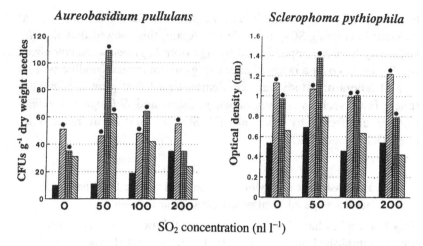

Fig. 7.2. Effect of fumigation with atmospheric sulphur dioxide $(0-200\,nl\,l^{-1})$ on the fungal colonization of needles of Scots pine seedlings pre-inoculated with either *Aureobasidium pullulans* or *Sclerophoma pythiophila* in closed fumigation chambers. Samples were taken just prior to inoculation (■), after 1 (▨), 3 (▦) and 5 (▧) days. Data are presented for total colony-forming units $(g^{-1}$ needles) for *A. pullulans*, and the fungal activity of *S. pythiophila* was assessed using the FDA (fluorescein diacetate) hydrolysis technique. Asterisks indicate significant differences $(P < 0.05)$ from controls.

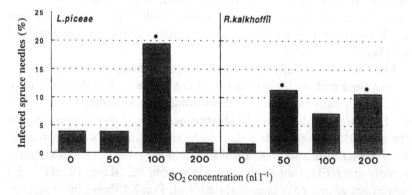

Fig. 7.3. Frequency of isolation of the endophytes *Lophodermium piceae* and *Rhizosphaera kalkhoffii* from pre-inoculated Sitka spruce seedlings exposed to $0-200\,nl\,l^{-1}$ atmospheric sulphur dioxide for a period of two weeks. Asterisks indicate significant differences $(P < 0.05)$ from controls.

low isolation level at $200\,nl\,l^{-1}$. This points to *L. piceae* perhaps being sensitive to highly elevated atmospheric SO_2 concentrations, while *R. kalkhoffii* was tolerant and able to colonize needles over a wider range of SO_2 exposure levels.

Effects of open-air fumigation with sulphur dioxide and ozone on phyllosphere and endophytic fungi

Over a three year period the impact of elevated SO_2 and O_3 on phyllosphere fungal populations of Scots pine, Sitka spruce and Norway spruce (*Picea abies*), and on isolation of *R. kalkhoffii* and *L. piceae* from green Sitka spruce needles, was determined in an open-air exposure system described by McLeod *et al.* (1992) (see Boddy *et al.*, Chapter 6). Detailed sampling using serial washing, direct plating of needle segments and measurement of total fungal activity by the FDA technique was done to assess the effect of SO_2 and O_3 treatments on the total and dominant phyllosphere fungi (Magan *et al.*, 1995). Table 7.3 summarizes the effects of SO_2 and O_3 on the total populations of fungi found over the whole experimental period. This shows that there was a significant decrease in fungal populations or fungal activity on some conifer species. Generally, there was a marked increase in fungal activity and total CFUs on needles in response to O_3. There were significant between–plot differences for a number of the dominant, component fungal species, including *A. pullulans*, *S. pythiophila*, *Epicoccum nigrum* and yeasts on needles of the three conifer species. However, analyses of treatment effects due to SO_2 (high, low and ambient) showed significant reductions in response to SO_2 of pink yeasts on Scots pine needles, and of *E. nigrum* and *S. pythiophila* on Sitka and Norway spruce, based on isolation frequencies from needle segments.

Previously, Magan & McLeod (1991*b*) showed that phyllosphere yeasts were more sensitive to gaseous SO_2 treatments than filamentous fungi in open-air fumigation experiments. These studies revealed that *A. pullulans* populations were significantly increased on leaves and ears of winter barley (*Hordeum vulgare*) exposed to $47\,nl\,l^{-1}$ SO_2. By contrast, populations of pink and white yeasts were significantly reduced, while those of *Cladosporium* spp. were unaffected. However, Fenn, Dunn and Durall (1989) reported that exposure of Valencia orange trees (*Citrus sinensis*) to $93\,nl\,l^{-1}$ SO_2 in open-top chambers resulted in a decrease in isolation of *A. pullulans*, *C. cladosporioides* and *A. alternata* from washed leaf segments.

Table 7.3. *Effects of atmospheric sulphur dioxide (SO_2) and ozone (O_3) fumigation on phyllosphere fungi. Mean total fungal populations (log-transformed total colony-forming units, CFU) and fungal activity using the fluorescein diacetate (FDA, optical density g^{-1} needles $g^{-1} h^{-1}$) over a three-year experimental period. Statistical analyses are of means using a one-way analysis of variance for SO_2 and O_3 treatment effects* (From Magan et al., 1995, with permission.)

Tree species			Scots pine		Sitka spruce		Norway spruce
Mean SO_2		O_3	FDA	CFU	FDA	CFU	CFU
Plot number	($nl l^{-1}$)						
1	4	25	0.4210	4.49	0.7617	4.36	4.91
2	22	25	0.2350	4.30	0.3072	4.30	4.82
3	13	30	0.3251	4.41	0.2884	4.04	4.89
4	5	30	0.4713	4.62	1.0462	4.38	4.90
5	22	29	0.1950	4.55	0.2703	4.51	4.83
6	12	25	0.2800	4.36	0.5276	3.87	4.84
7	4	25	0.2950	4.44	0.3839	4.30	4.87
ANOVA SO_2 $F(2,3)$			4.36	4.38	1.82	39.15**	6.74
ANOVA O_3 $F(1,3)$			0.75	10.26*	0.25	8.94	1.78

Note: $*P < 0.05$; $**P < 0.01$.

Dry deposition may modify the phyllosphere surface differently from wet deposition, influencing the ability of different species to colonize. Wet deposition has a significant effect on the leaf surface pH and can result in much higher actual pollutant concentrations in the phyllosphere. For example, recent studies on the phyllosphere fungi of pine and birch (*Betula*) leaves in the sub-arctic region of Finland demonstrated that populations of *A. pullulans* and *Hormonema* spp. were reduced in field irrigation experiments with acid rain (sulphuric acid or sulphuric + nitric acids) at $< pH$ 3.5 (Helander, Ranta & Neuvonen, 1993). Thus complex interactions may occur between dry and wet deposited SO_2 and this needs to be borne in mind when interpreting results from dry or wet deposition studies only, because often a combination of both types of pollution may impact on phyllosphere surfaces.

Of the two endophytes, *R. kalkhoffii* and *L. piceae*, the former was consistently isolated from the Sitka spruce needles in an open-air fumigation study (Magan *et al.*, 1995). The frequency of isolation was relatively low but increased with needle age. Generally, there was a higher level of isolation from both second and third year green needles on a number of sampling dates. This is similar to the results in a complementary field study where much higher colonization frequencies of first, second and third year needles by *R. kalkhoffii* than by *L. piceae* on Sitka spruce trees were obtained in poor growth stands that had been exposed to elevated pollution levels, compared to stands exhibiting good growth and exposed to low pollution levels (Smith, 1993).

The establishment of symptomless endophytic infection of green needles of young Sitka spruce stands could have important implications for tree vigour. For example, Tanaka (1980) suggested that trees under environmental stress were more susceptible to *R. kalkhoffii* and that SO_2 facilitated spread and reproduction of the endophyte in needle tissue of Japanese red pine (*Pinus densiflora*). This suggests that *R. kalkhoffii*, which is more tolerant than *L. piceae* of a range of environmental factors including SO_2, a_w and pH, may be at a competitive advantage once established in green needles. Interestingly, Smith (1993) found in interaction experiments on media containing between 250 and $500 \, nl \, l^{-1}$ dissolved SO_2 that *R. kalkhoffii* became dominant when paired with a range of *Lophodermium* spp., while in the absence of SO_2, *L. piceae* was dominant. Recent work in Sweden demonstrated that *R. kalkhoffii* was significantly linked with insect injury, while the opposite was true for *L. piceae* (Livsey & Barklund, 1992). Wet deposition of sulphuric and sulphuric + nitric acids has also been shown to influence the isolation of endophytes from Scots pine (Helander *et al.*, 1994). They found that *Cenangium ferruginosum* was the dominant endophyte (> 60% needles), with *Cyclaneusma minus* isolated less frequently (< 12% of needles). The colonization patterns were positively correlated with stand density, and infection was reduced by acidified sulphuric acid or sulphuric + nitric acid treatments but not by nitric acid alone. This may have been owing to a fertilizing effect on the trees. Thus, complex interactions may occur between biotic factors, such as phyllosphere and endophytic fungi and insects, and stresses due to different atmospheric pollutants, and the role of phyllosphere fungi in forest decline syndromes should be seen in this context.

Acknlowledgements

We are grateful to National Power Technology and Environment Centre and the Natural Environment Research Council for funding aspects of the work.

References

Anon. (1989). Reduced growth and bent top of Sitka spruce on the South Wales coalfield. Progress Report, Forestry Commission Northern Research Station, Bush Estate, Edinburgh, Scotland.

Babich, H. & Stotzky, G. (1982). Gaseous and heavy metal pollutants. In *Experimental Microbial Ecology*, ed. R. G. Burns & J. H. Slater, pp. 631–670. Blackwell Scientific Publications: Oxford, UK.

Butin, H. & Wagner, C. (1985). Mykologische untersuchungen zur nadelrote der fichte. *Forstwissenschaftliches Centralblatt*, **104**, 178–86.

Fenn, M. E., Dunn, P. H. & Durall, D. M. (1989). Effects of ozone and sulphur dioxide on phyllosphere fungi from three tree species. *Applied and Environmental Microbiology*, **55**, 412–18.

Helander, M. L., Ranta, H. & Neuvonen, S. (1993). Responses of phyllosphere microfungi to simulated sulphuric and nitric acid deposition. *Mycological Research*, **97**, 533–7.

Helander, M. L., Sieber, T. N., Petrini, O. & Neuvonen, S. (1994). Endophytic fungi in Scots pine needles: spatial variation and consequences of simulated acid rain. *Canadian Journal of Botany*, **72**, 1108–13.

Kandler, O. (1990). Epidemiological evaluation of the development of Waldsterben in Germany. *Plant Disease*, **74**, 5–12.

Kirkwood, I. A. (1991). Phyllosphere microflora of conifer needles and effects of atmospheric sulphur dioxide and ozone. PhD Thesis, Cranfield University, UK.

Livsey S. & Barklund, P. (1992). *Lophodermium piceae* and *Rhizosphaera Kalkhoffii* in fallen needles of Norway spruce (*Picea abies*). *European Journal of Forest Pathology*, **22**, 204–16.

Lockyer, D. R., Cowling, D. W. & Jones, L. H. P. (1976). A system for exposing plants to atmospheres containing low concentrations of sulphur dioxide. *Journal of Experimental Botany*, **27**, 397–409.

Magan, N. (1993). Tolerance of fungi to sulphur dioxide. In *Stress Tolerance of Fungi*, ed. D. H. Jennings, pp. 173–187. Marcel Dekker: New York.

Magan, N., Kirkwood, I. A., McLeod, A. R. & Smith, M. K. (1995). Effect of open-air fumigation with sulphur dioxide and ozone on phyllosphere and endophytic fungi of conifers. *Plant, Cell and Environment*, **18**, 291–302.

Magan, N. & McLeod, A. R. (1988). *In vitro* growth and germination of phylloplane fungi in atmospheric sulphur dioxide. *Transactions of the British Mycological Society*, **90**, 3–8.

Magan, N. & McLeod, A. R. (1991a). Effects of atmospheric pollutants on phyllosphere microbial communities. In *Microbiology of the Phyllosphere*, ed. J. H. Andrews & S. S. Hirano, pp. 379–400. Springer Verlag: New York.

Magan, N. & McLeod, A. R. (1991b). Effects of open-air fumigation with sulphur dioxide on the occurrence of phylloplane fungi on winter barley. *Agriculture, Ecosystems and Environment*, **33**, 245–62.

McLeod, A. R., Shaw, P. J. A., & Holland, M. R. (1992). The Liphook forest fumigation project: studies of sulphur dioxide and ozone effects on coniferous trees. *Forest Ecology and Management*, **51**, 121–7.

Rehfuess, K. E. & Rodenkirchen, H. (1985). Uber die nadelrote-erkrangun der fichte in SudDeutschland. *Forstwissenschaflisches Centralblatt*, **103**, 248–62.

Rinallo, C., Radii, P., Gellinin, R. & Di Lornardo, V. (1986). Effects of simulated acid deposition on the surface structure of Norway spruce and silver fir needles. *European Journal of Forest Pathology*, **16**, 440–6.

Schutt, P. & Cowling, E. B. (1985). Waldsterben, a general decline of forests in central Europe: symptoms, development and possible causes. *Plant Disease*, **69**, 548–58.

Shaw, P. J. A. & Johnston, A. (1993). Effects of SO$_2$ and O$_3$ on the chemistry and FDA activity of coniferous leaf litter in an open-air fumigation experiment. *Soil Biology and Biochemistry*, **25**, 897–908.

Smith, M. K. (1993). Effects of air pollution and environmental factors on endophytic fungi of Sitka spruce needles. PhD Thesis, Cranfield University, UK.

Tanaka, K. (1980). Studies on the relationship between air pollutants and microorganisms in Japan. *Forest and Forest Products Research Institute Report 1980*, pp. 110–116. USDA.

8

Influences of acid mist and ozone on the fluorescein diacetate activity of leaf litter

P. J. A. SHAW

One of the vital roles played by fungi in ecosystems lies in their being prime agents in the decomposition of plant material, thereby recycling nutrients (Swift, Heal & Anderson, 1979). It has been suggested that one insidious but important effect of pollution may be a reduction in the activity of decomposer communities (Ineson & Gray, 1980; Wookey, Ineson & Mansfield, 1991). Historically, the most significant atmospheric pollutant has been SO_2, whose adverse effects on fungal decomposers are well documented (Magan, 1993, also Magan, Chapter 7). In the UK the trend is for SO_2 to decline in importance as a pollutant (but see Boddy *et al.*, Chapter 6) while tropospheric ozone and nitrogenous pollutants are increasing (UK Terrestrial Effects Review Group, 1988; UK Review Group on the Impact of Atmospheric Nitrogen, 1994). The effects of these latter two classes of pollutants on fungi are still largely unresearched.

There have been several reports of decomposer activity in litter being adversely affected by treatment with simulated acid precipitation in which pH was controlled by sulphuric acid (Brown, 1985; Skiba & Cresser, 1986). By contrast, ambient rainwater has a significant component of nitric acid, amounting to approximately 30% of total acidity (Warren Spring Laboratory, 1990). Nitrogen has long been known as a promoter of the decay of plant material by fungi (Garrett, 1963), and this additional pollutant load might be expected partially to offset deleterious effects due to sulphurous pollutants. Additionally, there has been little work on the influence of O_3 on litter decomposition (Kickert & Krupa, 1990). Shaw and Johnston (1993) studied the early stages of coniferous leaf decomposition in an open-air fumigation system, and found no effects of pure ozone on decomposition processes but indications that N_2O_5-contaminated O_3 increased microbial activity.

Data are presented from an experiment in which decaying leaf material was exposed to acid mist containing a realistic mixture of sulphuric and nitric acids, as well as to controlled levels of O_3. This was part of a larger research programme on the influence of these factors on forest trees (Barnes & Brown, 1990; Barnes, Eamus & Brown 1990*a,b*).

The research described relied on the fluorescein diacetate (FDA) assay (Swisher & Carroll, 1980; Schnurer & Rosswall, 1982), which involves the hydrolysis of FDA into fluorescein by decomposer enzymes. Stubberfield and Shaw (1990) calibrated the FDA assay with other standard approaches, and concluded that it was a useful index of decomposer activity, although not of microbial biomass.

Fumigation and fluorescein diacetate assay

Samples were exposed in eight hemispherical greenhouses (solardomes) sited at the Central Elelctricity Research Laboratories, UK (National Grid Reference: TQ1657) and modified for prolonged fumigation. The domes are fully described in Barnes and Brown (1990), so only a brief description is given here. Each dome was approximately 4 m diameter and 2 m high, and took the form of a geodesic dome. Each dome received air containing an adjusted concentration of O_3, and also contained two sets of misting nozzles delivering either control (pH 5.6) or acidified (pH 3.3) mist – one misting treatment per half dome. Misting was applied automatically, with two spray cycles of 30 min each per day. Mist pH was controlled by use of 2 : 1 mixtures of H_2SO_4 : HNO_3.

O_3 was generated by high voltage discharge through air, with a water scrubber to remove N_2O_5 (Brown & Roberts, 1988). Concentrations were held constant by continuous computer monitoring, with feedback controls to the O_3 generating equipment. There were four O_3 treatments, each replicated twice. Table 8.1 gives the details of the ozone and acid misting treatments. Each dome contained potted saplings of silver fir (*Abies alba*) and beech (*Fagus sylvatica*) growing in soil collected from Fernworthy Forest, Dartmoor, UK (National Grid Reference: SX6684).

Sitka spruce needles were collected from heathland at Liphook, Surrey, UK in 1986 and dried at 70 °C prior to storage (McLeod, Shaw & Holland, 1992). They were prepared by autoclaving (120 °C for 20 min) in unsealed 0.5 mm mesh litter bags. Autoclaved bags were placed on the soil surface under silver fir saplings on 4 September 1987. Four litter bags were placed per dome as two replicates per misting treatment. At regular

Table 8.1. *Details of the experimental treatments used*

Ozone treatments (all concentrations were held constant for 24 h day^{-1} throughout the experiment)

Level 1: 20 nl l^{-1}
Level 2: 45 nl l^{-1}
Level 3: 70 nl l^{-1}
Level 4: 100 nl l^{-1}

Misting treatments

	pH	N deposition (kg ha^{-1} y^{-1})
Acid	3.3	4.3[a] 2.2[b]
Neutral	5.5	1.0[a] 0.5[b]

Note: a: 4 September 1987–5 November 1987.
　　　 b: 6 November 1987–end of experiment.

intervals subsamples of *c.* 0.2 g were removed from each litter bag, air dried for three days and assayed for FDA activity.

In a second experiment, the FDA activity of beech leaves from plants in the solardomes was assayed. Naturally fallen leaves were collected from pots in each half dome on 2 December 1987, air dried for 3 days, milled (< 1 mm) and assayed for FDA activity. The FDA assay is fully described by Stubberfield and Shaw (1990). Litter samples were air dried (21 °C) for three days, and *c.* 0.1 g placed in 19.5 ml pH 7.3 buffer in a sterile 25 ml universal tube, then 0.5 ml of a stock solution of FDA in acetone (320 mg l^{-1}) were added, making a final concentration of 8 μg ml^{-1}. Between experiments the stock solution was stored in the dark at −20 °C.

Each tube was shaken in a constant temperature facility at 21°C for approximately 90 min. An aliquot was transferred to a disposable plastic cuvette (path length 1 cm). Absorbance was measured at 490 nm on a Bausch & Lomb 601 spectrophotometer zeroed on a blank of FDA + buffer. Activity was calculated as μmol fluorescein produced g^{-1}h^{-1}.

Statistical analysis

Sequential harvests from the litter bags comprised a time series and could not be analysed as independent observations. Mean activities were therefore calculated for each litter bag as the integral of the time curve divided by the time period and analysed as individual observations. Mist and ozone effects were tested by an ANOVA model in which ozone was a main effect, but mist and the mist × ozone interaction were split domes 'subplots'.

ANOVA showed a weakly significant effect of mist ($F_{1,4}$ = 5.7, P < 0.10) on the FDA activity of Sitka spruce needles, while dome and dome × mist terms were non-significant, implying no effects of O_3. In Fig. 8.1 the activity of spruce litter is a function of misting treatment, showing that acidic mist increased litter FDA activity. For the beech leaf data (Fig. 8.2) ANOVA found a weakly significant effect of mist ($F_{1,4}$ = 5.0, P < 0.10), but dome and ozone effects were non-significant, again giving no indication that ozone affected FDA activity.

The stimulatory effect of acidic mist may reasonably be attributed to its nitrogen content, although the deposition rates were lower than those encountered in some areas of the UK (Warren Spring Laboratory, 1990). Killham, Firestone and McColl (1983) found that the principal mechanism by which simulated acid rain affected soil microbial activity was by changes in the nitrogen supply. Shaw and Johnston (1993) found the FDA activity of Scots pine (*Pinus sylvestris*) needles to be increased by exposure to N_2O_5 deposition, as well as by nitrogen-containing solutions.

Garrett (1963) described the nitrogen effect whereby decomposition of plant material was accelerated by application of nitrogenous materials, as being due to leaf material having a higher C:N ratio than decomposer microbes. This concept has been examined in greater detail by Fog (1988), who found it to be accurate only for cellulose-rich materials,

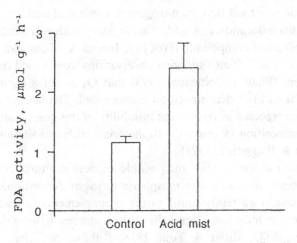

Fig. 8.1. Mean FDA activity of Sitka spruce needles during 6 months' exposure in an ozone/acid mist experiment as a function of misting treatment. Bars represent one treatment standard error.

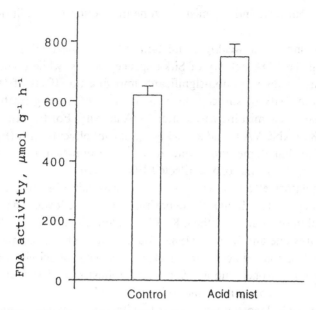

Fig. 8.2. Mean FDA activity of naturally fallen beech leaves collected from an ozone/acid mist experiment as a function of misting treatment. Bars represent one treatment standard error.

and that the decomposition of materials with a high lignin content could actually be inhibited by added nitrogen.

This study does suggest that the nitrogenous content of ambient pollution may stimulate decomposers, and offset in part the known deleterious effects of sulphurous compounds (Wookey, Ineson & Mansfield, 1991; Newsham et al., 1992). It also supports observations from a field fumigation experiment (Shaw & Johnston, 1993) that O_3 is not a significant pollutant as far as litter decomposition is concerned. The lack of ozone effects is not unexpected in view of the instability of this gas, resulting in its rapid decomposition in contact with materials such as soil and litter (Turner, Rich & Waggoner, 1973).

It is interesting to speculate on the possible impacts on fungal communities of projected increases in atmospheric nitrogen deposition. Field experiments have consistently found nitrogen supplements to lead to a reduction in mycorrhizal fungi, generally with an increase in litter decomposers (Meyer, 1985; Ritter & Tolle 1978; Ruhling & Tyler, 1991). Perhaps recent reports of serious declines in European macrofungi (Cherfas, 1992) may be attributable to increases in nitrogenous pollutants

(Skeffington & Wilson, 1988; UK Review Group on the Impact of Atmospheric Nitrogen, 1994), accompanied by unseen increases in populations of microfungi.

References

Barnes, J. D. & Brown, K. A. (1990). The influence of ozone and acid mist on the amount and wettability of the surface waxes of Norway spruce. *New Phytologist*, **114**, 531–5.

Barnes, J. D., Eamus, D. & Brown, K. A. (1990*a*). The influence of ozone, acid mist and soil nutrient status on Norway spruce. 1. Plant water relations. *New Phytologist*, **114**, 713–20.

Barnes, J. D., Eamus, D. & Brown, K. A. (1990*b*). The influence of ozone and acid mist on Norway spruce. 1. Photosynthesis, dark respiration and soluble carbohydrate status of trees in late autumn. *New Phytologist*, **115**, 149–56.

Brown, K. A. (1985). Acid deposition: effects of sulphuric acid at pH 3 on chemical and biochemical properties of bracken litter. *Soil Biology and Biochemistry*, **17**, 31–8.

Brown, K. A. & Roberts, T. M. (1988). Effects of ozone on foliar leaching in Norway spruce (*Picea abies* Karst.): confounding factors due to NO_x production during ozone generation. *Environmental Pollution*, **55**, 55–73.

Cherfas, J. (1992). Disappearing mushrooms: another mass extinction? *Science*, **245**, 1458.

Fog, K. (1988). The effect of added nitrogen on the rate of decomposition of organic matter. *Biological Review*, **63**, 433–62.

Garrett, S. D. (1963). *Soil Fungi and Soil Fertility*. Pergamon Press, Oxford, UK.

Ineson, P. & Gray, T. R. G. (1980). Monitoring the effects of acid rain and sulphur dioxide upon soil microorganisms. In *Microbial Growth and Survival in Extremes of Environment*, ed. G. W. Gould & J. E. L. Corry, pp. 21–26. Academic Press: London.

Kickert, R. N., & Krupa, S. V. (1990). Forest responses to tropospheric ozone and global climate change: an analysis. *Environmental Pollution*, **68**, 29–65.

Killham, K., Firestone, M. K. & McColl, J. G. (1983). Acid rain and soil microbial activity: effects and their mechanisms. *Journal of Environmental Quality*, **12**, 133–7.

McLeod, A. R., Shaw, P. J. A. & Holland, M. R. (1992). The Liphook forest fumigation project: studies of sulphur dioxide and ozone effects on coniferous trees. *Forest Ecology and Management*, **51**, 121–7.

Magan, N. (1993). Tolerance of fungi to sulphur dioxide. In *Stress Tolerance of Fungi*, ed. D. H. Jennings, pp. 173–187. Marcel Dekker Press: New York.

Meyer, F. H. (1985). Einfluss des Stickstoff-factors auf den Mykorrhizabesatz von Fichtelsamlingen im humus einer waldschadensflache. *Algemeine Forstzeitschrift*, **40**, 208–19.

Newsham,, K. K., Frankland, J. C., Boddy, L. & Ineson, P. (1992). Effects of dry-deposited SO_2 on fungal decomposition of angiosperm tree leaf litter. I, Changes in communities of fungal saprotrophs. *New Phytologist*, **122**, 97–110.

Ritter, G. & Tolle, H. (1978). Stickstoffdungung in Kiefernbestanden und ihre Wirkung auf Mykorrizabildung und Fruchtifikation der Symbiose-pilze. *Beitrage Forstwirtschaft*, **12**, 162–6.

Ruhling & Tyler (1991). Effects of simulated nitrogen deposition to the forest floor on the macrofungal flora of a beech forest. *Ambio*, **20**, 261–3.

Schnurer, J. & Rosswall, T. (1982). Fluroescein diacetate hydrolysis as a measure of total microbial activity in soil and litter. *Applied Environmental Microbiology*, **43**, 1256–61.

Shaw, P. J. A. & Johnston, J. P. N. (1993). Effects of SO_2 and O_3 on the chemistry and FDA activity of coniferous leaf litter in an open air fumigation experiment. *Soil Biology and Biochemistry*, **25**, 897–908.

Skeffington, R. A. & Wilson, E. J. (1988). Excess nitrogen deposition: issues for consideration. *Environmental Pollution*, **54**, 159–84.

Skiba, U. & Cresser, M. S. (1986). Effects of precipitation acidity on the chemistry and microbiology of Sitka spruce litter leachate. *Environmental Pollution*, **42**, 65–78.

Stubberfield, L. C. F. & Shaw, P. J. A. (1990). A comparison of tetrazolium reduction and FDA hydrolysis with other measures of microbial activity. *Journal of Microbiological Methods*, **12**, 151–62.

Swift, M. J., Heal, O. W. & Anderson, J. M. (1979). *Decomposition in Terrestrial Ecosystems*. Blackwell Scientific Publications: Oxford.

Swisher, R. & Carroll, G. C. (1980). Fluorescein diacetate hydrolysis as an estimator of microbial biomass on coniferous needle surfaces. *Microbial Ecology*, **6**, 217–26.

Turner, N. C., Rich, S. & Waggoner, P. E. (1973). Removal of ozone by soil. *Journal of Environmental Quality*, **2**, 259–65.

UK Terrestrial Effects Review Group (1988). *The Effects of Acid Deposition on the Terrestrial Environment in the UK*. Department of the Environment: London, UK.

UK Review Group on the Impact of Atmospheric Nitrogen (1994). *Impacts of Nitrogen Deposition in Terrestrial Ecosystems*. Department of the Environment: London, UK.

Warren Spring Laboratory (1990). *Acid Deposition in the United Kingdom*. Department of the Environment: London, UK.

Wookey, P. A., Ineson, P. & Mansfield, T. A. (1991). Effects of atmospheric SO_2 on microbial activity in decomposing forest litter. *Agriculture, Ecosystems and Environment*, **33**, 263–80.

9
Mycorrhizas and environmental stress

J. V. COLPAERT AND
K. K. VAN TICHELEN

The mycorrhizal symbiosis

The rhizosphere and the surface of plant roots are inhabited by populations of several microorganisms. Among these populations, a group of filamentous fungi can be present quite consistently on the root surfaces or in the tissues or cells of the roots, so that dual organs of consistent morphological and histological patterns are formed (Harley, 1989). In these 'mycorrhizas' the fungus and the host co-exist actively for long periods in a state called a mutualistic symbiosis. Most plant species in the greater part of the world's ecosystems are infected with these mycorrhizal fungi (Harley & Smith, 1983; Harley & Harley, 1987, Janos, Chapter 10).

Under natural conditions the vast majority of the fungi involved in this association appear to be obligate symbionts, with little or no ability for independent growth. This seems to be less true for the autobionts. Their degree of mycorrhizal dependency is somewhat more variable. Some plant species are infected occasionally while most others (70 % of the angiosperms according to Trappe, 1987) cannot complete their life cycles without mycorrhizas, at least in their natural environment. However, environmental factors as well as the presence of companion plants can determine whether a plant is mycotrophic or not in some settings (Miller, 1979; Molina, Massicotte & Trappe, 1992). Although mycorrhizal fungi have adapted to a wide range of environmental conditions (Stahl & Christensen, 1991), high levels of fertilizer, waterlogging, soil compaction (Sylvia & Williams, 1992) or soil disturbance (Jasper, Abbott & Robson, 1991) can restrict root colonization, but field observations show that many plants growing in a great variety of stressful situations are mycorrhizal. These 'extremophiles' live in extreme habitats conventionally characterized by low or high temperatures, low or high pH, drought, high salt or metal concentrations, excess nutrients and so forth (Jennings, 1993). In general, low nutrient availability favours mycorrhizal development, and mycorrhizal fungi can significantly enhance plant growth in such conditions (Read, Francis & Finlay, 1985; Daft, 1992). However, in some instances an increased uptake of P could not be found and sometimes no benefit at all is apparent, even in natural ecosystems (West, Fitter & Watkinson, 1993). The problem is that experimenters should be careful not to equate the beneficial effect of mycorrhizas only to an increase of a particular plant characteristic (height, biomass). It is often very difficult to demonstrate that the mycorrhizal symbiosis has a selective advantage to both host and fungus (Harley, 1989).

Two major groups of mycorrhizal fungi can be distinguished: ectomycorrhizal (ECM) fungi and vesicular-arbuscular mycorrhizal (VAM) fungi. General descriptions of both mycorrhizal associations can be found in excellent handbooks on mycorrhizas (Marks & Kozlowski, 1973; Harley & Smith, 1983; Norris, Read & Varma, 1991, 1992). A third class of mycorrhizas is typically found in heathland soils. In the distal parts of ericoid root systems, cortical cells are often strongly colonized with hyphae of the ascomycete *Hymenoscyphus ericae* or a few other fungal species. The ericoid mycorrhizal mycelium grows in habitats which can be seen as highly stressful for most plants and fungi. Read and his colleagues have shown clearly that ericoid mycorrhizas are involved in nutrient mobilization and detoxification processes (Read, 1992).

Environmental stress

How to define environmental stress for a mycorrhizal plant–fungus association

Stress can refer to those situations where a change in an environmental factor brings about clear and distinct changes in the metabolic state of both symbionts of the mycorrhizal association, regardless of whether this change is harmful to the system as a whole or to one of the partners. Normally it is assumed that essential nutrients are not limiting when studying a particular stress factor (Jennings, 1993). However, the mycorrhizal symbiosis is found most typically in those situations where mineral nutrients are growth limiting. It evolved in evolutionary history as a mechanism which overcame the general stress of low nutrient availability in terrestrial ecosystems. Therefore, as soon as mineral nutrients are no longer growth limiting for the autobiont, the fungus finds itself in an unfavourable condition as a consequence of the cost-benefit relationship between the symbionts. High nutrient availability reduces the carbon allocation to the roots and results in C deficiency for the fungus. Therefore it is difficult to give a clear definition of the normal, non-stress condition for a mycorrhizal plant. A normal situation is characterised by a mycorrhizal fungus and a plant that coexist in a mutualistic relationship over a relatively wide range of different, growth-limiting nutrient concentrations. However, small adjustments in their metabolism result in an optimal equilibrium in each of these conditions. An important practical consequence for mycorrhizal research is that mineral steady-state conditions should always be pursued when studying stress factors, even when this is not easy as plants have to be grown in solid substrata for long periods. Only in this condition, when it causes a clear change in the metabolic homeostasis of the symbiosis, can stress be recognized.

A model

Andersen and Rygiewicz (1991) presented a framework for studying responses of mycorrhizal plants to external stresses, including possible feedback effects which are likely to occur. In their conceptual model, the site of action of a stress factor varies and can be either the fungus or the plant (Fig. 9.1). The site of action will determine whether the primary response occurs in the fungus or the plant. The primary response to stress can be either positive or negative. The magnitude of the stress will be affected by external and internal factors, for example, stress avoidance mechanisms which limit the chemical or physical changes. Often a primary response may be followed by secondary and higher-order responses. Through a series of metabolic feedbacks, plant and fungus will eventually reach a new steady state.

Although the model has its theoretical value, investigators are confronted with many problems when they want to apply it in practice. The main reason is that many processes caused by stress occur in a solid substratum and only a few techniques and instruments are available for studying these belowground processes in a non-destructive manner. Minirhizotrons or root observation chambers, possibly provided with a hyphal compartment, have become useful tools for studying the growth of roots and mycorrhizal mycelium (Read *et al.*, 1985; Rygiewicz, Miller & Durall, 1988; Jakobsen & Rosendahl, 1990; Li, George & Marschner, 1991; Rygiewicz & Andersen, 1994).

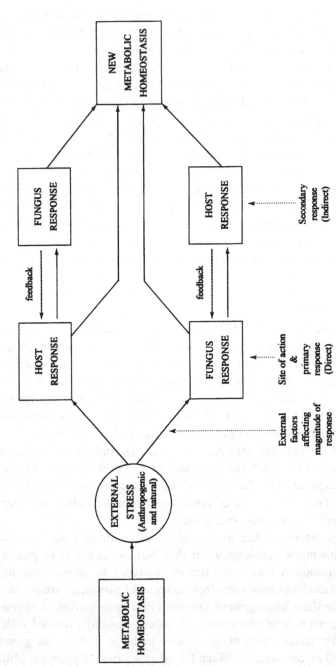

Fig. 9.1. Conceptual model showing how metabolic homeostasis could be disrupted by an external stress (After Andersen & Rygiewicz, 1991.)

Andersen and Rygiewicz (1991) argue that, once a stress factor causes a disturbance of the metabolic homeostasis, a new homeostasis will always be mediated by changes in carbohydrate allocation. However, the determination of biomass and carbon metabolism of a fungus in a substratum presents large problems. Only in recent years has some progress been made in this domain (Finlay & Söderström, 1992; Rygiewicz & Andersen, 1994). When considering the total higher plant-fungus system, knowledge of the plant's physiology will remain indispensable to the investigator.

The external mycelium of mycorrhizal fungi

Probably one of the best ways of studying the effect of an environmental stress factor on mycorrhizas is to focus on the growth of the external mycelium (Wallander & Nylund, 1992; Colpaert & Van Assche, 1992, 1993). For the mycobiont, growth of the external mycelium guarantees access to both mineral nutrients and carbohydrates as new roots become infected. Traditionally, researchers working with mycorrhizas used to focus on the growth and development of host plants. Differences between mycorrhizal and non-mycorrhizal control plants were stressed and the degree of root colonization was measured. Many studies have been carried out on the structure and functioning of mycorrhizal roots (Harley & Smith, 1983), but during the last decade more attention has been paid to the external mycelium of the mycorrhizas, in particular to its transport function (Read *et al.*, 1985; Li *et al.*, 1991; Abbott *et al.*, 1992; Jakobsen, Abbott & Robson, 1992; Frey & Schüepp, 1993). It is undoubtedly the most important structure of the mycorrhizal root for the nutrition of the host and the spread of the fungus. Despite being that part of the mutualism that is involved in the processes of mobilization and capture of nutrients from soil, it has frequently been ignored, probably because of the difficulties encountered when carrying out non-destructive investigations of the fragile mycelial systems (Read, 1992).

The external mycelium of ECM fungi can be of an extensive nature and highly differentiated (see also Janos, Chapter 10). Field studies show that the pattern of mycelial growth and differentiation in many basidiomycete species of ECM fungi is similar to that found in wood-rotting fungi the nutritional resources of which are also discontinuously distributed through the soil (Jennings, 1991; Read, 1992). In mycorrhizal fungi the resource units are the individual short roots and the pockets of soil and humus where nutrient mobilisation is most active (Read *et al.*, 1985;

Finlay & Read, 1986*a,b*). In many species these localized areas of nutrient enrichment by the fungus are interconnected by mycelial cords or strands. Cords also connect the mycelium with its fruit bodies. In some ectomycorrhizal systems, sheaths of mycorrhizas are very smooth and have few mycelial contacts with the soil around them. These mycorrhizas are found in soils which are unstable, such as the surface litter layers of beech forests. In such situations direct absorption of nutrients by the sheath may be important (Harley, 1989; Read, 1992). Our knowledge of the growth pattern and dynamics of the external mycelium is, however, still very meagre. Only a few species have been grown in root observation chambers and in a limited number of soils and growth conditions. For the majority of ectomycorrhiza-forming fungi no experimental data are available. Studies on the ecological function as well as of the ecological specificity of the ECM fungi are few in number (Molina *et al.*, 1992).

Although the VA mycorrhizal symbiosis is the oldest of the mycorrhizal associations with plants, the mycorrhizal mycelium forms less differentiated networks than the ECM mycelia. These networks facilitate the exploitation of the soil for nutrients, notably phosphate, and assure the infection of new, compatible roots, their unique resource for carbon compounds. VA mycorrhizal associations generally lack host specificity and the number of fungal species involved in the symbiosis appears to be much smaller than in the ectomycorrhizal associations (Harley & Smith, 1983; Harley, 1989). These elements illustrate the conservative but nevertheless very successful functioning of the VA mycorrhizal mycelium in evolutionary history.

An important functional difference between the mycelia of VAM and ECM fungi is that the latter seem to be more specialized in nutrient capture in soils with a high humus content. It becomes more and more evident that the external mycelium of some ECM fungi produces extracellular enzymes which interfere directly with nutrient mobilization from organic sources (Entry, Rose & Cromack, 1991; Read, 1992). Another difference between the mycelia of the two types of mycorrhizas is in the amount of the external biomass they produce. Quantifying this biomass is however difficult owing to a lack of adequate methods (Finlay & Söderström, 1992). Abbott and Robson (1985) found that different amounts of external hyphae were produced by four VAM fungi. Jakobsen and Rosendahl (1990) estimated that the biomass of external *Glomus fasciculatum* mycelium produced in less than one month on cucumber seedlings constituted 2.6% of root dry weight. From these and other studies, however, it is clear that species characteristics as well as many soil and plant factors determine the amount of hyphae produced (Wilson & Tommerup, 1992). It is worth noting that large amounts of mycelium are not synonymous with the largest positive plant response; the costs of maintaining the fungus can outcompete the benefit of a better nutrient acquisition.

Based on a field study, Vogt *et al.* (1982) estimated that 14% of net primary production may be diverted to the ectomycorrhizal fungi in a 23 year-old *Abies amabilis* ecosystem. The mycorrhizal fungal component was calculated to be 10.7% of the below-ground living biomass. Miller, Durall and Rygiewicz (1989) reported that between 25 and 30% of total root and fungal biomass was extramatrical hyphae in *Pinus ponderosa* seedlings grown in root microcosms for seven months. Colpaert, Van Assche and Luijtens (1992) estimated the biomass of some ECM mycelia produced by *Pinus sylvestris* seedlings in a semi-hydroponic cultivation system. Mycelial biomass varied from 2.9% of the total root dry weight for *Thelephora terrestris* to 25% for a *Suillus bovinus* strain. Wallander & Nylund (1992) found values of approximatelely 1 to 6% of living external mycelium on total root mass, produced in 6 to 7 weeks by *P. sylvestris* seedlings. Although these results cannot be directly extrapolated to the natural situation, they certainly demonstrate that the biomass production of mycorrhizal mycelium is highly variable between species. It appears also that in many members of the Pinaceae, at least, the major function of absorption from soil is fulfilled by the external mycelium which acts as a physiological extension of the root system (Read *et al.*, 1985). Soil stress factors (e.g. pH, metals) therefore will be met firstly by the fungus. This feature of the ECM fungi makes them capable of exerting a buffer function between the soil and the higher plant. Malloch, Pirozynski and Raven (1980) argued that ectomycorrhizal associations have a selective advantage in extreme environments. In VA mycorrhizal plants, the roots themselves are not fully screened from the soil environment. The effect of a soil stress factor on the plant root might therefore be greater in this association.

Environmental stress: three case studies

Excess nitrogen

In the Netherlands, field observations have revealed a significant impoverishment of the macromycete flora (Arnolds, 1988; see also Chapter 5). This decline is strongest among ECM fungi associated with coniferous trees. Arnolds (1988) argues that soil acidification due to wet and dry deposition of SO_2, NO_x and NH_3 is the main reason for the reduction. However, it is also known that these gaseous air pollutants directly affect plant physiology and reduce the transport of assimilates to the root system. A combined effect cannot be excluded.

Ammonia pollution may increase growth of epiphyllic algae, attack on leaves by fungal parasites and insects, and leaching of nutrients from leaves or needles, and may decrease frost hardiness (Nihlgård, 1985; Pearson & Stewart, 1993). Termorshuizen and Schaffers (1991) found strong negative correlations between the estimated nitrogen pollution and both the number of basidiomes and fruiting species in old pine forests. Nitrogen pollution leads to an increased uptake of nitrogen by the trees which might decrease carbon allocation to the mycorrhizas and consequently might result in deficiencies of other nutrients, especially in poor, acidic soils. Boxman and Roelofs (1988) showed that *P. sylvestris* seedlings had low uptake rates for other cations whenever ammonium was present in the culture solution. In normal conditions, the soil solution is soon depleted of ammonium. When ammonium is present in excess, a continuous ammonium load may lead to potassium and magnesium deficiencies. Pines infected with *Rhizopogon luteolus* were better protected against the negative effect of excess ammonium.

Wallander and Nylund (1992) studied the effect of excess nitrogen on the growth of the external mycelium of ectomycorrhizas of *P. sylvestris*. The external mycelium of *Laccaria bicolor* and *S. bovinus* was more sensitive to elevated ammonium concentrations than the mycelium in the mycorrhizas (Table 9.1). This decrease in biomass production could not be explained by the higher osmolarity or the change in pH, at least for the species they studied. When the plants were returned from the excess N solution to the normal solution, resumption of fungal growth took place. When excess nitrogen was combined with phosphorus starvation, a ten-fold increase in the production of external mycelium was observed; the increase was not observed when Mg was omitted from the nutrient solution. The stimulatory effect of P starvation indicates that the mechanisms of N-inhibition are indirect and mediated by carbon allocation.

The reduced carbon allocation to the root system can be the result of improved nitrogen nutrition and direct leaf damage (Gmur, Evans & Cunningham, 1983). When the reduced soil colonization by external mycorrhizal hyphae continues over long time periods, it is likely to

Table 9.1. *Fungal biomass, expressed as ergosterol per seedling, and concentration of nitrogen in needles of Pinus sylvestris grown for seven or eight weeks with or without excess nitrogen in the growth medium*

| Fungal associate | Ergosterol per seedling (µg) ± S.E. | | | | N concentration in needles (%) ± S.E. | |
| | Mycorrhizas | | External mycelium | | | |
	Control	Excess N	Control	Excess N	Control	Excess N
Laccaria bicolor	200 ±10	140 ±30	87 ±13	28 ±6	1.15 ±0.15	2.10 ±0.21
Suillus bovinus	19.5 ±3.4	4.2 ±2.0	9.8 ±7.7	0	1.12	2.33

Note: $n = 3$, the material of the three replicates of the *S. bovinus* condition was pooled in one sample for the nitrogen measurements.
Source: Wallander & Nylund, 1992.

lead to a secondary response which is reduced mineral availability; this might in turn affect tree vitality. Although the laboratory experiments showed no direct effect of N on the growth of the fungi, such effects might occur in certain species since up to now only a few quite common fungi have been studied.

Concerning VA mycorrhizas, Heijne *et al.* (1992) found no decrease in root colonisation of three heathland species treated with excess nitrogen in a nutrient-poor soil under natural conditions. VAM infection rate could not explain the observed decline of the test species. These herbs were probably outcompeted by grasses, which showed a greater growth response as a result of the increased input of nitrogen in the ecosystem.

Excess CO_2

CO_2 enrichment of the atmosphere is now well documented and its effect on plant growth in natural ecosystems is being questioned. Increased CO_2 assimilation leads to an increase in the biomass of the whole plant, especially in herbaceous crop plants. In tree species, root biomass very often increases proportionally more than shoot biomass (Mousseau & Saugier, 1992). In experiments where trees are grown in unfertilized soil, a substantial increase in the root/shoot ratio has been observed (Norby, O'Neill & Luxmore, 1986; El Kohen, Rouhier & Mousseau, 1992). Norby *et al.* (1987) found an increase in carbon allocation and root exudation in *Pinus echinata* seedlings grown under CO_2 enrichment. As a consequence mycorrhiza formation was clearly favoured.

Tree species grown in elevated CO_2 usually have smaller concentrations of nutrients in all plant parts except in fine roots compared to the controls (Mousseau & Saugier, 1992). Coûteaux *et al.* (1991) found C/N ratios of 40 and 75 in sweet chestnut (*Castanea sativa*) leaf litter for plants grown under ambient and enriched CO_2 concentration, respectively. This effect certainly will influence the decomposition of leaf litter, the communities of saprotrophic organisms and probably also the ECM fungal community. However, no firm conclusions can yet be drawn as further research is required.

If the effects of high CO_2 concentrations obtained in laboratory experiments are similar to the field situation, then mineral nutrient stress will increase in most natural ecosytems which might increase the mycorrhizal dependency of these ecosystems.

Heavy metal stress

Field observations

The detrimental effect of heavy metal pollution on soil processes and plant growth is a well-known phenomenon (Tyler *et al.*, 1989; Ernst, Verkleij & Schat, 1992; Gadd, 1993). Many plants growing on heavily polluted soils are mycotrophic. Field observations show that ECM fungi as well as VAM and ericoid species can occur (Bradley, Burt & Read, 1982; Gildon & Tinker, 1983; Rühling & Söderström, 1990). Ietswaart, Griffioen and Ernst (1992) found no significant differences in VAM infection percentages between *Agrostis capillaris* populations growing on polluted (Zn and Pb) and non-polluted soils. Gildon & Tinker (1983) also found high infection percentages in clover (*Trifolium*) plants surviving on a severely Zn- and Cd-polluted site. They found a positive correlation between infection and P concentration in the plant indicating that the VAM infection was effective in supplying P to the host. Rühling and Söderström (1990) showed that the ECM fungal flora was strongly affected by heavy metal pollution. Only a limited number of species was found on the most polluted sites.

In vitro *studies with ECM fungi*

Several authors have studied the *in vitro* tolerance of ECM fungi against heavy metals (see Morley *et al.*, Chapter 15). However, there is not necessarily a relationship between the *in vitro* tolerance of ECM fungi and their ability to increase metal tolerance of their host plants (Jones & Hutchinson, 1988a). *In vitro* studies with mycorrhizal fungi on agar media do not reflect the natural situation and can have only an indicative function for physiological and biochemical studies. As a second step, results should always be confirmed in a laboratory experiment with a mycorrhizal host and eventually in a field experiment. The low nutrient status of most natural, polluted or non-polluted environments never approaches that found in laboratory culture media. Many ECM fungi show signs of stress when grown under nutrient-rich conditions; secondary metabolite production, for example, can be seen as a means by which the fungus can cope with excess of a substrate that the organism never meets in such quantity in nature (Wainwright, 1993).

Experiments with entire plant-fungus associations

In laboratory experiments, an ameliorating effect of mycorrhizal fungi on metal phytotoxicity has been demonstrated in several investigations (Denny & Wilkins, 1987*a*; Jones & Hutchinson, 1988*b*; Colpaert & Van Assche, 1992, 1993). When different fungal species were included in such studies, large differences in the protective effect were observed. Jones and Hutchinson (1988*b*) found that birch seedlings infected with *Scleroderma flavidum* were better protected against Ni toxicity than seedlings inoculated with *Lactarius rufus*, which in turn did better than non-inoculated plants. Biomass of the shoots and roots indicated that the former fungus induced a larger carbon allocation to the roots than the latter species. The larger fungal biomass produced by *S. flavidum* confirmed this feature. Furthermore, *S. flavidum* continued to grow during the experiment contrary to *L. rufus* which failed to grow once exposed to Ni. Colpaert and Van Assche (1992) observed a similar reaction in an experiment concerned with Zn toxicity in mycorrhizal *Pinus sylvestris* seedlings. The external mycelium of *Paxillus involutus* stopped growing and died off once the seedlings were exposed to a high Zn concentration, although root growth was not inhibited (Table 9.2). The primary site of action in this case was clearly the fungus. The mycelium of a *S. bovinus* strain isolated from a Zn-contaminated soil was hardly affected by the treatment. Although both species reduced uptake of zinc in the shoots of their host (Table 9.2), it is evident that only *S. bovinus* would be effective in metal retention over a longer period of time. The fact that dead mycelium also retained high amounts of Zn suggests that the retention mechanism is, at least partly, of a passive nature. Denny and Wilkins (1987*b*) found that excess Zn was adsorbed in the hyphal cell walls and extrahyphal slime layers. Bradley *et al.* (1982) also suggested that the ericoid endophyte of ericaceous plants can adsorb large amounts of metals preventing translocation to the shoots. The retention of Zn in a perlite substratum, colonized by a dense mycelium of *S. bovinus* or a sparse *Laccaria laccata* mycelium is illustrated in Fig. 9.2. Pine seedlings were treated with a single dose of 33 mg Zn per plant. Most of the Zn had percolated through the substratum from the non-inoculated pine seedlings and those inoculated with *L. laccata*. A much higher percentage of the Zn was retained in the substratum colonized by *S. bovinus*. Despite this higher Zn concentration of the mycelium, less Zn was transported to the plant, mycorrhizas excluded. It is expected that the saturation of the metal retention sites

Table 9.2. *Effect of a high Zn substrate concentration (900 mg Zn l⁻¹ vermiculite) on water use and uptake of Zn in six-month-old Pinus sylvestris seedlings and on the growth of their mycobiont*

Fungal associate	Shoot Zn concentration ± S.E. ($\mu g\,g^{-1}$ dry wt)		% Substratum colonized by living external mycelium ± S.E.		Daily increase in water use per plant (ml) ± S.E.	
	control	Zn	before treatment	after treatment	control	Zn
Non-mycorrhizal plants	35 ± 3	197 ± 6	0	0	0.36 ± 0.01	0.34 ± 0.01
Thelephora terrestris	31 ± 3	240 ± 5	> 95	92 ± 3	0.33 ± 0.01	0.35 ± 0.01
Paxillus involutus	19 ± 2	106 ± 3	> 95	18 ± 2	0.30 ± 0.01	0.23 ± 0.02
Suillus bovinus	22 ± 2	139 ± 3	> 95	83 ± 5	0.22 ± 0.01	0.21 ± 0.01

Note: Treatment effects on plant biomass were not found after three months, $n = 4$ for control plants and $n = 12$ for Zn-treated plants.
Source: Colpaert & Van Assche, 1992.

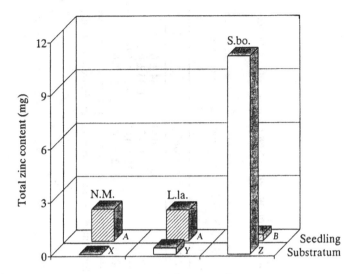

Fig. 9.2. Total amount of zinc in *Pinus sylvestris* seedlings and in their perlite substratum (including external mycelium), one month after a single Zn addition (33 mg per plant). N.M. = Non-mycorrhizal plants, L.la. = plants infected with a *Laccaria laccata* strain producing little external mycelium, S.bo. = plants infected with a *Suillus bovinus* strain with a dense external mycelium. Substratum or seedling bars which do not have the same letter as another substratum or seedling bar differ at $P < 0.05$ (Tukey groupings, $n = 6$) (unpublished observations).

might be influenced by the fungal biomass and the turnover rate of the mycelium or the mycorrhizal roots (ericoid type).

In mycorrhizas of *P. involutus* collected from heavily polluted sites, two types of electron-opaque granules have been localized in the vacuoles of the hyphae by means of electron spectroscopic imaging and electron energy loss spectroscopy (Turnau, Kottke & Oberwinkler, 1993). One type of granule, probably polyphosphate, contained high concentrations of P with S, Ca and Al. The other type of granule contained more N, S and Cd with low P. The accumulation of Cd in the fungal vacuoles may represent a detoxification mechanism by the fungus. However, Orlovich and Ashford (1993) argue that polyphosphate granules are formed during specimen preparation by an influx of divalent cations into the vacuole. In general, little is known about specific detoxification mechanisms in mycorrhizal fungi (Gadd, 1993).

Fungi producing little biomass are not very effective in metal retention and may even increase metal uptake in their hosts compared to non-inoculated plants (Jones & Hutchinson, 1988*c*; Colpaert & Van Assche, 1992). A comparable situation can be found in VA mycorrhizas. Gildon and Tinker (1981, 1983) found only slight reductions in transport of Zn and Cd to the shoots of mycorrhizal clover. However, they also showed that a VAM strain from a polluted site was much more tolerant of heavy metals than a strain from an unpolluted soil. Although the Zn-tolerant fungus could not prevent metal uptake in the host, its presence was indispensable to ensure sufficient P uptake from the very poor soil. The authors did not examine the external mycelium of the mycorrhizas, but they assumed that the growth of the mycelium of the sensitive VAM strain was suppressed by the high soil metal concentrations. When the metal retention capacity of mycorrhizal fungi is limited, higher amounts of metals will be transported to their autobiont. From that moment, metal tolerance mechanisms in the host plant become important. If these mechanisms reduce carbon allocation to the roots, mycorrhizal development may diminish and further negative responses can be expected.

Heavy metals can also change the growth habit of mycorrhizal fungi. Cd induces increased hyphal density in *P. involutus* mycelia, before the biomass is affected, by both increasing the number of laterals at a branch point and decreasing the distance between branch points (Darlington & Rauser, 1988). When hyphal density increases around mycorrhizal roots, a larger retention capacity per unit soil will be obtained, although the uptake of essential elements will decrease. As a consequence of their growth habit, starting from a point source, fungi are able to change their immediate environment, such as by excretion of protons and seques-tration of toxic metals (Jennings, 1993). Heavy metal stress can also lead to a faster replacement of mycelium (Colpaert & Van Assche, 1993). Such an increase in mycelial turnover will reduce metal supply to the host but also implies an additional cost. Replacement of external mycelium has also been observed in mycorrhizal Sitka spruce (*Picea sitchensis*) subjected to waterlogging (Coutts & Nicoll, 1990).

Conclusions

When the impact of environmental stress factors on mycorrhizas are studied, the response of both the fungus and the host should be moni-tored carefully. Furthermore, experiments are only ecologically relevant when they are carried out, as far as possible, in a natural environment. The wide variation in stress tolerance observed among ECM fungi is often the result of the different growth characteristics of the wide array of mycorrhizal fungi. In VAM fungi stress tolerant ecotypes are adapted to a variety of stress factors such as heavy metals, salt, excess fertilizer (P) and others (Sylvia & Williams, 1992 for a review). ECM fungi appear to have a greater impact on the amount of a soil stress factor that reaches their autobiont than VAM fungi. This hypothesis seems not to be true when the primary action site of stress is located in the plant.

References

Abbott, L. K. & Robson, A. D. (1985). Formation of external hyphae in soil by four species of vesicular–arbuscular mycorrhizal fungi. *New Phytologist*, **99**, 245-55.

Abbott, L. K., Robson, A. D., Jasper, D. A. & Gazey, C. (1992). What is the role of VA mycorrhizal hyphae in soil? In *Mycorrhizas in Ecosystems*, ed. D. J. Read, D. H. Lewis, A. H. Fitter & I. J. Alexander, pp. 37-41. CAB International: Wallingford.

Andersen, C. P. & Rygiewicz, P. T. (1991). Stress interactions and mycorrhizal plant response: understanding carbon allocation priorities. *Environmental Pollution*, **73**, 217-44.

Arnolds, E. (1988). The changing macromycete flora in the Netherlands. *Transactions of the British Mycological Society*, **90**, 391-406.

Boxman, A. W. & Roelofs, J. G. M. (1988). Some effects of nitrate versus ammonium nutrition on the nutrient fluxes in *Pinus sylvestris* seedlings. Effects of mycorrhizal infection. *Canadian Journal of Botany*, **66**, 1091-7.

Bradley, R., Burt, A. J. & Read, D. J. (1982). The biology of mycorrhiza in the Ericaceae. VIII. The role of mycorrhizal infection in heavy metal resistance. *New Phytologist*, **91**, 197-209.

Colpaert, J. V. & Van Assche, J. A. (1992). Zinc toxicity in ectomycorrhizal *Pinus sylvestris*. *Plant and Soil*, **143**, 201-11.

Colpaert, J. V. & Van Assche, J. A. (1993). The effects of cadmium on ectomycorrhizal *Pinus sylvestris*. *New Phytologist*, **123**, 325-33.

Colpaert, J. V., Van Assche, J. A. & Luijtens, K. (1992). Relationship between the growth of the extramatrical mycelium of ectomycorrhizal fungi and the growth response on *Pinus sylvestris* plants. *New Phytologist*, **120**, 127-35.

Coûteaux, M-M., Mousseau, M., Célérier, M-L. & Bottner, P. (1991). Increased atmospheric CO_2 and litter quality: decomposition of sweet chestnut leaf litter with animal food webs of different complexities. *Oikos*, **61**, 54-64.

Coutts, M. P. & Nicoll, B. C. (1990). Waterlogging tolerance of roots of sitka-spruce clones and of strands from *Thelephora terrestris* mycorrhizas. *Canadian Journal of Forest Research*, **20**, 1896-9.

Daft, M. J. (1992). Use of VA mycorrhizas in agriculture: problems and prospects. In *Mycorrhizas in Ecosystems*, ed. D. J. Read, D. H. Lewis, A. H. Fitter & I. J. Alexander, pp. 198-201. CAB International: Wallingford.

Darlington, A. B. & Rauser, W. E. (1988). Cadmium alters the growth of the ectomycorrhizal fungus *Paxillus involutus*: a new growth model accounts for changes in branching. *Canadian Journal of Botany*, **66**, 225-9.

Denny, H. J. & Wilkins, D. A. (1987a). Zinc tolerance in *Betula* spp. III. Variation in response to zinc among ectomycorrhizal associates. *New Phytologist*, **106**, 535-44.

Denny, H. J. & Wilkins, D. A. (1987b). Zinc tolerance in *Betula* spp. IV. The mechanism of ectomycorrhizal amelioration of zinc toxicity. *New Phytologist*, **106**, 545-53.

El Kohen, A., Rouhier, H. & Mousseau, M. (1992). Changes in dry weight and nitrogen partitioning induced by elevated CO_2 depend on soil nutrient availability in sweet chestnut (*Castanea sativa* Mill). *Annales Sciences Forestiéres*, **2**, 83-90.

Entry, J. A., Rose, C. L. & Cromack, K., Jr. (1991). Litter decomposition and nutrient release in ectomycorrhizal mat soils of a douglas fir ecosystem. *Soil Biology and Biochemistry*, **23**, 285-90.

Ernst, W. H. O., Verkleij, J. A. C. & Schat, H. (1992). Metal tolerance in plants. A Review. *Acta Botanica Neerlandica*, **41**, 229-48.

Finlay, R. D. & Read, D. J. (1986a). The structure and function of the vegetative mycelium of ectomycorrhizal plants. I. Translocation of ^{14}C-labelled carbon between plants interconnected by a common mycelium. *New Phytologist*, **103**, 143-56.

Finlay, R. D. & Read, D. J. (1986b). The structure and function of the vegetative mycelium of ectomycorrhizal plants. II. The uptake and distribution of phosphorus by mycelial strands interconnecting host plants. *New Phytologist*, **103**, 157-65.

Finlay, R. D. & Söderström, B. (1992). Mycorrhiza and carbon flow to the soil. In *Mycorrhizal Functioning. An Integrative Plant–Fungal Process*, ed. M. J. Allen, pp. 134-160. Chapman & Hall: New York.

Frey, B. & Schüepp, H. (1993). Acquisition of nitrogen by external hyphae of arbuscular mycorrhizal fungi associated with *Zea mays* L. *New Phytologist*, **124**, 221-30.

Gadd, G. M. (1993). Interactions of fungi with toxic metals. Tansley Review 47. *New Phytologist*, **124**, 25-60.

Gildon, A. & Tinker, P. B. (1981). A heavy metal-tolerant strain of a mycorrhizal fungus. *Transactions of the British Mycological Society*, **77**, 648-9.

Gildon, A. & Tinker, P. B. (1983). Interactions of vesicular–arbuscular mycorrhizal infection and heavy metals in plants. I. The effects of heavy metals on the development of vesicular-arbuscular mycorrhizas. *New Phytologist*, **95**, 247-61.

Gmur, N. F., Evans, L. S. & Cunningham, E. A. (1983). Effects of ammonium sulphate aerosols on vegetation. II. Mode of entry and responses of vegetation. *Atmospheric Environment*, **17**, 715-21.

Harley, J. L. (1989). The significance of mycorrhiza. *Mycological Research*, **92**, 129-39.

Harley, J. L. & Harley, E. L. (1987). A check-list of mycorrhiza in the British flora. *New Phytologist*, supplement to Vol. **105**, 1-102.

Harley, J. L. & Smith, S. E. (1983). *Mycorrhizal Symbiosis*. Academic Press: London and New York.

Heijne, B., Hofstra, J. J., Heil, G. W., Van Dam, D. & Robbink, R. (1992). Effect of the air pollution component ammonium sulphate on the VAM infection rate of three heathland species. *Plant and Soil*, **144**, 1-12.

126   *J. V. Colpaert and K. K. Van Tichelen*

Ietswaart, J. H., Griffioen, W. A. J. & Ernst, W. H. O. (1992). Seasonality of VAM infection in three populations of *Agrostis capillaris* (Gramineae) on soil with or without heavy metal enrichment. *Plant and Soil*, **139**, 67-73.

Jakobsen, I., Abbott, L. K. & Robson, A. D. (1992). External hyphae of vesicular-arbuscular mycorrhizal fungi associated with *Trifolium subterraneum* L. 2. Hyphal transport of ^{32}P over defined distances. *New Phytologist*, **120**, 509-16.

Jakobsen, I. & Rosendahl, L. (1990). Carbon flow into soil and external hyphae from roots of mycorrhizal cucumber plants. *New Phytologist*, **115**, 77-83.

Jasper, D. A., Abbott, L. K. & Robson, A. D. (1991). The effect of soil disturbance on vesicular-arbuscular mycorrhizal fungi, in soils from different vegetation types. *New Phytologist*, **118**, 471-6.

Jennings, D. H. (1991). Techniques for studying the functional aspects of rhizomorphs of wood-rotting fungi: some possible applications to ectomycorrhiza. In *Methods in Microbiology, volume 23*, ed. J. R. Norris, D. J. Read & A. K. Varma, pp. 309-329. Academic Press: London.

Jennings, D. H. (1993). Understanding tolerance to stress: laboratory culture versus environmental actuality. In *Stress Tolerance of Fungi*, ed. D. H. Jennings, pp. 1-12. Marcel Dekker, Inc.: New York.

Jones, M. D. & Hutchinson, T. C. (1988*a*). The effects of nickel and copper on the axenic growth of ectomycorrhizal fungi. *Canadian Journal of Botany*, **66**, 119-24.

Jones, M. D. & Hutchinson, T. C. (1988*b*). Nickel toxicity in mycorrhizal birch seedlings infected with *Lactarius rufus* or *Scleroderma flavidum*. I. Effects on growth, photosynthesis, respiration and transpiration. *New Phytologist*, **108**, 451-9.

Jones, M. D. & Hutchinson, T. C. (1988*c*). Nickel toxicity in mycorrhizal birch seedlings infected with *Lactarius rufus* or *Scleroderma flavidum*. II. Uptake of nickel, calcium, magnesium, phosphorus and iron. *New Phytologist*, **108**, 461-470.

Li, X.-L., George, E. & Marschner, H. (1991). Extension of the phosphorus depletion zone in VA-mycorrhizal white clover in a calcareous soil. *Plant and Soil*, **136**, 41-8.

Malloch, D. W., Pirozynski, K. A. & Raven, P. H. (1980). Ecological and evolutionary significance of mycorrhizal symbioses in vascular plants (a review). *Proceedings of the North Atlantic Academy of Science*, **77**, 2113-18.

Marks, G. C. & Kozlowski, T. T. (1973). *Ectomycorrhizae. Their Ecology and Physiology*. Academic Press: London and New York.

Miller, R. M. (1979). Some occurrences of vesicular-arbuscular mycorrhiza in natural and disturbed ecosystems of the Red Desert. *Canadian Journal of Botany*, **57**, 619-23.

Miller, S. L., Durall, D. M. & Rygiewicz, P. T. (1989). Temporal allocation of ^{14}C to extramatrical hyphae of ectomycorrhizal ponderosa pine seedlings. *Tree Physiology*, **5**, 239-49.

Molina, R., Massicotte, H. & Trappe, J. M. (1992). Specificity phenomena in mycorrhizal symbioses: community–ecological consequences and practical implications. In *Mycorrhizal Functioning. An Integrative Plant-Fungal Process*, ed. M. J. Allen, pp. 357-423. Chapman & Hall: New York.

Mousseau, M. & Saugier, B. (1992). The direct effect of increased CO_2 on gas exchange and growth of forest tree species. *Journal of Experimental Botany*, **43**, 1121-30.

Nihlgård, B. (1985). The ammonium hypothesis – an additional explanation to the forest dieback in Europe. *Ambio*, **14**, 2-8.

Norby, R. J., O'Neill, E. G., Hood, W. G. & Luxmore, R. B. (1987). Carbon allocation, root exudation and mycorrhizal colonization of *Pinus echinata* seedlings grown under CO_2 enrichment. *Tree Physiology*, **3**, 203-10.

Norby, R. J., O'Neill, E. G. & Luxmore, R. B. (1986). Effects of atmospheric CO_2 enrichment on the growth and mineral nutrition of *Quercus alba* seedlings in nutrient poor soil. *Plant Physiology*, **82**, 83-9.

Norris, J. R., Read, D. J. & Varma, A. K. (1991). *Methods in Microbiology. Volume 23*. Academic Press: London.

Norris, J. R., Read, D. J. & Varma, A. K. (1992). *Methods in Microbiology. Volume 24*. Academic Press: London.

Orlovich, D. A. & Ashford, A. E. (1993). Polyphosphate granules are an artefact of specimen preparation in the ectomycorrhizal fungus *Pisolithus tinctorius*. *Protoplasma*, **173**, 91-102.

Pearson, J. & Stewart, G. R. (1993). Tansley Review N°. 56. The deposition of atmospheric ammonia and its effects on plants. *New Phytologist*, **125**, 283-305.

Read, D. J. (1992). The mycorrhizal mycelium. In *Mycorrhizal Functioning. An Integrative Plant–Fungal Process*, ed. M. J. Allen, pp. 102-133. Chapman & Hall: New York.

Read, D. J., Francis, R. & Finlay, R. D. (1985). Mycorrhizal mycelia and nutrient cycling in plant communities. In *Ecological Interactions in Soil: Plants, Microbes and Animals*, ed. A. H. Fitter, D. Atkinson, D. J. Read & M. B. Usher, pp. 193-217. Blackwell Scientific Publications: Oxford.

Rühling, A. & Söderström, B. (1990). Changes in fruitbody production of mycorrhizal and litter decomposing macromycetes in heavy metal polluted coniferous forests in north Sweden. *Water, Air and Soil Pollution*, **49**, 375-87.

Rygiewicz, P. T. & Andersen, C. P. (1994). Mycorrhizae alter quality and quantity of carbon allocated below ground. *Nature*, **369**, 58-60.

Rygiewicz, P. T., Miller, S. L. & Durall, D. M. (1988). A root-mycocosm for growing ectomycorrhizal hyphae apart from host roots while maintaining symbiotic integrity. *Plant and Soil*, **109**, 281-4.

Stahl, P. D. & Christensen, M. (1991). Population variation in the mycorrhizal fungus *Glomus mosseae*: Breadth of environmental tolerance. *Mycological Research*, **95**, 300-7.

Sylvia, D. M. & Williams, S. E. (1992). Vesicular-arbuscular mycorrhizae and environmental stress. In *Mycorrhizae in Sustainable Agriculture*, ed. G. J. Bethlenfalvay & R. G. Linderman, pp. 101-124. ASA Special Publication N° 54, Madison, Wisconsin, USA.

Termorshuizen, A. J. & Schaffers, A. (1991). The decline of carpophores of ectomycorrhizal fungi in stands of *Pinus sylvestris* L. in The Netherlands: possible causes. *Nova Hedwigia*, **53**, 267-89.

Trappe, J. M. (1987). Phylogenetic and ecologic aspects of mycotrophy in the angiosperms from an evolutionary standpoint. In *Ecophysiology of VA Mycorrhizal Plants*, ed. G. R. Safir, pp. 2-25. CRC Press: Boca Ratan, FL.

Turnau, K., Kottke, I. & Oberwinkler, F. (1993). *Paxillus involutus – Pinus sylvestris* mycorrhizae from heavily polluted forest. I. Element localization using electron energy loss spectroscopy and imaging. *Botanica Acta*, **106**, 213-19.

Tyler, G., Balsberg Påhlsson, A.-M., Bengtsson, G., Bååth, E. & Tranvik, L. (1989). Heavy-metal ecology of terrestrial plants, microorganisms and invertebrates. A review. *Water, Air and Soil Pollution*, **47**, 189-215.

Vogt, K. A., Grier, C. C., Meier, C. E. & Edmonds, R. L. (1982). Mycorrhizal role in net primary production and nutrient cycling in *Abies amabilis* ecosystems in western Washington. *Ecology*, **63**, 370-80.

Wainwright, M. (1993). Oligotrophic growth of fungi – stress or natural state? In *Stress Tolerance of Fungi*, ed. D. H. Jennings, pp. 127-144. Marcel Dekker, Inc.: New York.

Wallander, H. & Nylund, J-E. (1992). Effects of excess nitrogen and phosphorus starvation on the extramatrical mycelium of ectomycorrhizas of *Pinus sylvestris* L. *New Phytologist*, **120**, 495-503.

West, H. M., Fitter, A. H. & Watkinson, A. R. (1993). Response of *Vulpia ciliata* ssp. *ambigua* to removal of mycorrhizal infection and to phosphate application under natural conditions. *Journal of Ecology*, **81**, 351-8.

Wilson, J. M. & Tommerup, I. C. (1992). Interactions between fungal symbionts: VA mycorrhizae. In *Mycorrhizal Functioning. An Integrative Plant–Fungal Process*, ed. M. J. Allen, pp. 199-248. Chapman & Hall: New York.

10

Mycorrhizas, succession, and the rehabilitation of deforested lands in the humid tropics

D. P. JANOS

Introduction

Once covered by seemingly limitless forest, tropical lands today suffer escalating deforestation to satisfy the food, fuel and fibre demands of burgeoning human populations. Although disease and inaccessibility long protected the humid tropics – those tropical regions where annual precipitation exceeds potential evapotranspiration – deforestation is now rampant. Year-round warm temperatures and an excess of precipitation combine to exacerbate decline of soil fertility after deforestation. Diminished fertility and unacceptably low productivity, in turn, lead to evermore land clearing, such that, except for highly inaccessible areas of rugged terrain, strict national parks, and forest reserves, little undisturbed forest will remain by the end of this century (see Myers, 1991).

Utilization and conversion of humid tropical forests produces an extreme range of vegetations from largely intact forest matrix disturbed only by extraction of a few valuable timber species to barren, abandoned wastelands. As the latter predominate, rehabilitation of these highly degraded lands becomes a necessity. 'Rehabilitation' is used here to indicate the return of productive capacity to degraded land. With some types of land use, diminished primary production may be a consequence of disruption of mycorrhizal associations. In such instances, rehabilitation requires restoration of mycorrhizas.

This chapter examines the role of mycorrhizas in rehabilitation of degraded, low elevation, humid tropical lands. The analysis offered here is applicable in general outline to other ecosystems. This chapter: (i) describes the reliance of humid tropical plants on mycorrhizas, (ii) examines the characteristics of mycorrhiza inocula, the assessment of inoculum potential, and the likelihood of mycorrhizal fungus loss with deforestation, and (iii) considers the restoration of mycorrhizal fungi.

Mycorrhiza effects on plants

In the lowland humid tropics, pervasive soil infertility makes mycorrhizas essential to a wide variety of plant species for mineral nutrient uptake (Janos, 1983, 1987). The acid, red-yellow clays that comprise two-thirds of tropical soils (Oxisols & Ultisols; Sanchez, 1976) typically immobilize phosphorus, so that phosphorus limitation of plant growth is widespread (Sanchez, 1976; Vitousek, 1984). Because mycorrhizas can greatly enhance phosphorus supply to plants (Harley & Smith, 1983), the insurance of abundant, rapid mycorrhiza formation can be a key to the maintenance of productivity of humid tropical lands.

The types of mycorrhizas considered here are those that occur on the majority of plant species exploited for production: vesicular-arbuscular mycorrhizas (VAM) and ectomycorrhizas (EM). See also Colpaert *et al.*, Chapter 9. 'Vesicular-arbuscular mycorrhiza' is a term used generically to encompass all mycorrhizas formed by fungi in the Glomales (Zygomycetes; Morton & Benny, 1990), whether or not they produce vesicles. In the humid tropics, vesicular-arbuscular mycorrhizas formed by Glomineae seem more common than strictly arbuscular mycorrhizas involving Gigasporineae (Sieverding, 1991). Both trees and herbs in all parts of the tropics form VAM (see Janos, 1980*a*, 1987). Ectomycorrhizas are formed exclusively by trees and lianas in the lowland, humid tropics (Janos, 1983), predominantly in association with basidiomycetes (Alexander, 1989). Ectomycorrhizas are common where dipterocarps dominate Asian forests (Smits, 1992), are locally abundant, especially with caesalpinioid legumes in the tribe Amhersteae, in humid tropical Africa (Alexander, 1989), and are relatively rare among plant species native to Neotropical humid lowlands except on extremely infertile, sandy soils (Janos, 1983; Moyersoen, 1993). Ectomycorrhizal fungi, however, are potentially introduced and spread where pines, some *Eucalyptus* species (Lapeyrie *et al.*, 1992), *Casuarina* spp. (Theodorou & Reddell, 1991), and *Acacia* spp. (Bâ, Balaji & Piché, 1994) are imported for reforestation.

Plant growth and survival

In the humid tropics, reliance on mycorrhizas for mineral nutrient uptake, growth and survival by trees of many species can be absolute (VAM: Janos, 1977, 1980*a*; EM: see Redhead, 1980), but not all plant species fail to establish if they lack mycorrhizas. Janos (1980*a*) called those species unable to grow without mycorrhizas in the most fertile of their natural habitats 'obligate mycotrophs', and those which can grow without mycorrhizas in relatively fertile natural soils 'facultative mycotrophs'. In studies of 32 humid tropical plant species, Janos (1975, 1977, 1980*a*) found that VAM significantly improved the growth of 27 species, including the majority of mature forest tree species, and significantly increased the survival of eight species. Several early successional shrub and small tree species could grow without VAM, however, and one pioneer species proved to be 'non-mycorrhizal' (Janos, 1980*a*). Subsequent research with additional tropical plant species by other investigators generally has confirmed these observations (see Sieverding, 1991). Late successional and climax forest plant species tend to be obligately mycotrophic, but early successional species, and perhaps the majority of herbs, tend to be facultatively mycotrophic (see Trappe, 1987). Some pioneer, weedy ruderals are non-mycorrhizal, neither requiring mycorrhizas for nutrient uptake, nor forming significant numbers of functional mycorrhizas (see Janos, 1980*b*; Tester, Smith & Smith, 1987).

Several caveats pertain to generalizations concerning the reliance of plant species upon mycorrhizas. The first is that the correlation between early seral status and independence of mycorrhizas is loose. For example, in the Neotropics the genera *Cecropia* and *Ochroma* are aggressive colonizers of disturbed areas, but, while *Cecropia* species are facultatively mycotrophic as expected, *Ochroma pyramidale* is greatly dependent on VAM (D.P. Janos, unpublished observations). A second caveat is that the terms 'obligately mycotrophic' and 'facultatively mycotrophic' refer to dependence upon mycorrhizas, not to responsiveness to mycorrhizas (see Janos, 1988, 1993; Sieverding, 1991). For example, a second-growth Neotropical tree, *Stemmadenia donnell-smithii*, fails to grow without VAM, but does not grow rapidly even when mycorrhizal (that is, it is dependent but not responsive; Janos, 1980*a*). The final caveat is that one cannot precisely infer mycorrhiza dependence from field observations of the occurrence of mycorrhizas. For example, some *Piper* species may consistently form mycorrhizas in terrestrial field soils but be capable of growth without mycorrhizas (see Maffia, Nadkarni & Janos, 1993). On the other hand, some species may lack mycorrhizas as slow-growing adults (for example, some Lecythidaceae, St John, 1980*a*) but be incapable of survival without mycorrhizas as seedlings, hence are obligate mycotrophs.

Some obligate mycotrophs are capable of growth without mycorrhizas when artificially fertilized, but others, such as some palms (Janos, 1977), may have completely abdicated ability to take up mineral nutrients without mycorrhizas. Facultative mycotrophs, on the other hand, typically suppress mycorrhiza formation at high fertility (see Allen, 1991), thereby avoiding the carbon cost of sustaining mycorrhizal fungi that are not needed. By allocating all photosynthate to their own growth, facultative mycotrophs may achieve higher growth rates than obligate mycotrophs at high fertility. Fertilization of obligate mycotrophs, in contrast, may not preclude the formation of mycorrhizas (Janos, 1985; Alexander, Ahmad & Lee, 1992; Peng et al., 1993).

Plant species independence of mycorrhizas is a consequence of low nutrient requirements, or of ability of roots alone to take up mineral nutrients. Mineral nutrient requirements may be low because of slow growth, vegetative maturity combined with infrequent reproduction, or 'cheap, throw-away' tissue with low concentrations of nutrients (see Janos, 1983, 1987). Roots may have substantial ability to take up mineral nutrients without mycorrhizas if well endowed with root hairs (see Baylis, 1975), if highly branched with small diameter ultimate rootlets (see St John, 1980b), or if capable of excreting substantial mineral-solubilizing compounds such as organic acids (see Woolhouse, 1975). Several of these attributes typify early successional plant species, and are probably responsible for their non-mycorrhizal or facultatively mycotrophic status. Most herbaceous crop plants are facultatively mycotrophic, especially if they have been bred for responsiveness to fertiliser. Among lowland, humid tropical crops, bananas and papaya are facultatively mycotrophic (D. P. Janos, personal observation), but avocado, cacao, cassava, citrus and mango (see Sieverding, 1991; Read, 1993) are strongly dependent on VAM.

All known lowland, humid tropical ectomycorrhizal species are woody, and most are likely to be obligately mycotrophic. Janos (1985) argued that EM generally represent a higher carbon cost to host plants than do VAM, and, therefore, that host reliance on EM may be correspondingly great. Notwithstanding, some EM plant species can be facultatively mycotrophic (for example, *Neea laetevirens*, an early-successional, nyctaginaceous, small tree; Janos, 1985 and unpublished observations), but facultative mycotrophism among EM species is probably uncommon. An alternative to facultative mycotrophism is for young plants with few leaves or small crowns and high requirement of photosynthate for their own growth to initially depend upon VAM, and later in life to switch to EM. A switch from VAM to EM may occur in some tropical nyctaginaceous species (Moyersoen, 1993) and eucalypts (Lapeyrie et al.,

1992). The ecosystem consequence of the supposed higher cost of EM than of VAM is that a higher proportion of net primary production goes below ground in EM than in VAM-dominated communities. Those who would advocate planting pines and eucalypts indiscriminately throughout the tropics should consider that vegetation conversion to EM-dominated stands ultimately may sacrifice harvestable aboveground production to ectomycorrhizal fungi.

Over the full range of plant reliance upon mycorrhizas, that of lowland, humid tropical mature-forest canopy trees tends to be extreme, even though some herbaceous perennials such as warm-season, temperate-zone grasses also can be highly dependent and highly responsive to VAM. Mycorrhizas are crucial for tight closure of lowland, humid tropical nutrient cycles (Janos, 1983), and so, in degraded sites of greatly diminished mineral nutrient availability, mycorrhizas can prove indispensable for tree survival and growth.

Community effects

Because of their dramatic effects on individual plant performance, mycorrhizas influence the productivity of plant communities, and can affect community composition, succession, and within-habitat species diversity. Mycorrhizas influence community composition by affecting the competitive abilities of plant species in consequence of soil fertility and the availability of mycorrhiza inocula.

Succession

Janos (1980*b*) presented a general model of mycorrhiza influence on plant communities and succession that provides a conceptual framework for understanding the potential role of mycorrhizas in rehabilitation. Although the model can describe temperate-zone successions (see Allen, 1991), it pertains especially to tropical seres because of prevalent soil infertility which makes mycorrhizas important, and because lowland, humid tropical soils typically contain few mycorrhizal fungus spores.

Abundant mycorrhizal fungus spores buffer soils against failure of mycorrhizas to form, but humid tropical soils are not so buffered (Janos, 1992). When soils are not buffered by spores, their mycorrhizal fungus populations are especially sensitive to the extent to which mycelia are sustained by vegetation. Neither VAM fungi nor EM fungi appear to have significant free-living ability in competition with soil saprotrophs (see Harley & Smith, 1983; Allen, 1991), and hence, they are nutritionally dependent upon host photosynthate. If a plant community that is not buffered by spores comprises many non-mycorrhizal species or facultative mycotrophs growing without mycorrhizas, mycorrhizal fungi will be correspondingly rare, and mycorrhiza inocula will be lacking for the subsequent sere.

When inocula are lacking, obligately mycotrophic species cannot establish, so plant communities consist of non-mycorrhizal and facultatively mycotrophic species (Janos, 1980b). To the extent that occurrence of these species is determined by competitive ability, soil fertility may influence performance and hence relative abundances of facultatively mycotrophic and non-mycorrhizal species. Non-mycorrhizal species are the only ones likely to be successful at low fertility in the absence of mycorrhizal inocula. Nevertheless, non-mycorrhizal species occur at high fertility, although they may be poor competitors because of the costs of their adaptations alternative to mycorrhizas for mineral nutrient uptake. Facultative mycotrophs probably take best advantage of high fertility because they can reject mycorrhizas which they do not need, and they do not suffer the costs of alternative adaptations of non-mycorrhizal species. At low fertility, however, provided that mycorrhiza formation is assured, obligate mycotrophs are most successful because of the efficiency of mycorrhizas in the uptake of scarce mineral nutrients.

Throughout succession, as soil fertility declines with the mineral nutrient capital of the system shifting to biomass (Odum, 1969), the potential benefits of mycorrhizas increase, and obligately mycotrophic species are favoured. Obligate mycotrophs, in their turn, are most likely to sustain populations of mycorrhizal fungi, insuring the availability of inocula subsequently derived from them. However, if mycorrhizal fungus populations are diminished by succession-initiating events, or by predominance of non-mycorrhizal pioneer colonizers, the fungi require mycotrophic hosts in order to rebound. Especially when fertility is moderate, facultative mycotrophs are more likely to establish and persist with very sparse mycorrhiza inocula than are obligate mycotrophs (see Cuenca & Lovera, 1992). Establishment of obligately mycotrophic species requires adequate mycorrhiza formation within a time limit imposed by the survival capacity of seedlings without mycorrhizas. This is the decisive factor for the importance of mycorrhizas in the rehabilitation of degraded lands. Facultative mycotrophs are not likely to colonize degraded sites of low fertility and few mycorrhizal fungi, and thus are not present to facilitate restoration of mycorrhizal fungus populations. When the probability of obligately mycotrophic seedling establishment in such sites is very low, natural succession and site rehabilitation may be greatly retarded.

Species diversity

Plant communities typically are mixtures of non-mycorrhizal, facultative and obligately mycotrophic species, each of which has its own pattern of response to soil fertility and mycorrhizas. In habitats where mycorrhiza formation is heterogeneous, each mycotrophic seedling experiences a likelihood of mycorrhiza formation that depends upon its capacity for survival and root spread (see Janos, 1992). These patterns and probabilities for species and habitats may influence community composition. This potential of mycorrhizas to influence community composition implies that mycorrhizas may also influence plant species diversity.

Janos (1983, 1985) suggested that mycorrhizas influence plant species richness by affecting the abilities of species to coexist. He argued that VAM enhance coexistence of many tropical tree species by minimizing differences among them in ability to compete for phosphorus, which is likely to be the limiting nutrient. Climax forest tree species probably depend on the same, relatively few VAM fungus species (see Janos, 1980a) for virtually all phosphorus uptake, so no host species is likely to be competitively much superior to others, nor able to exclude them. A multiple species, tropical field experiment in which VAM more than doubled the number of individuals of four obligately mycotrophic tree species surviving in mixed stands (Janos, 1985) supports this argument. Grime *et al.* (1987) obtained similar results for mixed communities of temperate-zone grassland plants in growth-chamber microcosms. Gange and coworkers (Gange, Brown & Farmer, 1990; Gange, Brown & Sinclair, 1993) showed that reduction of VAM of herbs and grasses in the field at Silwood Park, Berkshire, UK reduced vascular plant species diversity in seral communities. Recent research in the lowland humid tropics of Costa Rica has shown that, twenty years after a one-time elimination of mycorrhizal inocula from 10×20 m field plots, tree species richness was lower than in plots from which mycorrhizal fungi were not removed (D. P. Janos & G. S. Hartshorn, unpublished observations).

Janos (1983, 1985, 1988) indicated that, where EM plant species occur in the humid tropics, they tend to be among the most common species, thereby limiting the abundances of subsidiary species. Consistent with this suggestion, Newberry *et al.* (1988) found that EM plants dominated groves associated with especially infertile soils in Cameroon. Connell and

Lowman (1989), arguing that EM might contribute to the unusual occurrence of monospecific tree stands in the tropics, extended suggestions of Baylis (1975), Malloch, Pirozynski and Raven (1980), and Janos (1980*b*) that optimal, exclusive mycorrhizal fungus associates could assist a host species to dominance. Connell and Lowman (1989), however, seemed to impute more specificity to EM associations than generally exists (see Molina, Massicotte & Trappe, 1992 but also Smits, 1992). Newman and Reddell (1988) used multiple regression to examine whether or not species richness in the Great Smoky Mountains of the eastern United States was correlated with the relative abundance of EM hosts. They did not find a significant relationship between tree species richness and EM host abundance relative to that of hosts with other types of mycorrhizas. They did find that the richness of all plant species and of herbs alone was significantly related to the relative abundance of trees capable of forming VAM. The latter observation is consistent with data from the manipulative experiments on community composition discussed above, and underscores the potential of type, rapidity and extent of mycorrhiza formation to influence plant communities.

Inocula and inoculum potential

The presence of inoculum is best understood as a population process that depends on the balance of *in situ* production and 'immigration' (deposition) versus 'emigration' and death. The results of these competing processes partially determine the inoculum potential of a site. Mycorrhiza inoculum potential (MIP) can be defined as *the capacity of the inocula existing at a site to produce mycorrhizas sufficient to affect plant performance (positively or negatively)*. This is an unconventional (see Tommerup, 1992) but ecologically relevant definition. Defining MIP in this way encompasses the quantity and virulence of fungus propagules, the propensity of a host genotype to form mycorrhizas, and the site-dependent physiological receptivity of the host. It also incorporates the rapidity of root colonization to the point at which plant performance is affected. The magnitude of effects on host performance, called mycorrhizal fungus 'effectiveness' (see Janos, 1993), is not included.

A corollary of the definition of MIP suggested here is that different host species in the same place may experience very different inoculum potentials. Plant species differ in the length of time they can survive without mycorrhizas (especially when stressed by drought, herbivory, or competition), and in the maximum extent and density of root spread prior to mycorrhiza formation. These temporal and spatial differences among plant species in the way they 'census' their environment for inocula may cause large-seeded VAM species (see Janos, 1980*b*) to 'perceive' a habitat as providing uninterrupted inocula while other, less robust, mycorrhiza-dependent species 'sense' it as patchy (Janos, 1992). EM hosts may be less likely to suffer inoculum limitation than VAM hosts, because the spores, mycelial strands (or cords) and rhizomorphs of EM fungi may be more uniformly and broadly distributed than spores and hyphae of VAM fungi. However, even in environments where EM plants predominate, such as the Pacific Northwest of the United States, a temporally narrow opportunity for seedling establishment between spring thaw and summer drought can contribute to 'patchy' ectomycorrhiza inoculum potential (Trappe, 1977).

Types of inocula

Mycorrhizas can be initiated by hyphae emanating from spores, from live roots attached to shoots, and from root fragments (VAM: see Tommerup, 1992; EM: see Marx, Ruehle & Cordell, 1991). The large spores of glomalean fungi are regularly used to establish VAM 'pot cultures', and basidiospores of several EM fungi have been used to inoculate EM seedlings in nurseries (Marx *et al.*, 1991). Hyphae of both VAM and EM fungi can spread along and between roots, even linking roots of different plants via 'hyphal bridges' (Read, Francis & Finlay, 1985). Excised mycorrhizal roots too can serve as inoculum for experiments, implying that naturally occurring root fragments can give rise to infective hyphae. Because each type of mycorrhizal inoculum has different 'population' characteristics, MIP can be difficult to predict *a priori*.

In intact, humid tropical forest which typically has relatively few VAM fungus spores (Janos, 1992) and where viable EM fungus spores may not persist for long, most spread of mycorrhizas is probably by root-to-root growth of hyphae (Alexander *et al.*, 1992). From rates of spread of introduced fungi (see Janos, 1988), it seems that VAM hyphae are most likely to extend several centimetres to colonize roots but that the mycelial strands and rhizomorphs of some EM fungi can extend several metres from established mycorrhizal roots. This ability of EM strands and rhizomorphs to spread probably depends upon their capacity to transport nutrients from an established food base. Although mycorrhizal fungi attached to live host roots may be most competent to colonize new roots (especially because VAM fungi when spreading from a host may colonize the roots of species that do not normally form mycorrhizas, as observed by Ocampo, Martin and Hayman, 1980), Janos (1992) showed that fewer VAM hyphal 'propagules' may emanate from live host roots than from roots detached from shoots. Cessation of photosynthate supply to excised VAM roots may increase the ratio of 'runner hyphae' (see Friese & Allen, 1991), which are infective, to 'absorptive network hyphae', which are not. Ectomycorrhizal fungi do not appear to have such differentiated hyphae. Hyphal, mycelial strand, and rhizomorph spread are constrained by physical breakage, desiccation, and predation by mycophagous soil mesofauna such as collembola and nematodes (see references in Allen, 1991).

Frequent, relatively long-distance dispersal of mycorrhizal fungi depends upon sporulation and spore vectors. Spores of epigeous, ecto-mycorrhizal basidiomycetes are predominantly dispersed by wind, although mammalian mycophagists may be important vectors especially of hypogeous fungi (see Allen, 1991). Wind may transport spores for very long distances, but because spore concentrations in a volume of air diminish in proportion to a power of distance, low spore concentration makes substantial, rapid EM formation a rare event beyond several kilometres from a spore source (Bowen & Theodorou, 1973). A regular direction of prevailing winds, and regular patterns of habitat use by vertebrate vectors contribute to patchy spore dispersal. Wind erosion from broad expanses of bare, dry soil can move the large spores of VAM fungi (see Allen, 1991), but this is not likely to be of much importance in the humid tropics. There, flowing water (Lodge, 1990; Maffia et al., 1993), and small mammals (Janos, 1992) can move VAM fungus sporocarps and spores hundreds of metres. On a much more limited scale of perhaps a few metres, earthworms and various arthro-pods (see Allen, 1991; Janos, 1993) may transport VAM spores, but the latter may also prey upon them. Birds infrequently might transport a few spores in soil on their bodies but may make a significant contribution especially to vertical inoculum dispersal (see Janos, 1993) if spores, root fragments, or EM fungus rhizomorphs are associated with or used as nest material.

Spores can be eliminated as viable inocula by predation, by 'leaching' to deep soil horizons to which feeder rootlets do not penetrate, and by occlusion in organo-mineral complexes with clay colloids. The latter two physical processes are most likely to affect the smallest VAM fungus spores and EM fungus spores. Among EM fungus spores, those species with hydrophobic spores (see Borchers & Perry, 1987) are most likely to be occluded, and those with hydrophilic spores are most likely to be leached.

Specific germination cues and directed germ tube growth toward roots would maximise the MIP to be derived from spores. Once spores germinate, hyphal growth and competence to colonize roots may be influenced by spore energy reserves and the capacity of a fungus to gain additional energy saprophytically. Some EM fungi seem to have the ability to be partially saprophytic, but VAM appear to lack this ability (Janos, 1983). The large spores of VAM fungi, however, are likely to have greater energy reserves than the typically much smaller spores of EM fungi. The spores of EM fungi are haploid, and even if their germ tubes are competent to colonize roots prior to somatogamy and dikaryotization, dikaryotic mycelia are likely to be more vigorous and phenotypically plastic than monokaryons (Trappe & Molina, 1986).

Although gradual root death in response to seasonal change may stimulate sporulation by mycorrhizal fungi, rapid root death by tree fall or animal predation produces root fragment inocula. Powell (1976) showed that hyphae from VAM root fragments can colonize new roots without intervening sporulation. Under natural conditions, root fragments may be moved by water or vertebrates. Koske and Gemma (1990) have found that fragments of roots and other plant materials in the drift line on a Hawaiian beach could produce VAM, but accompanying spores may have been involved.

The energy reserves of root fragment inocula are likely to be limited to stores in roots and vesicles (VAM) or in roots and hyphal mantles (EM). Under humid tropical conditions of warmth and moisture, high respiration rates of roots and hyphae may quickly exhaust these energy stores (see Perry, Molina & Amaranthus, 1987). In highly seasonal sites, however, drought may slow respiration to the extent that a small percentage of 'dormant' root fragment inocula remains viable for many months (Alexander *et al.*, 1992).

Humans may intentionally or unwittingly transport mycorrhizal inocula. Ectomycorrhizal fungi frequently have been imported to inoculate pine plantations where pines are not native (Mikola, 1973). However, our present inability to cultivate VAM fungi *en masse* has limited their intentional transport and introduction of non-indigenous strains (but see Sieverding, 1991 for a description of the commercial 'Manihotina' VAM inoculant). Inadvertent transport of mycorrhizal fungi is likely with bare-root transplant stock that has formed mycorrhizas prior to transplant, and with movement of soil that contains mycorrhizal propagules.

Overall, EM fungi seem to have a capacity to spread farther and faster than VAM fungi. EM fungi that are dispersed by wind and that have saprophytic ability should be especially capable of rapid spread. An apparent, nearly complete lack of host specificity on the part of VAM fungi, however, compensates for their relatively low mobility and lack of free-living ability. Most VAM hosts associate promiscuously with VAM fungus species, and VAM fungi seem to have nearly unrestricted host ranges (Harley & Smith, 1983). Some EM fungi can be quite restricted in the host species with which they will associate (Molina *et al.*, 1992; Smits, 1992), although others, such as *Pisolithus tinctorius* which is widely used as a plantation inoculant (Marx *et al.*, 1991), have nearly unrestricted host ranges among EM host species.

Notwithstanding their putative capacity for distance dispersal, whether or not EM fungi are any more or less patchy than VAM fungi, at the scale of centimetres which is important in determining the acquisition of mycorrhizas by seedlings, probably depends upon the identities of the EM fungus species. Some, such as *Hysterangium* species (see Allen, 1991), form dense 'mats' of hyphae over areas of square metres which may comprise areas of homogeneous high MIP. Other EM fungi with diffuse mycelia, and VAM fungi are likely to be locally patchy with respect to MIP (Janos, 1992).

Assessing inoculum potential

The amount of mycorrhizal inoculum existing at a site has been assessed in various ways that fall into two major categories: (i) direct enumeration of propagules, and (ii) indirect assessment by 'bioassay' that involves baiting mycorrhizal fungi with a receptive host. All methods of inoculum assessment are problematic. Direct counts of propagules suffer two problems: difficulty of determining propagule viability, and uncertainty as to what constitutes the unit to be enumerated. Spores of VAM fungi can be extracted from soil, identified, and counted relatively readily, but, because thick walls of dead spores are persistent, total spore counts can be poorly correlated with mycorrhiza formation (for example, Alexander *et al.*, 1992; Fischer *et al.*, 1994). No quantitative relationship between measures of extramatrical hyphae and consequent mycorrhiza formation is known for either VAM or EM fungi. Bioassay techniques, because they include root colonization, are more ecologically interpretable than direct counts, and may be the only practical approach for studying EM inocula. Typically, bioassays such as the commonly used 'most probable number' technique (Porter, 1979) are performed upon soil samples removed from the field. The inevitable problem of extractive bioassays is that removal from the field may alter inoculum potential. Janos (1992) presented evidence that severed VAM roots gave rise to mycorrhizas more quickly than roots still attached to shoots.

The most ecologically relevant bioassay is likely to be one conducted in the field that measures MIP in units of time instead of measuring it as either the proportion of roots colonized in a fixed time or the most probable number of propagules. How quickly mycorrhizas are formed is likely to be most relevant to seedling persistence in a competitive milieu. Janos (1992) scored MIP in units of time by using the growth performance of seedlings incapable of growth without mycorrhizas to infer mycorrhiza formation. By planting bait seedlings on a regular grid throughout a site and monitoring their growth rates at frequent intervals by non-destructive morphometric measurement, MIP can be assessed for each planting position as *time until a seedling attains a threshold growth rate indicative of having adequate mycorrhizas to affect growth.* Such spatially arrayed point data can be analysed with geostatistical techniques for mapping within-site patterns of MIP. Within-site heterogeneity is likely to be more indicative than site average MIP of potential mycorrhiza limitation of plant establishment which may constrain rehabilitation.

Land use effects on inoculum potential

At the locations of individual seedlings, and across whole sites, mycorrhizal inocula can be diminished by land use in two ways which are complementary. Inoculum potential is reduced when patches of inoculum-free soil are exposed, and it is also reduced by vegetation that sustains few mycorrhizas. Additionally, various methods of disinfection of seedbed and nursery soils can eliminate mycorrhizal fungi (VAM: Sieverding, 1991; EM: Smits, 1992), and tillage that disrupts mycelia in the soil (see Read, 1993 and references therein) can lower MIP. In disturbed sites, even when mycorrhizas continue to form, the fungus species involved may be ineffective, and rehabilitation is again required.

Reduction of inoculum potential

Deforestation can markedly affect lowland, humid tropical MIP because hyphae are vulnerable to predation, root fragments may have inadequate energy stores to meet the demands of high respiration rates, and viable spores seem not to persist for long in this habitat. Although Alexander *et al.* (1992) state that VAM hyphae can survive in dead root fragments for at least 12 months, such survival may depend upon a marked dry season which minimizes biological activity (see Miller, 1987). In a continuously moist soil, Janos (1992) found that, after four months in pots in the presence of a non-mycorrhizal plant species, root fragment inocula had drastically reduced capacity to form mycorrhizas in comparison to freshly excised VAM. Sieverding (1991) noted that VAM spore and root inocula stored in a cold room may markedly attenuate within two years if the soil in which they are stored has as little as 5% humidity. For dipterocarp EM 'soil inoculum' containing root fragments, Smits (1992) noted that fewer than 50% of inoculated plants form mycorrhizas if the inoculum is stored for more than eight days.

Data on spore turnover in humid tropical soils are sparse but suggest that even in tropical forests with indistinct seasonality there can be large variations in VAM spore populations (Louis & Lim, 1987; Musoko, Last & Mason, 1994). Little information is available, however, concerning the fruiting of EM fungi and dissemination of their spores in the lowland humid tropics. Although rain facilitates fruiting by EM fungi, heavy rain can wash spores out of air (Trappe, 1977). The processes that remove EM and VAM fungus spores from the inoculum pool can be expected to operate frequently in tropical soils. Arthropods with potential as spore predators are abundant, high rainfall may rapidly leach EM spores and the smallest VAM spores, and spores may be occluded by abundant clay colloids.

Overall, mycorrhizal fungus populations in lowland, humid tropical soils are probably highly dynamic, and in the absence of live mycorrhizal roots MIP may attenuate rapidly. In the strongly seasonal tropics, however, a distinct dry season might slow mycorrhizal fungus population decline by limiting biological activity, and might stimulate VAM (but not EM) spore production (Mosse & Bowen, 1968).

Inoculum-free soil exposure

Land clearing methods that remove mycorrhiza-rich surface soil, exposing subsoil, or that result in the death of mycorrhizal fungi in surface soil can be expected to lower MIP or to eliminate inocula entirely (Alexander, 1989; Sieverding, 1991; Michelsen, 1992). The majority of fine roots (see references in Janos, 1984) and mycorrhizas (Rose & Paranka, 1987) in lowland, humid tropical forests occur within the top 10 centimetres depth of the mineral soil. Mycorrhizal fungi are obligately aerobic (Harley & Smith, 1983), and so are unlikely to function far below the surface, especially in clayey tropical soils, unless deep roots are in proximity to large channels open to the surface such as those excavated by 'leaf-cutter' ants (*Atta* spp.). Spores may be leached downward through the soil profile by water or be carried by arthropods, and so can be found deep in the soil, but their numbers are very low, below 50 to 60 cm (Habte, 1989; Sieverding, 1991). Surface (strip) mining, clearing by bulldozer, gullying by erosion (especially in pastures), skidding of logs, and craters resulting from warfare are likely to expose inoculum-free substrata.

There have been few studies in the tropics of the effects on mycorrhizal inocula of subsoil exposure by human activity. Habte (1989) showed that simulated erosion left very few VAM propagules in residual tropical soils, and Alexander *et al.* (1992) showed that compaction and erosion which resulted from logging had similar effects. Subsoil is naturally exposed by landslides, which can be extensive (for example, Garwood, Janos & Brokaw, 1979). Lodge (1990) and Simmons and Lodge (1990) showed that VAM inoculum was lacking for the most part in a large landslide in Puerto Rico. Inoculum was present and VAM did form on the slide, however, in proximity to its vegetated edges, and where level places accumulated material, eroded from upslope forest. In the temperate zone, Amaranthus and Trappe (1993) similarly found that eroded topsoil was an important source of inoculum for VAM plants but not for EM. They speculated that this was because the very small spores of EM fungi were less likely than VAM fungus spores to be trapped in positions of gentle slope or by accumulations of plant debris.

The consequences for subsequent site productivity of mechanised land clearing in the tropics can vary greatly depending upon whether or not a straight or floating 'shear' blade is used on a bulldozer, the skill of the operator, and whether or not large woody litter, stumps, and surface roots are pushed into windrows (Cassel & Alegre, 1994). Cuenca and Lovera (1992) studied the effects of bulldozer clearing of topsoil in the 'Gran Sabana' of Venezuela. Two years after clearing, disturbed sites had only 13 VAM fungus spores 100 g^{-1} soil compared to 200 spores 100 g^{-1} soil in undisturbed savanna. The authors concluded that this greatly retarded plant establishment on disturbed sites.

Even if topsoil is not removed, land clearing can elevate surface soil temperature to levels inimical to the survival of EM fungi, thereby reducing EM inocula (Smits, 1983). Smits (1992) indicated that dipterocarp EM inocula can be eliminated from soil by letting the sun heat soil under agricultural (black) plastic for at least one hour which will raise its temperature to more than 50 °C. However, similar treatment of soil in Colombia was ineffective in removing VAM inocula (Howeler, Sieverding & Saif, 1987). A bare soil in Zaire reached a record 86 °C surface temperature, likely to be sufficient to kill EM fungi, but temperature attenuated rapidly with depth such that this soil was a nearly normal 30 °C at 10 cm (Sanchez, 1976). At the same site, surface soil temperature was 25 °C in forest. Clearing may elevate tropical surface soil temperatures by seven to eleven degrees (Sanchez, 1976), so usually should not produce temperatures lethal to all types of EM and VAM fungus propagules. Elevated temperatures, however, and their secondary adverse effects on organic matter content, soil structure, aeration porosity, and water percolation (see Sanchez, 1976) may be inimical to mycorrhiza function even if not all propagules are killed (VAM: see Janos, 1987; EM: see Smits, 1983). Becker (1983) and Smits (1983) suggested that failure of dipterocarp reforestation after severe logging partly results from inability of EM fungi to tolerate altered soil conditions. If soil compaction increases surface runoff and erosion, then, if not killed, mycorrhizal fungus propagules may be exported downslope.

It is difficult to generalize on the effects of burning slash on mycorrhizal fungus survival and function because fires can differ greatly in intensity. Although Sieverding (1991) suggested that 'slash and burn' agriculture has little detrimental effect on VAM propagules, several investigations in the temperate zone have reported reduction of VAM propagules after fire (see references in Bellgard, Whelan & Muston, 1994). Fire in temperate-zone clearcuts has been reported to reduce EM inoculum potential (see Amaranthus & Trappe, 1993 and references therein). In the tropics, where many plants resprout from stumps after clearing, their surviving root systems are likely to maintain mycorrhizal fungi (see Bellgard *et al.*, 1994).

In addition to potential direct sterilizing effects, fire may indirectly affect mycorrhizal fungi through the 'liming' and fertilizing effects of ash (Janos, 1987). Soil fertility does not appear to be elevated sufficiently by 'slash and burn' techniques to suppress VAM formation (Sieverding, 1991), although it is well known that high levels of artificial fertilizer applied to facultatively mycotrophic host species can preclude mycorrhiza formation. Fertilization of obligate mycotrophs, however, may not suppress mycorrhiza formation (for example, Alexander *et al.*, 1992; Peng *et al.*, 1993), and would not reduce subsequent MIP. Fertilization of extremely infertile tropical soils may actually enhance

VAM formation by contributing to improved establishment and growth of facultatively mycotrophic species, thereby providing more root sites for mycorrhizas, and increasing the probability of roots encountering inoculum (Cuenca & Lovera, 1992). Nevertheless, over the long term, fertilization may be detrimental to mycorrhizas even though not sufficient to preclude mycorrhiza formation (Johnson, 1993). Ruiz (1987) studied mycorrhiza formation by maize in soil from a high input site that was continuously cropped for 14 years at Yurimaguas, Peru. Although he found that, in soil receiving for one year only the same high fertilizer input as the continuously cropped site, maize had abundant VAM (reported as 100%), in the 14-year continuously cropped soil, maize formed very few mycorrhizas (10%). Ruiz (1987) suggested that this might be a consequence of the fertilization regime favouring non-mycorrhizal weed species, presence of which reduced MIP.

Forest conversion

Land use that produces vegetation which sustains few mycorrhizas can exacerbate lack of mycorrhizal inoculum. For example, heavily fertilized plantations of facultative mycotrophs such as bananas (Cuenca, Herrera & Meneses, 1991; D. P. Janos personal observation) and overgrazed, sedge-filled pastures (Janos, 1988, 1992) have limited inoculum potential because their plant communities sustain few mycorrhizal fungi. Such sites, after abandonment to secondary succession, with their limited inocula tend to favour colonization by plant species that do not require mycorrhizas, thereby further retarding restoration of mycorrhizal fungus communities.

Few studies have examined the effects of tropical vegetation on MIP, although many counts of VAM fungus spores under lowland, humid tropical tree plantations and herbaceous crops have been published. When both MIP and total number of spores have been assessed, they are often not correlated (An *et al.*, 1990; Johnson *et al.*, 1991; Fischer *et al.*, 1994). This is not surprising because: hyphae and root fragment inocula may be more important than spores; it is difficult to distinguish viable spores from dead ones, and propagule counts cannot indicate either activation (germination) rates of propagules or rates of spread of infective hyphae through soil.

Studies that have examined the effects of vegetation on mycorrhizal inocula to be derived from it have tended to confirm the importance of maintenance of active mycorrhizas by vegetation. In tropical India, Harinikumar and Bagyaraj (1988) showed that growth of a non-mycorrhizal crop (mustard) significantly reduced VAM of a subsequent host crop, and they suggested that it would take at least two sequential crops of mycorrhizal hosts to rebuild inocula to the level that preceded the non-mycorrhizal crop. Where roots are sparse, as in semi-arid savanna, mycorrhizal inocula are low (Michelsen & Rosendahl, 1990; Sieverding, 1991). Dodd et al. (1990) showed that by growing VAM host plants in a savanna soil inocula were enriched and the yield of a subsequent crop was improved in comparison with savanna sown directly to the crop. Where EM roots occupy the soil, VAM inocula can be sparse (Tobiessen & Werner, 1980; Kovacic, St John & Dyer, 1984; Benjamin, Anderson & Liberta, 1989, Johnson et al., 1991; Moyersoen, 1993).

Cattle pasture is an anthropogenic vegetation which has become common, especially in the Neotropics (for example, Nepstad, Uhl & Serrão, 1991). Great variation in the management and intensity of use of pastures confounds generalization concerning pasture effects on MIP, but, where overgrazing has been severe, reduction of MIP should be considered as a major factor potentially impeding regeneration of woody vegetation. Janos (1988, 1992) showed that little VAM inoculum in an overgrazed pasture dominated by several species of non-mycorrhizal sedges (Cyperaceae) retarded growth of obligately mycotrophic trees. Fertilized or annually burned pastures might also have low MIP if facultatively mycotrophic grasses reject mycorrhizas. On the other hand, moderately grazed, unburned grass pastures can have relatively large populations of VAM spores (Fischer et al., 1994), although they contain neither as many spores as agricultural fields, nor as many fungus species as native forest vegetation (see Sieverding, 1991).

Alteration of mycorrhizal fungus community composition

When inocula are not entirely eliminated, alteration of the composition of a mycorrhizal fungus community is another potential effect of land use which should be considered. Although research in both the temperate zone and in the tropics has documented apparent simplification of mycorrhizal fungus communities as a result of various land uses or cropping sequences, it is difficult to predict what the consequences may be of reduction in number of mycorrhizal fungus species and changes in their relative abundance. Simplification of the composition of a mycorrhizal fungus community might diminish potential site productivity, and might diminish the resilience of vegetation to recover from perturbation or to accommodate environmental change.

Several investigators have compared the richness of VAM fungus species in natural tropical vegetation with that in agroecosystems, and have usually found spores of fewer species in modified than in natural vegetation. Schenck, Siqueira and Oliveira (1989) found 18 VAM fungus species in Brazilian 'Cerrado' but a range of seven to 15 species in various monoculture and mixed crops. Sieverding (1991) summarized four studies that recorded 16 to 21 species in native vegetation in Brazil, Colombia and Zaire, but just ten to 15 species under low-input agriculture (including mixed crops), and only six to nine species in high-input, intensive agriculture (including monocultures) in the same locations. Cuenca *et al.* (1991) reported four to 26 species in natural vegetations ranging from very dry forest, through sclerophyllous scrub, to cloud forest and evergreen rainforest in Venezuela. They reported just one to five species in cacao plantations that were as much as 60 years old. In contrast, Musoko *et al.* (1994) reported 16 species in forest in Cameroon, and a similar number in *Terminalia* plantations that had been manually cleared (Mason, Musoko & Last, 1992). Clearing, however, had occurred just 11 months prior to sampling. These authors remarked that mechanical clearing caused large changes in the relative abundances of fungus species which they associated with rapid invasion by an herbaceous weed. A study that apparently contradicts the aforementioned is that of Wilson *et al.* (1992) in Côte d'Ivoire in which, although a comparable 16 spore types were found in forest, 26 to 35 spore types were found in *Terminalia* plantations from one to 23 years of age. These data do not necessarily suggest a doubling of species richness from forest to plantations, however, because two plantations cleared and planted in the year of sampling had 26 and 30 spore types. It is most likely that clearing simply stimulated simultaneous sporulation of more species than in forest. This underscores an important caveat for interpreting fungus community composition based upon the presence of VAM fungus spores and EM fungus basidiomes: not all species present may sporulate every year.

Mechanisms of mycorrhizal fungus community change

Even though the absence of spores and sporocarps need not indicate absence of vegetative mycelia, it is likely that over a period of many years land use involving vegetation change does alter the composition of a mycorrhizal fungus community. The mechanisms of fungus community change are not well understood, but two phenomena must be considered: the appearance of species that were not originally present, and disappearance of species that were. For VAM fungi, appearance of species not originally present is probably a rare event. The observed declines in VAM fungus species richness with land use, already cited, support this contention. True introductions require both a nearby source, and vectors moving between source and introduction sites. Moreover, introduced fungi must be able to tolerate climatic and edaphic conditions at the introduction site. In many cases, the apparent occurrence in a disturbed site of a VAM fungus species of which spores were not found in pre-disturbance vegetation most likely reflects disturbance favouring sporulation of mycelium already present. When exotic mycorrhizal fungi are introduced, humans are the agent most likely deliberately or inadvertently to have transported them. However, ectomycorrhizal fungi may be more likely than VAM fungi to be introduced naturally, because of their potential long-distance dispersal.

Three mechanisms have been postulated to cause disappearance of mycorrhizal fungus species after disturbance: host species discrimination against fungus associates, inability of fungi to tolerate an altered soil environment, and lack of resistant propagules. The latter two mechanisms are more likely than discrimination by hosts to cause simplification of VAM fungus communities, but all three might cause change in EM fungus communities.

The idea that host species might directly and selectively favour certain VAM fungus species periodically resurfaces in the literature because of observations of great spore abundance of some species within host rhizospheres (for example, Musoko et al., 1994). A problem with such inference from observations made in tropical forest is that at any point in the forest the feeder rootlets of a median of six different plant species may be found intermixed (Janos, 1984), and the mycorrhizal rootlets of any one tree are not necessarily located closer to its bole than to the boles of other individuals. Its own leaf and branch litter are most likely to surround an individual, however, and soil properties beneath the crown reflect the influence of species-specific characteristics of this litter. Neither is greater sporulation of some VAM fungus species than of others in crop monocultures sufficient evidence of active host 'preference', because of the possible lack of correlation between mycelium presence and sporulation. Although a mycorrhizal fungus may require substantial carbon from a host in order to form many spores, a fungus receiving adequate carbon need not sporulate. Experiments investigating preferential host association with VAM fungus species have failed to provide conclusive evidence of this phenomenon (Lopez-Aguillon & Mosse, 1987; Sanders, 1993). Nevertheless, it is possible that a host species can indirectly favour increases in mycelial abundance of some VAM species by providing soil conditions favourable for sporulation, provided that spores are the primary means of mycorrhiza persistence or transmission.

Soil physical, chemical and biological characteristics are well known to affect spore germination and mycelial spread of VAM and EM fungi (Harley & Smith, 1983; Allen, 1991), so inability of some fungus species to tolerate soil conditions altered by land use is highly likely to contribute to their disappearance. After land clearing, if vegetative mycelia cannot survive in bare ground, only those fungus species present as resistant propagules will be available to form mycorrhizas with new vegetation. Seasonal differences in sporulation might result in the disappearance of species that are not sporulating at the time of disturbance. With repeated sampling, Louis and Lim (1987) and Musoko *et al.* (1994) observed large fluctuations in total numbers of VAM fungus spores in lowland, humid tropical soils, but these fluctuations were not correlated with season. In a study of rodent mycophagy in Amazonian Peru, however, D. P. Janos, C. T. Sahley and L. H. Emmons (unpublished observations) found that the presence of spores of four species of *Glomus* in rodent faeces was associated with the dry season. Moreover, broad peaks of abundance in faeces of *Sclerocystis coremioides* sporocarps versus all *Glomus* species occurred at different times of year. D. P. Janos and C. A. Lumio (unpublished observations) observed a similar difference in seasonal abundance of *Acaulospora foveata* versus two *Glomus* species in a Costa Rican lowland, wet forest soil. Differences in time of sporulation among mycorrhizal fungus species could therefore lead to differential loss after land clearing.

All three mechanisms mentioned above may cause change, but not necessarily simplification, in EM fungus communities. In the temperate zone, as host stands age, a succession of different EM fungus species are present both as basidiomes and as mycorrhizas (see Keizer & Arnolds, 1994, and references therein). This succession is most likely to be caused by host and soil changes with stand development (see Janos, 1988). Ectomycorrhizal fungus species differ in their abilities to tolerate soil stresses such as high temperature, low pH, and high metal concentrations (see Harley & Smith, 1983), so soil changes are highly likely to alter EM fungus communities. As already noted, EM fungi may be less capable of surviving land clearing than are the large, resistant spores of VAM fungi, but high mortality may be ameliorated by the rapid recolonizing ability of EM fungus species that fruit near cleared sites.

Ectomycorrhizal fungus community change may be caused by hosts selectively favouring fungus species, but evolutionary considerations argue against this except in invariant, climax environments. Janos (1985) suggested that EM hosts may have an ability to differentially favour fungus species mycorrhizal on different root tips in a way unlikely for VAM hosts. In order to exercise selectivity, however, hosts require a means of 'comparing' the benefit they receive from different fungus species. A mechanism for direct physiological comparison by hosts among fungus species is difficult to envisage for fungi among which different types of benefit accrue, for example, between a fungus species that enhances nitrogen nutrition and another that enhances water supply without affecting nitrogen uptake. Nevertheless, in a stable, predictable environment, EM hosts might physiologically 'discriminate' among fungus species from which the same suite of benefits are obtained. However, in a temporally variable environment in which the relative importance of different benefits change, simultaneous comparison among fungus associates could lead to disadvantageous discrimination against fungus species with potential for future benefit. In a fluctuating environment, or among seral host species, natural selection might disfavour hosts that discriminate among fungus associates in ecological time. Over evolutionary time EM hosts certainly gain the ability to discriminate absolutely among fungus species as demonstrated by many pure culture, synthesis studies (see Molina *et al.*, 1992), and by failures of native tropical EM host species to provide suitable fungus associates for introduced pines (Redhead, 1980).

Consequences of mycorrhizal fungus community change

Especially because of the 'soil specificity' of both VAM and EM fungi, it would be remarkable if land use that changes soil properties did not also change the composition of mycorrhizal fungus communities. Predicting the potential consequences of such change requires that the quality (effectiveness as mutualists) and quantity of mycorrhizal fungus species colonizing a host be considered. This consideration raises a fundamental question concerning the evolution of mycorrhizal mutualisms: can ineffective mycorrhizal fungi persist in nature? If they can, then land use that increases the abundance of ineffective fungi is likely to be detrimental to mycotrophic hosts (see Johnson, 1993; Blal & Gianinazzi-Pearson, 1992).

An ineffective mycorrhizal fungus can be defined as one that has no net positive effect on the fitness of its host. Fungus 'effectiveness' (see Janos,

1988, 1993) is most often perceived as improved growth, although size and fitness need not be correlated. That mycorrhizal fungus species can be ineffective with respect to host growth for particular artificial host and soil combinations has been demonstrated with crops. In an extreme case, a VAM fungus was shown to function as a parasite of cotton (Modjo & Hendrix, 1986). Ineffectiveness results when the carbon cost to the host of a mycorrhizal fungus exceeds (or equals) the benefits gained from that fungus, so that the net effect of association on host fitness is negative (or zero). The cost per ectomycorrhiza or VAM infection unit is probably relatively fixed for a fungus species by its physiology, but benefit is highly situation dependent. For example, imagine a fungus that serves only to protect its host from root pathogens. In a fumigated soil this fungus, diminishing host growth because of its carbon cost, might be considered to be a 'parasite' (net negative effect on host fitness), but, in the presence of pathogens, such a fungus would increase host fitness by improving survival. In the appropriate context, the fungus is a mutualist, and natural selection will favour continued association of host and fungus. If host fitness benefits are not tightly coupled with material supply (water and mineral nutrients) or growth, hosts will have no basis for physiological comparison and immediate discrimination among fungus associates. Such inability to discriminate might allow the persistence of wholly ineffective fungi.

On the other hand, evolutionary discrimination against the persistence of ineffective mycorrhizal fungi is probably strong. Prolonged, great interdependence between hosts and fungi, and consistently high encounter probability tend to strengthen natural selection against ineffective associates (see Soberon & Martinez del Rio, 1985), and to improve mutualistic function of effective strains (Law, 1985). Because both EM (see Marks & Foster, 1973) and VAM fungi (see Wilson & Tommerup, 1992) compete for root sites by exploitation, mycorrhizal fungi on the same plant affect one another negatively, although all might be favoured by improvement of host performance (that is, an increase in the host 'carrying capacity' for mycorrhizal fungi). In order for an ineffective fungus species to proliferate under such conditions (a 'Type II *aprovechado*' of Soberon & Martinez del Rio, 1985), generally it must be a facultative, non-specific associate with weak effects on the other mutualistic partners (Soberon & Martinez del Rio, 1985). Among root mutualists, this describes *Rhizobium* and *Bradyrhizobium*, ineffective strains of which may be relatively common. Most mycorrhizal fungi, however, are obligate mutualists. Because the strength of effects of ineffective mycorrhizal fungus species will increase in proportion to their abundance and to host dependence on mycorrhizas, natural selection is expected to minimize the overall abundance of ineffective species. This should be especially true in lowland, humid tropical forests, where host dependence on mycorrhizas is extreme.

These evolutionary considerations suggest that ineffective mycorrhizal fungus species are least likely to occur in lowland, humid tropical forests, so that poor function of simplified mycorrhizal fungus communities in deforested lands is most likely to be the consequence of altered soil conditions and novel host species. Diminished mycorrhiza function could result because hosts have few mycorrhizal fungus species with which to associate. Although hosts exposed to mixed inoculum, containing several species of mycorrhizal fungi, may not perform as well as when inoculated with the most effective one of those species in relatively short-term experiments under invariant conditions (see Janos, 1988), this is not always the case (VAM: Hurtado & Sieverding, 1986; EM: Mikola, 1973). Janos (1988) suggested that, as field soil conditions change seasonally, or during the development of tree plantations, multiple mycorrhizal fungus associates are likely to be more beneficial to a host than are single mycorrhizal fungus associates. If so, simplification of mycorrhizal fungus communities by land use could retard succession on fallowed or abandoned sites.

Rehabilitation and conservation

Rehabilitation of sites in which mycorrhiza function has been diminished may require both the quantitative and qualitative re-establishment of mycorrhizal fungus populations. That is, not only may MIP need to be enhanced, as in sites from which inocula have been substantially lost, but mycorrhizal fungus species may have to be replaced to restore a diverse fungus community. Although extreme land degradation may eliminate mycorrhizal fungi from entire sites, leading to highly depauperate vegetation, patchy mycorrhiza formation that limits establishment of only some seedlings is more common than the extreme. Patchy lack of mycorrhizas may not be suspected as a cause of failures of plant establishment, because not all plant species are similarly affected. Facultative mycotrophs and species capable of great root spread prior to mycorrhiza formation are least sensitive to patchy deficiency of mycorrhiza inoculum. Even when inocula are uniformly available, if the composition of mycorrhizal fungus communities is altered by land use, especially if the fungus communities are impoverished of species, productivity may be jeopardized (for example, see Johnson *et al.*, 1992).

Recently, numerous authors have considered production and utilization of mycorrhizal fungus inoculants in agriculture, tree nurseries, and plantation forestry (for example, VAM: Abbott, Robson & Gazey, 1992; Bagyaraj, 1992; Feldmann & Idczak, 1992; EM: Marx *et al.*, 1991), and this copious literature may be consulted for procedures. Two conclusions of relevance to rehabilitation of deforested lands emerge. The first is that field-scale, direct inoculation with mycorrhizal fungi is unlikely to be economic because of the difficulty of producing large quantities of VAM inocula (see Sieverding, 1991; Bagyaraj, 1992; Read, 1993), and because field inoculation is unnecessary for EM tree hosts that are customarily produced in nurseries before outplanting (see Mikola, 1973; Trappe, 1977; Marx *et al.*, 1991). The second is that, although transplanted mycorrhizal seedlings can successfully establish EM fungi in outplanting sites, persistence and spread of VAM fungi from transplanted inoculated seedlings is problematic (Abbott *et al.*, 1992). Part of the latter difference between EM and VAM fungi is attributable to the widespread utilization of *Pisolithus tinctorius*, which has great tolerance of high soil temperatures and acidity, as an EM inoculant fungus. No VAM fungus equivalent to *P. tinctorius* has emerged, which may be fortunate if indeed mycorrhizal fungus monocultures lack ecological resilience.

Rehabilitation of mycorrhizal fungi and plant communities on deforested sites can require amelioration of soil compaction, high soil temperature, aluminum toxicity, and acidity (see Janos, 1987, 1988). Establishment of vegetation cover, increase in root density, and augmentation of soil organic matter will ameliorate inhibitory site factors, and favour the proliferation of mycorrhizal fungi. However, such biotic improvements are themselves influenced by mycorrhizas, compounding the difficulty of rehabilitation. The solution to this conundrum may lie in the management of plant secondary succession to facilitate restoration of mycorrhizas.

The general model of mycorrhiza influence on plant succession of Janos (1980*b*) suggests two novel approaches to site rehabilitation in the lowland humid tropics. Where MIP is low but mycorrhizal fungi are not completely absent, the model indicates that favouring facultative mycotrophs should foster a rebuilding of mycorrhizal fungus populations. In the tropics, the key to establishing facultative mycotrophs to sustain mycorrhizas, contrary to convention in the temperate zones, may be to fertilize. Fertilization of tropical nurseries (Michelsen, 1993) and field sites (Cuenca & Lovera, 1992) need not suppress mycorrhizas. The second novel approach is to mimic the functional role of native tropical pines in secondary succession by employing EM transplant stock to establish an initial vegetative cover on extremely impoverished sites (especially sites lacking non-occluded, light-fraction organic matter and nitrogen). Ectomycorrhizal hosts may effectively 'pump' nitrogen from heavy-fraction soil organic matter into a new, accessible light-fraction formed by their accumulating litter (see Janos, 1988), thereby improving conditions for intercropped VAM facultative mycotrophs. Ectomycorrhizal *Acacia* species, especially if also capable of forming VAM and of nodulating with nitrogen-fixing bacteria, could be a highly appropriate choice for such usage because their leaf litter is less likely than that of *Casuarina* and *Eucalyptus* species to contain compounds that are inhibitory to other plants.

Where mycorrhizal fungi are completely absent from rehabilitation sites, their reintroduction may be accomplished most practically by transplanting stock previously inoculated in the nursery. Outplanting probably will have to be accompanied by site treatment to mitigate inhibitory factors (for example, Cassel & Alegre, 1994), and direct-seeded or non-inoculated transplanted 'companion' plants also may be needed to facilitate VAM fungus spread (see Janos, 1988). Recent advice of Abbott *et al.* (1992) concerning the selection of inoculant fungus species should be heeded. Capacity to improve host growth in the nursery should be neither the sole nor the primary criterion for inoculant fungus selection. Instead, inoculant fungus impact upon host survival after outplanting, ability to persist and spread, resilience under aperiodic stresses, and provision of food for vertebrates (which, in their turn, may foster sporocarpic mycorrhizal fungus spread; see Allen, 1991) should be considered. In view of the potential importance of a diverse mycorrhizal fungus community, it may be most prudent, and technologically the simplest (see Michelsen, 1993), to use field-collected, mycorrhizal roots from native vegetation as inoculum to employ as wide a range of fungus species as possible.

As the availability of forested tropical lands diminishes, and the consequent need to rehabilitate deforested lands increases, forest remnants may take on increasing importance as mycorrhizal fungus germplasm reservoirs. They are potential sources of VAM fungi and late-stage EM fungi that are best adapted to local climatic and edaphic conditions under forest cover (see Michelsen, 1993, and references in Abbott *et al.*, 1992). These are the species most likely to be lost as VAM fungus communities are simplified by land use, and are the EM fungi perhaps least likely to be cultivable. The need to conserve them provides a strong rationale for retention of patches of native vegetation in the landscape. Consequences of extinction of these species for ecosystem productivity and stability are not known but may be severe. Concern for loss of biodiversity too often ignores diversity belowground, but in lowland, humid tropical forests it may be mycorrhizal fungus diversity upon which conservation of aboveground biodiversity depends.

Acknowledgements

Completion of this chapter was supported by a Charles Bullard Fellowship in Forest Research from Harvard University which is gratefully acknowledged. The author thanks Zita de Andrade, Mohamed Bakarr, Ralph E. J. Boerner, Blase Maffia, Anders Michelsen, Bernard Moyersoen, Pathmaranee Nadarajah, and Chris Picone for sharing publications and work in progress, and the editors of this volume for encouragement. This is contribution 445 from the Program in Tropical Biology, Ecology, and Behavior of the Department of Biology, University of Miami.

References

Abbott, L. K., Robson, A. D. & Gazey, C. (1992). Selection of inoculant vesicular-arbuscular mycorrhizal fungi. In *Methods in Microbiology, Volume 24*, ed. J. R. Norris, D. J. Read & A. K. Varma, pp. 1-21. London: Academic Press.

Alexander, I. (1989). Mycorrhizas in tropical forests. In *Mineral Nutrients in Tropical Forest and Savanna Ecosystems*, ed. J. Proctor, pp. 169-188. Oxford: Blackwell Scientific Publications.

Alexander, I., Ahmad, N. & Lee, S. S. (1992). The role of mycorrhizas in the regeneration of some Malaysian forest trees. *Philosophical Transactions of the Royal Society London B*, **335**, 379-88.

Allen, M. F. (1991). *The Ecology of Mycorrhizae*. Cambridge: Cambridge University Press.

Amaranthus, M. P. & Trappe, J. M. (1993). Effects of erosion on ecto- and VA-mycorrhizal inoculum potential of soil following forest fire in southwest Oregon. *Plant and Soil*, **150**, 41-9.

An, Z.-Q., Hendrix, J. W., Hershman, D. E. & Henson, G. T. (1990). Evaluation of the 'most probable number' (MPN) and wet-sieving methods for determining soil-borne populations of endogonaceous mycorrhizal fungi. *Mycologia*, **82**, 576-81.

Bá, A. M., Balaji, B. & Piché, Y. (1994). Effect of time of inoculation on *in vitro* ectomycorrhizal colonization and nodule initiation in *Acacia holosericea* seedlings. *Mycorrhiza*, **4**, 109-19.

Bagyaraj, D. J. (1992). Vesicular-arbuscular mycorrhiza: application in agriculture. In *Methods in Microbiology, Volume 24*, ed. J. R. Norris, D. J. Read & A. K. Varma, pp. 359-373. London: Academic Press.

Baylis, G. T. S. (1975). The magnolioid mycorrhiza and mycotrophy in root systems derived from it. In *Endomycorrhizas*, ed. F. E. Sanders, B. Mosse & P. B. Tinker, pp. 373-389. London: Academic Press.

Becker, P. (1983). Ectomycorrhiza on *Shorea* (Dipterocarpaceae) seedlings in a lowland Malaysian rainforest. *The Malaysian Forester*, **46**, 146-70.

Bellgard, S. E., Whelan, R. J. & Muston, R. M. (1994). The impact of wildfire on vesicular-arbuscular mycorrhizal fungi and their potential to influence the re-establishment of post-fire plant communities. *Mycorrhiza*, **4**, 139-46.

Benjamin, P. K., Anderson, R. C. & Liberta, A. E. (1989). Vesicular-arbuscular mycorrhizal ecology of little bluestem across a prairie-forest gradient. *Canadian Journal of Botany*, **67**, 2678-85.

Blal, B. & Gianinazzi-Pearson, V. (1992). Interactions between indigenous VAM fungi and soil ecotype in *Terminalia superba* in the wet tropics (Ivory Coast). In *Mycorrhizas in Ecosystems*, ed. D. J. Read, D. H. Lewis, A. H. Fitter & I. J. Alexander, pp. 372-373. Wallingford: CAB International.

Borchers, J. G. & Perry, D. A. (1987). Mycorrhizal inoculum potential in size fractions of southwest Oregon forest soils. In *Mycorrhizae in the Next Decade*, ed. D. M. Sylvia, L. L. Hung & J. H. Graham, p. 188. Gainesville: Institute of Food and Agricultural Sciences.

Bowen, G. D. & Theodorou, C. (1973). Growth of ectomycorrhizal fungi around seeds and roots. In *Ectomycorrhizae*, ed. G. C. Marks & T. T. Kozlowski, pp. 107-150. New York: Academic Press.

Cassel, K. & Alegre, J. C. (1994). *Land Clearing and Reclamation of Ultisols and Oxisols* (Soil management CRSP Bulletin No. 94-01). Raleigh: Soil Management Collaborative Research Support Program.

Connell, J. H. & Lowman, M. D. (1989). Low-diversity tropical rainforests: some possible mechanisms for their existence. *American Naturalist*, **134**, 88-119.

Cuenca, G., Herrera, R. & Meneses, E. (1991). Las micorrizas vesiculo arbusculares y el cultivo del cacao en Venezuela. *Acta Cientifica Venezolana*, **42**, 153-9.

Cuenca, G. & Lovera, M. (1992). Vesicular–arbuscular mycorrhizae in disturbed and revegetated sites from La Gran Sabana, Venezuela. *Canadian Journal of Botany*, **70**, 73-9.

Dodd, J. C., Arias, I. Koomen, I & Hayman, D. S. (1990). The management of populations of vesicular-arbuscular mycorrhizal fungi in acid-infertile soils of a savanna ecosystem. I. The effect of pre-cropping and inoculation with VAM fungi on plant growth and nutrition in the field. *Plant and Soil*, **122**, 229-40.

Feldmann, F. & Idczak, E. (1992). Inoculum production of vesicular-arbuscular mycorrhizal fungi for use in tropical nurseries. In *Methods in Microbiology, Volume 24*, ed. J. R. Norris, D. J. Read & A. K. Varma, pp. 339-357. London: Academic Press.

Fischer, C. R., Janos, D. P., Perry, D. A., Linderman, R. G. & Sollins, P. (1994). Mycorrhiza inoculum potentials in tropical secondary succession. *Biotropica*, **26**, 369-77.

Friese, C. F. & Allen, M. F. (1991). The spread of VA mycorrhizal fungal hyphae in the soil: inoculum types and external hyphal architecture. *Mycologia*, **83**, 409-18.

Gange, A. C., Brown, V. K. & Farmer, L. M. (1990). A test of mycorrhizal benefit in an early successional plant community. *New Phytologist*, **115**, 85-91.

Gange, A. C., Brown, V. K. & Sinclair, G. S. (1993). Vesicular-arbuscular mycorrhizal fungi: a determinant of plant community structure in early succession. *Functional Ecology*, **7**, 616-22.

Garwood, N. C., Janos, D. P. & Brokaw, N. (1979). Earthquake-caused landslides: a major disturbance to tropical forests. *Science*, **205**, 997-9.

Grime, J. P., Mackey, J. M., Hillier, S. H. & Read, D. J. (1987). Floristic diversity in a model system using experimental microcosms. *Nature*, **328**, 420-2.

Habte, M. (1989). Impact of simulated erosion on the abundance and activity of indigenous vesicular-arbuscular mycorrhizal endophytes in an Oxisol. *Biology and Fertility of Soils*, **7**, 164-7.

Harinikumar, K. M. & Bagyaraj, D. J. (1988). Effect of crop rotation on native vesicular arbuscular mycorrhizal propagules in soil. *Plant and Soil*, **110**, 77-80.

Harley, J. L. & Smith, S. E. (1983). *Mycorrhizal Symbiosis*. London: Academic Press.

Howeler, R. H., Sieverding, E. & Saif, S. R. (1987). Practical aspects of mycorrhizal technology in some tropical crops and pastures. *Plant and Soil*, **100**, 249-83.

Hurtado V., M. A. & Sieverding, E. (1986). Estudio del efecto de hongos formadores de micorriza vesiculo arbuscular (MVA) en cinco especies latifoliadas regionales en la zona geográfica del Valle del Cauca, Colombia. *Suelos Ecuatoriales*, **16**, 109-15.

Janos, D. P. (1975). Effects of vesicular-arbuscular mycorrhizae on lowland tropical rainforest trees. In *Endomycorrhizas*, ed. F. E. Sanders, B. Mosse & P. B. Tinker, pp. 437-446. London: Academic Press.

Janos, D. P. (1977). Vesicular-arbuscular mycorrhizae affect the growth of *Bactris gasipaes*. *Principes*, **21**, 12-18.

Janos, D. P. (1980a). Vesicular-arbuscular mycorrhizae affect lowland tropical rain forest plant growth. *Ecology*, **61**, 151-62.

Janos, D. P. (1980b). Mycorrhizae influence tropical succession. *Biotropica*, **12** (Supplement), 56-64.

Janos, D. P. (1983). Tropical mycorrhizas, nutrient cycles and plant growth. In *Tropical Rain Forest: Ecology and Management*, ed. S. L. Sutton, T. C. Whitmore & A. C. Chadwick, pp. 327-345. Oxford: Blackwell Scientific Publications.

Janos, D. P. (1984). Methods for vesicular-arbuscular mycorrhiza research in the lowland wet tropics. In *Physiological Ecology of Plants of the Wet Tropics (Tasks for Vegetation Science: 12)*, ed. E. Medina, H. A. Mooney, & C. Vasquez-Yanes, pp. 173-187. The Hague: Junk.

Janos, D. P. (1985). Mycorrhizal fungi: agents or symptoms of tropical community composition? In *Proceedings of the Sixth North American Conference on Mycorrhizae*, ed. R. Molina, pp. 98-103. Corvallis: Forest Research Laboratory.

Janos, D. P. (1987). VA mycorrhizas in humid tropical ecosystems. In *Ecophysiology of VA Mycorrhizal Plants*, ed. G. R. Safir, pp. 107-134. Boca Raton: CRC Press.

Janos, D. P. (1988). Mycorrhiza applications in tropical forestry: are temperate-zone approaches appropriate? In *Trees and Mycorrhiza*, ed. F.S.P. Ng, pp. 133-188. Kuala Lumpur: Forest Research Institute, Malaysia.

Janos, D. P. (1992). Heterogeneity and scale in tropical vesicular-arbuscular mycorrhiza formation. In *Mycorrhizas in Ecosystems*, ed. D. J. Read, D. H. Lewis, A. H. Fitter & I. J. Alexander, pp. 276-282. Wallingford: CAB International.

Janos, D. P. (1993). Vesicular-arbuscular mycorrhizae of epiphytes. *Mycorrhiza*, **4**, 1-4.

Johnson, N. C. (1993). Can fertilization of soil select less mutualistic mycorrhizae? *Ecological Applications*, **3**, 749-57.

Johnson, N. C., Copeland, P. J., Crookston, R. K. & Pfleger, F. L. (1992). Mycorrhizae: possible explanation for yield decline with continuous corn and soybean. *Agronomy Journal*, **84**, 387-90.

Johnson, N. C., Zak, D. R., Tilman, D. & Pfleger, F. L. (1991). Dynamics of vesicular-arbuscular mycorrhizae during old field succession. *Oecologia*, **86**, 349-58.

Keizer, P. J. & Arnolds, E. (1994). Succession of ectomycorrhizal fungi in roadside verges planted with common oak (*Quercus robur* L.) in Drenthe, The Netherlands. *Mycorrhiza*, **4**, 147-59.

Koske, R. E. & Gemma, J. N. (1990). VA mycorrhizae in strand vegetation of Hawaii: evidence for long-distance codispersal of plants and fungi. *American Journal of Botany*, **77**, 466-74.

Kovacic, D. A., St John, T. V. & Dyer, M. I. (1984). Lack of vesicular-arbuscular mycorrhizal inoculum in a ponderosa pine forest. *Ecology*, **65**, 1755-59.

Lapeyrie, F., Garbaye, J., de Oliveira, V. & Bellei, M. (1992). Controlled mycorrhization of eucalypts. In *Mycorrhizas in Ecosystems*, ed. D. J. Read, D. H. Lewis, A. H. Fitter & I. J. Alexander , pp. 293-299. Wallingford: CAB International.

Law, R. (1985). Evolution in a mutualistic environment. In *The Biology of Mutualism*, ed. D. H. Boucher, pp. 145-170, New York: Oxford University Press.

Lodge, D. J. (1990). Patterns of mycorrhizal symbiosis following landslides and hurricanes. In *Luquillo Experimental Forest – Long-term Ecological Research Site Annual Report*, pp. 34-40. San Juan: Center for Energy and Environmental Research.

Lopez-Aguillon, R. & Mosse, B. (1987). Experiments on competitiveness of three endomycorrhizal fungi. *Plant and Soil*, **97**, 155-70.

Louis, I. & Lim, G. (1987). Spore density and root colonization of vesicular-arbuscular mycorrhizas in tropical soil. *Transactions of the British Mycological Society*, **88**, 207-12.

Maffia, B., Nadkarni, N. M. & Janos, D. P. (1993). Vesicular-arbuscular mycorrhizae of epiphytic and terrestrial Piperaceae under field and greenhouse conditions. *Mycorrhiza*, **4**, 5-9.

Malloch, D. W., Pirozynski, K. A. & Raven, P. H. (1980). Ecological and evolutionary significance of mycorrhizal symbiosis in vascular plants (a review). *Proceedings of the National Academy of Sciences, USA*, **77**, 2113-18.

Marks, G. C. & Foster, R. C. (1973). Structure, morphogenesis, and ultrastructure of ectomycorrhizae. In *Ectomycorrhizae*, ed. G. C. Marks & T. T. Kozlowski, pp. 1-41. New York: Academic Press.

Marx, D. H., Ruehle, J. L. & Cordell, C. E. (1991). Methods for studying nursery and field response of trees to specific ectomycorrhiza. In *Methods in microbiology, Volume 23*, ed. J. R. Norris, D. J. Read & A. K. Varma, pp. 383-411. London: Academic Press.

Mason, P. A., Musoko, M. O. & Last, F. T. (1992). Short-term changes in vesicular-arbuscular mycorrhizal spore populations in *Terminalia* plantations in Cameroon. In *Mycorrhizas in Ecosystems*, ed. D. J. Read, D. H. Lewis, A. H. Fitter & I. J. Alexander, pp. 261-267. Wallingford: CAB International.

Michelsen, A. (1992). Mycorrhiza and root nodulation in tree seedlings from five nurseries in Ethopia and Somalia. *Forest Ecology and Management*, **48**, 335-44.

Michelsen, A. (1993). Growth improvement of Ethiopian acacias by addition of vesicular-arbuscular mycorrhizal fungi or roots of native plants to non-sterile nursery soil. *Forest Ecology and Management*, **59**, 193-206.

Michelsen, A. & Rosendahl, S. (1990). Propagule density of VA-mycorrhizal fungi in semi-arid bushland in Somalia. *Agriculture, Ecosystems and Environment*, **29**, 295-301.

Mikola, P. (1973). Application of mycorrhizal symbiosis in forestry practice. In *Ectomycorrhizae*, ed. G. C. Marks & T. T. Kozlowski, pp. 383-411. New York: Academic Press.

Miller, R. M. (1987). The ecology of vesicular-arbuscular mycorrhizae in grass- and shrublands. In *Ecophysiology of VA Mycorrhizal Plants*, ed. G. R. Safir, pp. 135-170. Boca Raton: CRC Press.

Modjo, H. S. & Hendrix, J. W. (1986). The mycorrhizal fungus *Glomus macrocarpum* as a cause of tobacco stunt disease. *Phytopathology*, **76**, 688-91.

Molina, R., Massicotte, H. & Trappe, J. M. (1992). Specificity phenomena in mycorrhizal symbioses: community–ecological consequences and practical implications. In *Mycorrhizal Functioning*, ed. M. F. Allen, pp. 357-423. New York: Chapman & Hall.

Morton, J. B. & Benny, G. L. (1990). Revised classification of arbuscular mycorrhizal fungi (Zygomycetes): a new order, Glomales, two new suborders, Glomineae and Gigasporineae, and two new families, Acaulosporaceae and Gigasporaceae, with an emendation of Glomaceae. *Mycotaxon*, **37**, 471-91.

Mosse, B. & Bowen, G. D. (1968). *Endogone* spores in some Australian & New Zealand soils, and in an experimental field soil at Rothamsted. *Transactions of the British Mycological Society*, **51**, 485-92.

Moyersoen, B. (1993). Ectomicorrizas y micorrizas vesiculo-arbusculares en Caatinga Amazónica del Sur de Venezuela. *Scientia Guaianae*, **3**, 1-82.

Musoko, M., Last, F. T. & Mason, P. A. (1994). Populations of spores of vesicular-arbuscular mycorrhizal fungi in undisturbed soils of secondary semideciduous moist tropical forest in Cameroon. *Forest Ecology and Management*, **63**, 359-77.

Myers, N. (1991). Tropical forests: present status and future outlook. *Climatic Change*, **19**, 3-32.

160   D. P. Janos

Nepstad, D. C., Uhl, C. & Serrão, E. A. S. (1991). Recuperation of a degraded Amazonian landscape: forest recovery and agricultural restoration. *Ambio*, **20**, 248-55.

Newberry, D. M., Alexander, I. J., Thomas, D. W. & Gartlan, J. S. (1988). Ectomycorrhizal rain-forest legumes and soil phosphorus in Korup National Park, Cameroon. *New Phytologist*, **109**, 433-50.

Newman, E. I. & Reddell, P. (1988). Relationship between mycorrhizal infection and diversity in vegetation: evidence from the Great Smoky Mountains. *Functional Ecology*, **2**, 259-62.

Ocampo, J. A., Martin, J. & Hayman, D. S. (1980). Influence of plant interactions on vesicular-arbuscular mycorrhizal infections. I. Host and non-host plants grown together. *New Phytologist*, **84**, 27-35.

Odum, E. P. (1969). The strategy of ecosystem development. *Science*, **164**, 262-70.

Peng, S, Eissenstat, D. M., Graham, J. H., Williams, K. & Hodge, N. C. (1993). Growth depression in mycorrhizal Citrus at high-phosphorus supply. *Plant Physiology*, **101**, 1063-71.

Perry, D. A., Molina, R. & Amaranthus, M. P. (1987). Mycorrhizae, mycorrhizospheres, and reforestation: current knowledge and research needs. *Canadian Journal of Forest Research*, **17**, 929-40.

Porter, W. M. (1979). The 'Most Probable Number' method for enumerating infective propagules of vesicular arbuscular fungi in soil. *Australian Journal of Soil Research*, **17**, 515-19.

Powell, C. Ll. (1976). Development of mycorrhizal infection from *Endogone* spores and infected root segments. *Transactions of the British Mycological Society*, **66**, 439-45.

Read, D. (1993). Mycorrhizas. In *Tropical Soil Biology and Fertility*, ed. J. M. Anderson & J. S. I. Ingram, pp. 121-131. Wallingford: CAB International.

Read, D. J., Francis, R. & Finlay, R. D. (1985). Mycorrhizal mycelia and nutrient cycling in plant communities. In *Ecological Interactions in Soil: Plants, Microbes, and Animals*, ed. A. H. Fitter, D. Atkinson, D. J. Read & M. B. Usher, pp. 193-217. Oxford: Blackwell Scientific Publications.

Redhead, J. F. (1980). Mycorrhiza in natural tropical forests. In *Tropical Mycorrhiza Research*, ed. P. Mikola, pp. 127-142. Oxford: Clarendon Press.

Rose, S. L. & Paranka, J. E. (1987). Root and VAM distribution in tropical agricultural and forest soils. In *Mycorrhizae in the Next Decade*, ed. D. M. Sylvia, L. L. Hung & J. H. Graham, p. 56. Gainesville: Institute of Food and Agricultural Sciences.

Ruiz, P. O. (1987). Occurrence of VAM in tropical soils under different management practices. In *Mycorrhizae in the Next Decade*, ed. D. M. Sylvia, L. L. Hung & J. H. Graham, p. 57. Gainesville: Institute of Food and Agricultural Sciences.

Sanchez, P. A. (1976). *Properties and Management of Soils in the Tropics*. New York: John Wiley & Sons.

Sanders, I. R. (1993). Temporal infectivity and specificity of vesicular-arbuscular mycorrhizas in co-existing grassland species. *Oecologia*, **93**, 349-55.

Schenck, N. C., Siqueira, J. O. & Oliveira, E. (1989). Changes in the incidence of VA mycorrhizal fungi with changes in ecosystems. In *Interrelationships between Microorganisms and Plants in Soil*, ed. V. Vancura & F. Kunc, pp. 125-129. Amsterdam: Elsevier.

Sieverding, E. (1991). *Vesicular-Arbuscular Mycorrhiza Management in Tropical Agrosystems*. Eschborn: Deutsche Gesellschaft für Technische Zusammenarbeit (GTZ) GmbH.

Simmons, N. & Lodge, J. (1990). Colonization of a large landslide in Puerto Rico by VA mycorrhizae. *Belowground Ecology*, **1**, 15.

Smits, W. Th. M. (1983). Dipterocarps and mycorrhiza. *Flora Malesiana Bulletin*, **36**, 3926-37.

Smits, W. T. M. (1992). Mycorrhizal studies in dipterocarp forests in Indonesia. In *Mycorrhizas in Ecosystems*, ed. D. J. Read, D. H. Lewis, A. H. Fitter & I. J. Alexander, pp. 283-292. Wallingford: CAB International.

Soberon M., J. & Martinez del Rio, C. (1985). Cheating and taking advantage in mutualistic associations. In *The Biology of Mutualism*, ed. D. H. Boucher, pp. 192-216, New York: Oxford University Press.

St John, T. V. (1980a). A survey of mycorrhizal infections in an Amazonian rain forest. *Acta Amazonica*, **10**, 527-33.

St John, T. V. (1980b). Root size, root hairs, and mycorrhizal infection: a re-examination of Baylis's hypothesis with tropical trees. *New Phytologist*, **84**, 483-7.

Tester, M., Smith, S. E. & Smith, F. A. (1987). The phenomenon of 'nonmycorrhizal' plants. *Canadian Journal of Botany*, **65**, 419-31.

Theodorou, C. & Reddell, P. (1991). *In vitro* synthesis of ectomycorrhizas in Casuarinaceae with a range of mycorrhizal fungi. *New Phytologist*, **118**, 279-88.

Tobiessen, P. & Werner, M. B. (1980). Hardwood seedling survival under plantations of Scotch pine and red pine in Central New York. *Ecology*, **61**, 25-9.

Tommerup, I. C. (1992). Methods for the study of the population biology of vesicular-arbuscular mycorrhizal fungi. In *Methods in Microbiology, Volume 24*, ed. J. R. Norris, D. J. Read & A. K. Varma, pp. 23-51. London: Academic Press.

Trappe, J. M. (1977). Selection of fungi for ectomycorrhizal inoculation in nurseries. *Annual Review of Phytopathology*, **15**, 203-22.

Trappe, J. M. (1987). Phylogenetic and ecologic aspects of mycotrophy in the angiosperms from an evolutionary standpoint. In *Ecophysiology of VA Mycorrhizal Plants*, ed. G. R. Safir, pp. 5-25. Boca Raton: CRC Press.

Trappe, J. M. & Molina, R. (1986). Taxonomy and genetics of mycorrhizal fungi: their interactions and relevance. In *Physiological and Genetical Aspects of Mycorrhizae*, ed. V. Gianinazzi-Pearson & S. Gianinazzi, pp. 133-146. Paris: Institut National de la Recherche Agronomique.

Vitousek, P. M. (1984). Litterfall, nutrient cycling, and nutrient limitation in tropical forests. *Ecology*, **65**, 285-98.

Wilson, J., Ingleby, K., Mason, P. A., Ibrahim, K. & Lawson, G. J. (1992). Long-term changes in vesicular-arbuscular mycorrhizal spore populations in *Terminalia* plantations in Côte d'Ivoire. In *Mycorrhizas in Ecosystems*, ed. D. J. Read, D. H. Lewis, A. H. Fitter & I. J. Alexander, pp. 268-275. Wallingford: CAB International.

Wilson, J. M. & Tommerup, I. C. (1992). Interactions between fungal symbionts: VA mycorrhizae. In *Mycorrhizal Functioning*, ed. M. F. Allen, pp. 199-248. New York: Chapman & Hall.

Woolhouse, H. W. (1975). Membrane structure and transport problems considered in relation to phosphorus and carbohydrate movements and the regulation of endotrophic mycorrhizal associations. In *Endomycorrhizas*, ed. F. E. Sanders, B. Mosse & P. B. Tinker, pp. 209-239. London: Academic Press.

11

Potential effects on the soil mycoflora of changes in the UK agricultural policy for upland grasslands

R. D. BARDGETT

Introduction

Hill and upland areas, defined as land typically above 150 m altitude, within designated agriculturally less favoured areas, and composed predominately of dwarf shrub heaths, grassland and peat bogs, make up more than one-third (5.8 M ha) of the total UK land surface (Ratcliffe & Thompson, 1988). Agriculture, in particular sheep farming, is the primary industry of such hill and upland areas, and its expansion over the last seven–eight centuries has resulted in the development of vast areas of sheep walk composed of indigenous *Agrostis-Festuca*, *Nardus* and *Molinia* dominant grassland. These indigenous grasslands or rough grazings presently constitute some 70% of the land within the hills and uplands (Newbould, 1985).

Traditionally, indigenous hill grasslands have been lightly grazed by pure bred sheep. Over the last few decades, however, intensities of sheep farming have increased dramatically, due largely to improvements in grassland productivity from use of fertilizer and lime. These agricultural improvements have caused considerable concern amongst conservationists, particularly for the loss of indigenous flora and fauna of the hills and uplands (Sydes & Miller, 1988; Bardgett & Marsden, 1992; Bardgett, Marsden & Howard, 1995). Recent changes in UK agricultural policy have reflected these concerns and addressed others, such as the over-production of meat. Consequent reductions in financial subsidies based on output, and the introduction of monetary incentives for hill farmers to manage their land in such a way as to preserve or improve the environment (for example, environmentally sensitive areas) are likely to result in an overall reduction in liming and the use of fertilizer, and hence of the intensity of sheep grazing on upland grasslands.

163

There is a wealth of information concerning the influence of management practices on grassland composition and productivity (Milton, 1940; Newbould, 1985; Hill, Evans & Bell, 1992). However, in order to understand the functional mechanisms driving vegetational change in upland grasslands, information on the soil biota which mediates processes of organic matter decomposition and nutrient turnover are required. This chapter discusses the implications of reductions in the intensity of upland grassland management on the soil microflora, with special reference to fungi.

Changes in soil conditions

In order to understand the effects of agricultural practices on the soil mycoflora, it is necessary to consider the soil physico-chemical environment and its response to management change. Upland soils developed under high rainfall conditions and low temperatures are generally acidic and have high organic matter contents and low nutrient availability. Agricultural practices, such as the application of fertilizers and lime, have been widely used to overcome these limitations and to increase plant productivity and sheep stocking density. Such improvements and increase in sheep grazing pressure have led to enhanced decomposition of organic matter and more efficient recirculation of soil nutrients through the animal excreta pathway, leading to the development of fertile brown earth soils (Floate, 1972; Floate et al., 1973).

The effects of reducing the intensity of management on the physico-chemical conditions of upland grassland soils were highlighted in a recent study by Bardgett, Frankland & Whittaker (1993) of sites in Cumbria (Nat. Grid Ref. SD 6590). Their study sites represented a gradient of decreasing intensity of sheep management and ranged from a heavily grazed (5-8 ewe ha^{-1}), fertilized and limed *Agrostis-Festuca* grassland with a brown earth soil to an ungrazed grassland with a ferric stagno-podzol soil. Intermediate sites, which were moderately (3–5 ewe ha^{-1}) or lightly grazed (1 ewe ha^{-1}) and received no fertilizers or lime were also studied. Analysis of soil samples taken from these sites demonstrated trends of increasing organic matter content, total N, P and moisture content, and decreasing soil pH and bulk density in the surface soil, along the gradient towards less intensive management (Table 11.1). Similarly, the short-term (two-year) removal of sheep grazing by fenced exclosures within the three grazed sites, resulted in an increase in moisture content and acidity in the surface soil (Table 11.1). However, in agreement with the findings of Rawes (1981) and Marrs, Rizand & Harrison (1989) from longer-term studies, the short-term (two-year) removal of sheep grazing from the grazed grasslands had no effect on soil nutrient status (Table 11.1; Bardgett, Frankland & Whittaker, 1993).

Similar findings were also derived from a study of a grazed and 'improved' (limed and fertilized) upland grassland site in the Brecon Beacons, south Wales (Nat. Grid Ref. SN 8832) by Bardgett and Leemans (1995). This study demonstrated that short-term cessation of fertilizer applications and liming (three years) with or without the removal of sheep resulted in a decrease in soil pH from 5.4 to 4.7 but no change in soil moisture content.

It would appear, therefore, that the principal effects of reducing the intensity of sheep farming on 'improved' upland grasslands are an increase in moisture content and acidity, and a consequent longer-term increase in organic matter content in the surface soil. Such changes are prerequisite to the development of stagnopodzol soils which are prevalent under lightly grazed *Nardus* grasslands and are characterised by the presence of a mat of decaying plant material on the soil surface (Askew, Payton & Sheil, 1985).

Changes in fungal biomass

Bardgett *et al.* (1993), in their study of upland grassland sites in Cumbria, found that the abundance of fungal mycelium (live and dead), measured using a modified membrane filter technique (Bardgett, 1991), increased along a gradient of decreasing intensity of sheep management (Fig. 11.1). In the three grazed grassland sites, mycelial abundance was greatest in October, whereas in the ungrazed site it was greatest in January (Bardgett *et al.*, 1993) (Fig. 11.1). In contrast to total mycelial abundance (live and dead), the proportion of fluorescein diacetate-active (FDA-active), live fungal mycelium declined along the same gradient of reduced management intensity (Fig. 11.2(a),(b)), suggesting a trend of increasing accumulation of inactive and dead hyphae in the wetter, more acidic and organic soils (Bardgett *et al.*, 1993). This latter finding was in agreement with Latter, Cragg and Heal (1967), who in a study of grazed upland grassland sites at Moor House in the North Pennine uplands showed that the abundance of mycelium stained with phenolic analine blue (assumed active), expressed as a percentage of the total, was greatest (56%) in a heavily grazed *Agrostis-Festuca* grassland with a freely drained, relatively basic soil, and least (14%) in a lightly grazed mixed moor with a poorly

Table 11.1. *Mean values (± SE) of soil chemical/physical properties in the upper 0–3 cm and lower 3–6 cm soil, in enclosed (UG) (2 years) and grazed (G) plots within three grazed upland grasslands, representing a gradient of sheep management intensity in Cumbria*

		Heavily grazed		Moderately grazed		Lightly grazed	
		G	UG	G	UG	G	UG
Loss on ignition (%)	0–3 cm	25 ± 2	26 ± 3	29 ± 1	26 ± 1	43 ± 3	40 ± 3
	3–6 cm	13 ± 1	14 ± 1	18 ± 0	18 ± 1	24 ± 1	26 ± 2
Total N(%)	0–3 cm	0.79 ± 0.05	0.90 ± 0.09	1.04 ± 0.03	0.92 ± 0.02	1.41 ± 0.06	1.32 ± 0.06
	3–6 cm	0.51 ± 0.04	0.55 ± 0.06	0.78 ± 0.02	0.77 ± 0.03	0.94 ± 0.04	0.96 ± 0.07
Total P(%)	0–3 cm	0.10 ± 0.01	0.11 ± 0.01	0.14 ± 0.01	0.13 ± 0.00	0.15 ± 0.01	0.13 ± 0.01
	3–6 cm	0.08 ± 0.01	0.08 ± 0.01	0.12 ± 0.01	0.12 ± 0.00	0.12 ± 0.01	0.11 ± 0.01
Extractable PO_4-P(mg 100g^{-1})	0–3 cm	1.69 ± 0.30	1.65 ± 0.38	1.15 ± 0.17	0.86 ± 0.09	1.43 ± 0.22	1.16 ± 0.21
	3–6 cm	0.33 ± 0.03	0.32 ± 0.02	0.30 ± 0.04	0.45 ± 0.11	0.27 ± 0.05	0.32 ± 0.03
Extractable NH_4-N(mg 100g^{-1})	0–3 cm	4.9 ± 0.3	7.3 ± 1.3	7.8 ± 1.2	7.8 ± 1.7	14.1 ± 3.8	8.8 ± 1.7
	3–6 cm	1.3 ± 0.2	1.9 ± 0.3	6.4 ± 3.3	2.7 ± 0.4	4.2 ± 0.6	3.8 ± 0.6
Extractable NO_3-N(mg 100g^{-1})	0–3 cm	0.03 ± 0.00	0.11 ± 0.04	0.23 ± 0.08	0.06 ± 0.01	0.16 ± 0.06	0.16 ± 0.10
	3–6 cm	0.08 ± 0.05	0.09 ± 0.04	0.21 ± 0.06	0.05 ± 0.01	0.18 ± 0.07	0.10 ± 0.05
Mineralizable N(mg 100g^{-1}14d^{-1})	0–3 cm	41 ± 3	47 ± 6	58 ± 7	36 ± 3	58 ± 6	51 ± 5
	3–6 cm	–	–	–	–	–	–
pH	0–3 cm	5.1	4.8	5.0	4.8	4.7	4.6
Moisture content (%)	0–3 cm	55	58	50	52	59	62
Bulk density (g cm^{-3})	0–3 cm	0.54	0.53	0.49	0.53	0.40	0.38

Source: After Bardgett, Frankland & Whittaker, 1993.

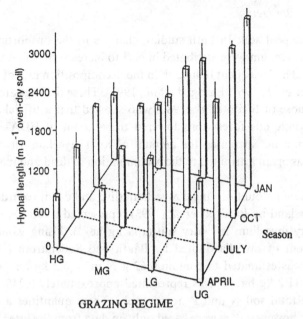

Fig. 11.1. Mean seasonal abundance of total hyphae (m g^{-1} oven-dry soil ± SE) in the surface soil of four upland grasslands in Cumbria (HG, heavily grazed; MG, moderately grazed; LG, lightly grazed; and UG, ungrazed); $n = 5$. (After Bardgett, Frankland & Whittaker, 1993.)

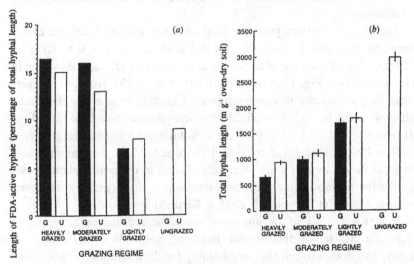

Fig. 11.2. The effects of removing sheep grazing for 2 years (G = grazed, UG = ungrazed) on (a) the abundance of fluorescein diacetate-active (FDA-active) mycelium (percentage of total hyphal length), and (b) total abundance of fungal hyphae (m g^{-1} oven-dry soil ± SE), in the surface soil of four upland grasslands; $n = 5$. (After Bardgett, Frankland & Whittaker, 1993.)

drained, acidic peat soil. In both studies, changes in the proportion of active fungal mycelium were attributed in part to increases in soil wetness and acidity and a consequent reduction in the decomposition rate of dead hyphae (Latter et al., 1967; Bardgett, et al., 1993). These findings are also in line with those of Bååth et al. (1980) who showed that artificial acidification of coniferous forest soils lowered the amount of FDA-active fungal mycelium but increased the amount of total mycelium (live and dead). This was again attributed to the accumulation of dead and inactive hyphae.

The maximum amount of fungal mycelium (live and dead) recorded in ungrazed grassland by Bardgett et al. (1993) represented a total biomass of 16.83 g dry mycelium m^{-1} dry soil (168 kg ha^{-1}). Using values of nutrient content of mycelia quoted by Bååth and Söderström (1979), this biomass was estimated to contain 0.62 g N m^{-2} (6.2 kg ha^{-1}) and 0.12 g P m^{-2} (1.2 kg ha^{-1}). This represented approximately 0.24% and 0.34% of the total soil N and P, respectively. These quantities appear relatively low; however, they were based only on data from the litter layer and on standing crop of mycelium rather than turnover. In view of this, meaningful comparisons could not be made with data from other studies (for example, Kjøller & Struwe, 1982) and of their significance as a source of nutrients.

The short-term (two-year) removal of sheep grazing from the exclosures on the upland grassland sites in Cumbria was shown to have no effect on the abundance of total (live and dead) or FDA-active hyphae in the surface soil (Fig. 11.2(a),(b); Bardgett et al., 1993). However, longerterm (approximately 30 years) removal of grazing from a heavily grazed (five–six ewe ha^{-1}), unfertilized Agrostis-Festuca grassland at Moor House resulted in a 45% increase in total fungal hyphal length from 1460 ± 159 to $2111 \pm 152 \, m \, g^{-1}$ soil (Bardgett, 1991). Similar increases in total fungal hyphal length were also found in different upland grassland sites in Snowdonia, North Wales, where sheep grazing had been removed for over 30 years (Fig. 11.3; Bardgett, 1990).

In a recent study of an 'improved', sheep-grazed upland grassland at Bronydd Mawr, in the Brecon Beacons, South Wales, Bardgett & Leemans (1995) showed that withholding fertilizer and lime applications for three grazing seasons resulted in reductions in both total microbial biomass (fumigation–extraction) and activity (basal respiration and dehydrogenase activity) in the surface soil (Table 11.2). The short-term cessation of fertilizer applications, liming and grazing resulted in even greater reductions in total soil microbial biomass and activity (Table

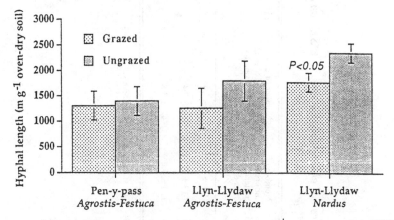

Fig. 11.3. Length of fungal hyphae (mean m g^{-1} oven-dry soil ± SE) in the surface 3 cm soil of three grazed and adjacent ungrazed grassland exclosure plots at different locations in Snowdonia, North Wales; $n = 5$. (After Bardgett, 1990.)

11.2). Such changes in microbial biomass and activity were attributed largely to observed increases in soil acidity following the cessation of liming and grazing. This suggestion was supported by the finding that dehydrogenase activity, an index of total soil microbial activity, declined linearly with increasing soil acidity in the upland grasslands (Fig. 11.4). In addition, it is well known that low microbial biomass values are associated with acid soils (see Wardle, 1992). Reductions in microbial biomass and activity were also attributed in part to a reduced input of 'potentially available' nutrients from sheep excreta and urine in the less heavily grazed and ungrazed grassland (Bardgett & Leemans, 1995).

Changes in microbial community structure

The above findings of Bardgett & Leemans (1995) appeared contrary to those of earlier studies by Bardgett (1991) and Bardgett *et al.* (1993), who as already mentioned showed that reductions in sheep management intensity and the long-term removal of grazing, resulted in an increase in total fungal biomass (live and dead) in the surface soil. It was suggested by Bardgett & Leemans (1995), therefore, that the changes observed in both total microbial and total fungal biomass (live and dead), at the different sites, were in fact a reflection of shifts in microbial community structure, that is an increase in the proportion of fungi relative to bacteria concurrent with a reduced total microbial biomass. Indeed, it is generally

Table 11.2. *Effect of cessation of fertilizer applications, liming and grazing on microbial characteristics (mean $\pm$ SE) and pH of the surface soil of an improved upland grassland at Bronydd Mawr, South Wales*

Treatment	Microbial biomass-C (μg C g^{-1} oven-dry soil)	Microbial biomass-N (μg N g^{-1} oven-dry soil)	Microbial C:N ratios	Basal respiration (μg CO$_2$-C g^{-1} oven-dry soil h^{-1})	Dehydrogenase (nmol INTF g^{-1} oven-dry soil 2 h^{-1})	Number of culturable bacteria (no. g^{-1} oven-dry soil $\times 10^{-7}$)	Number of c.f.u. of culturable fungi (no. g^{-1} oven-dry soil $\times 10^{-8}$)	Soil pH
Fertilized, limed and grazed	1035 ± 170	104 ± 7	6.2	0.54 ± 0.04	890 ± 140	9.1 ± 2.3	3.2 ± 2.4	5.4
Unfertilized, unlimed and grazed	847 ± 173*	102 ± 173*	5.8	0.51 ± 0.07	708 ± 49**	10.3 ± 7.2	3.2 ± 2.4	4.7
Unfertilized, unlimed and ungrazed	579 ± 258***	93 ± 9*	4.4	0.46 ± 0.04*	616 ± 101***	11.1 ± 3.0	4.9 ± 3.8	4.5

Note: Significant differences from the fertilized, limed and grazed treatment are shown as *, **, *** for $P < 0.05$, $P < 0.01$ and $P < 0.001$, respectively.
c.f.u., colony-forming units.
INTF, idonitrotetrazolium formazan.
Source: After Bardgett & Leemans, 1995.

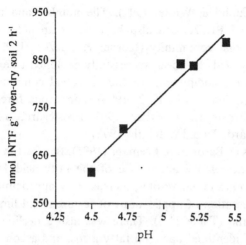

Fig. 11.4. Relationship between soil pH and dehydrogenase activity (nmol idonitrotetrazolium formazan [INTF] g^{-1} oven-dry soil 2 h^{-1}) in an improved upland grassland soil, at Bronydd Mawr, South Wales (r^2 = 0.97; $P < 0.01$). (After Bardgett & Leemans, 1995.)

accepted that the proportion of eukaryotes to prokaryotes rises at lower pH owing to the greater tolerance of the former to acidity (Alexander, 1977).

Bardgett & Leemans (1995) had shown that reductions in management intensity and associated increases in acidity of upland grassland soils resulted in a fall in the ratio of microbial biomass-C : biomass-N (Table 11.2). This finding may be evidence of an increase in the abundance of fungi relative to bacteria, since the N content of bacteria is considerably less than that of fungi (Alexander, 1977). Despite this, the change in agricultural practice did not affect the relative abundance of culturable bacteria and fungi, measured using selective plating techniques (Table 11.2; Bardgett & Leemans, 1995). However, it was suggested that some of the microorganisms developing under the changed soil physico-chemical conditions may have been unculturable or not obtainable on the selective media used (Bardgett & Leemans, 1995).

In a study by Bardgett & Hobbs (1994) changes in microbial community structure in upland grasslands under different intensities of management were further evaluated by analysing the ester-linked phospholipid fatty acid (PLFA) composition of the soils. The method relies on the fact that different groups of microorganisms (for example, fungi, actinomycetes, Gram-positive and Gram-negative bacteria) have different 'signature' PLFAs (Tunlid & White, 1992). The number and relative abundance of identified PLFAs can also be used as an index of the diversity of the microbial community (Korner & Laczkó, 1992). The use of 'signature' fatty acids has been successfully exploited to examine structural and functional components of the microbial community of various agricultural (Zelles et al., 1992, 1994), coniferous forest and arable soils (Bååth, Frostegård & Fritze, 1993; Frostegård, Bååth & Tunlid, 1993; Frostegård, Tunlid & Bååth, 1993).

Using the same sites as Bardgett & Leemans (1995), Bardgett & Hobbs (1994) showed that the relative abundance of the fatty acid 18:2ω6, which is commonly found in eukaryotes, increased by approximately 5 times following the cessation of applications of fertilizer and lime on a grazed upland grassland (Table 11.3). There was also a decline in the relative abundance of identified bacterial fatty acids, and a consequent increase in the ratio of fungal:bacterial fatty acids (Table 11.3). If the relative abundance of fungal and bacterial fatty acids extracted from soil can be used as indicators of changes in biomass, this suggests that the proportion of fungi relative to bacteria increases when lime is withheld from grazed grassland and soils become more acidic.

Surprisingly, the cessation of fertilizer applications, liming and grazing had no effect on the relative abundance of the fungal fatty acid 18:2ω6 (Bardgett & Hobbs, 1994). This appears to be contrary to the finding of Bardgett et al. (1993) that the removal of sheep grazing from different upland grasslands resulted in increases in the abundance of total fungal biomass. This discrepancy, however, is likely to be due partly to that fact that the phospholipid component of cell membranes can be detected only in live cells, whereas the membrane filter technique measures both live and dead fungal hyphae. In view of this, there is support for the earlier suggestion that increases in the abundance of fungal mycelia observed in ungrazed grassland are in fact a reflection of an accumulation of dead and inactive fungal hyphae in the surface soil.

The increase in the relative abundance of the fungi-specific fatty acid 18:2ω6 only when sheep grazing was maintained and fertilizer and lime withheld suggests that the input of animal faeces to soil had an important role in encouraging the growth of fungi in these acid soils, and may have been partly due to the presence of coprophilous species on sheep dung. In the absence of coprophilous species, the increase in the ratio of fungal:bacterial fatty acids in the ungrazed grassland (with no fertilizer or lime) (Table 11.3) supports the suggestion that the proportion of fungi relative to bacteria increases as upland grassland soils become more acidic. All

Table 11.3. *Effect of cessation of fertilizer applications, liming and grazing on the mean (±SE) relative abundance (% total) of phospholipid fatty acids (PLFAs) specific to bacteria and fungi, and the ratio of fungal : bacterial PLFAs in the surface soil of an improved upland grassland at Bronydd Mawr, South Wales*

Treatment	Bacterial PLFAs (% total)	Fungal PLFAs (% total)	Ratio fungal : bacterial PLFAs
Fertilized, limed and grazed	74.3 ± 3.1	6.2 ± 1.3[a]	0.08[a]
Unfertilized, unlimed and grazed	66.9 ± 1.3[b]	30 ± 4.1[b]	0.16[b]
Unfertilized, unlimed and ungrazed	69.8 ± 1.4[b]	6.7 ± 0.4[a]	0.10[a]

Note: Significant differences ($P < 0.05$) are shown by values without the same letter.
Source: After Bardgett & Hobbs, 1994.

the above changes were found to be concurrent with a reduced microbial biomass (Bardgett & Leemans, 1995).

Increases in the proportion of fungi relative to bacteria in the surface 10 cm soil were also observed following the long-term removal of sheep grazing (37 years) from a range of *Agrostis* and *Nardus* upland grasslands in Snowdonia, North Wales (Bardgett & Leemans, unpublished observations). As in the study of upland grassland soils in South Wales (Bardgett & Leemans, 1995), these changes were concurrent with a reduced total microbial biomass and activity. However, unlike the short-term effects already discussed, the long-term removal of sheep grazing had no effect on soil pH. This would suggest that long-term changes in soil microbial characteristics following the cessation of sheep grazing are related to factors other than soil pH, such as the feeding activities of soil animals.

Changes in species composition of the fungal community

Using a washed litter technique (Harley & Waid, 1955), Bardgett *et al.* (1993) isolated five frequently occurring Deuteromycotina fungi from the three adjacent and differently managed upland grassland sites in Cumbria. The relative frequency of the five species isolated in each of the grassland sites is shown in Fig. 11.5. In addition, two members of the Zygomycotina, *Mortierella ramanniana* and *Mucor hiemalis* were also commonly found on litter placed in damp chambers (*sensu* Keyworth, 1951) from these sites. All the fungi isolated were species common on plant litter and several have been recorded in upland soils by other authors. For example, *M. ramanniana*, *M. hiemalis* and *Trichoderma viride* Pers. were found to be common in soils along an altitudinal gradient at Moor House (Latter *et al.*, 1967; Widden, 1987), and all three species have been shown to occur in tussock soils in New Zealand (Thornton, 1958). Species of *Phoma* and *Fusarium* are commonly isolated from grassland soils (Warcup, 1951), and *Epicoccum nigrum* is a known primary colonizer of many types of decaying plant material (Webster, 1957).

The relative abundance of the five commonly isolated Deuteromycotina fungi was found to vary in the three grassland sites (Fig. 11.5; Bardgett *et al.*, 1993). However, a general trend of decreasing abundance along the gradient of less intensive sheep management was shown by the species *E. nigrum* and *Fusarium lateritium*, whereas for *Phoma exigua*, *Cladosporium cladosporioides* and possibly *T. viride*, the reverse was true. Similar trends, in the distribution of *T. viride* were also

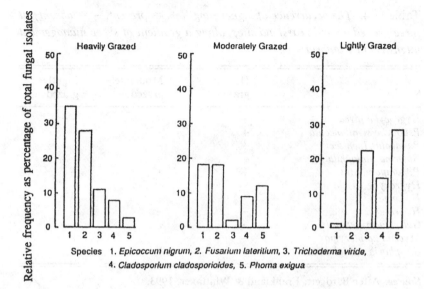

Fig. 11.5. Relative frequency of fungal species isolated from litter taken from three upland grasslands, expressed as percentage of total isolates. (After Bardgett, Frankland & Whittaker, 1993.)

shown by Latter *et al.* (1967) who found that this species was more common in poorly drained peat soils than in more fertile, freely drained limestone grassland soils at Moor House.

In addition to the above, various species of *Penicillium* were isolated from litter in damp chambers, and the fruit bodies of 11 basidiomycetes were observed on the sites (Table 11.4); however, none of them occurred on all three grassland sites (Bardgett *et al.*, 1993).

Fungal-faunal interactions

Populations of fungal-feeding soil microarthropods, in particular the Collembola have been shown to be sensitive to changes in upland grass-land management (Bardgett *et al.*, 1993). In contrast to fungal biomass, the abundance of total Collembola and the fungal-feeding species *Onychiurus procampatus* in the surface soil decreased dramatically along a gradient toward less intensive sheep management (Fig. 11.6(*a*),(*b*)). In addition, the two-year removal of sheep grazing from the three differently managed grasslands resulted in a decline in the abundance of total Collembola and the species *O. procampatus*, particularly in the heavily grazed *Agrostis-Festuca* grassland where the effects of

Table 11.4. *The occurrence of higher fungi (+ = present, − = absent) at three grazed upland grassland sites, along a gradient of sheep management intensity, in Cumbria*

Species	Heavily grazed	Moderately grazed	Lightly grazed
Hygrocybe nivea	+	−	−
Psilocybe semilanceata	+	−	−
Panaeolus foenisecii	+	−	−
Galerina mniophila	+	+	−
P. campanulatus	+	+	−
Hygrocybe coccineus	−	+	−
H. langei	−	+	−
H. pratensis	−	−	+
Galerina mycenopsis	−	−	+
Mycena epipterygia	−	−	+
Stropharia semiglobata	−	−	+

Source: After Bardgett, Frankland & Whittaker, 1993.

removing grazing on soil conditions were more pronounced. Reductions in the abundance of Collembola were attributed to increases in soil acidity and soil wetness under less intensive management.

It was hypothesized that simultaneous changes in the relative abundance of Collembola and fungal mycelium may have been owing in part to interactions between the two biota. Indeed, it is well established that field populations of soil arthropods feed preferentially on fungal hyphae (for example, Anderson & Healey, 1972) and the collembolan, *O. procampatus*, a known fungal-feeder (Healey, 1965) was found to be dominant in these upland grassland sites. Moreover, it fed preferentially on fungal species in the laboratory, the most preferred species being *P. exigua* (Bardgett, Whittaker & Frankland, 1993a). In view of this finding, it was suggested that grazing of fungal mycelium in particular by the high number of the collembolan *O. procampatus* in the heavily grazed *Agrostis-Festuca* grassland (Fig. 11.6(b)) might have resulted in a lower mycelial biomass. In addition, grazing by Collembola may have stimulated fungal activity and hyphal branching (Hanlon, 1981), which could in turn explain the higher proportion of FDA-active fungal mycelium in this grassland (Fig. 11.2).

A laboratory microcosm study was also conducted (Bardgett, Whittaker & Frankland, 1993b) to assess the impact of grazing by the collembolan *O. procampatus* on the abundance, nutrient release and respiration of the saprotrophic fungus *P. exigua* grown on sterile, irradiated grass litter from the three differently managed sites in Dentdale. Different numbers of Collembola, reflecting changes in mean field densities (Fig. 11.6(*a*),(*b*)), were added to microcosms (Anderson & Ineson, 1982) which were then analysed over a period of 12 weeks. The results of the study showed firstly that *P. exigua* was highly efficient at immobilising N and P (for example, 1–2 mg NH_4-N g^{-1} oven-dry litter), but nutrient release was not affected by the fauna (Fig. 11.7), and secondly that fungal biomass and its activity were reduced only when Collembola were in excess of mean field densities (i.e. > *ca* 50 000 m^{-2}). For example, the abundance of total and FDA-active fungal hyphae on grass litter from the heavily grazed grassland (*Agrostis-Festuca*) was reduced only when Collembola numbers equivalent to double the mean field density were added to the microcosms (Fig. 11.8). Also, the addition of increasingly higher numbers of Collembola, far in excess of mean field densities, to similar microcosms containing sterilized and *Phoma*-inoculated grass litter from the moderately grazed grassland (*Agrostis-Festuca*) resulted in a linear decrease in fungal activity, measured as CO_2 evolution (Fig. 11.9). Both the above findings suggested that localized grazing activities

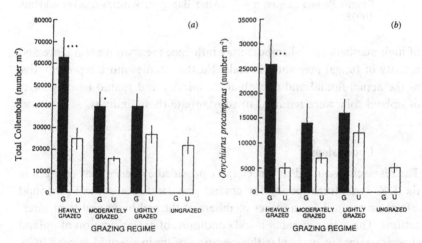

Fig. 11.6. The effects of removing sheep grazing for two years (G = grazed, UG = ungrazed) on (*a*) the abundance (mean number m^{-2} ± SE) of total Collembola, and (*b*) the abundance (mean number m^{-2} ± SE) of the fungal-feeding collembolan *Onychiurus procampatus*, in the surface soil of four upland grassland sites, representing a gradient of sheep management intensity, in Cumbria; *, *** = $P < 0.05$, $P < 0.001$, respectively; $n = 5$. (After Bardgett, Frankland & Whittaker, 1993.)

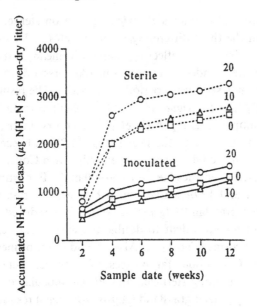

Fig. 11.7. Effect of Collembola (*Onychiurus procampatus*) grazing activities (numbers added per microcosm; 0, 10 and 20) on accumulated release of NH_4-N from sterilized grass litter taken from a heavily grazed *Agrostis-Festuca* upland grassland, in the presence and absence of the fungus *Phoma exigua*; $n = 3$. (After Bardgett, Whittaker & Frankland, 1993b.)

of high numbers of Collembola may influence the spatial abundance and activity of fungal mycelium in soils. Further studies more representative of the actual faunal and microbial complexity and spatial heterogeneity of upland soils were required to substantiate these findings.

Conclusions

The research discussed in this chapter provided evidence that changes in the UK agricultural policy for upland grasslands could have profound effects on the soil fungi, other members of the soil biota and their interactions. The major impact on soil conditions of extensification of upland sheep farming, in particular the cessation of liming, would appear to be an increase in acidity and moisture content. These changes in soil conditions, coupled with pertubations in the soil food-web structure (for example, fungal-feeding microarthropods), are likely to result in shifts in microbial community structure. In general, the proportion of fungi increases relative to bacteria as soils become more acidic. These changes

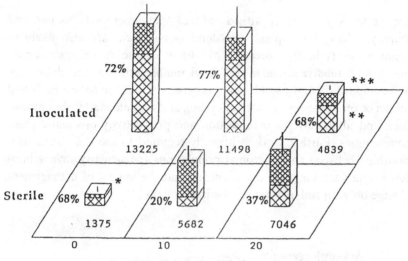

Fig. 11.8. Abundance (m g^{-1} oven-dry litter ± SE) and fluorescein diacetate (FDA) active (percentage of total) fungal hyphae on sterilized grass litter taken from a heavily grazed *Agrostis-Festuca* upland grassland, at the end of a 12-week experimental period in microcosms to which Collembola had been added, in the presence and absence of the fungus *Phoma exigua*; ▨ FDA-active hyphae, ▓ total hyphae; *,**,***, $P < 0.05$, $P < 0.01$, $P < 0.001$, respectively; $n = 3$. (After Bardgett, Whittaker & Frankland, 1993*b*.)

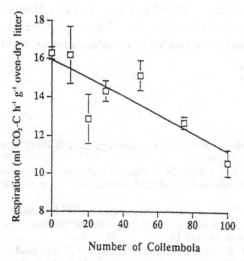

Fig. 11.9. Relationship between number of Collembola added to microcosms containing sterilized grass litter taken from a grazed *Agrostis-Festuca* upland grassland, inoculated with the fungus *Phoma exigua*, and respiration rate (CO_2 evolution) ($r^2 = 0.703$; $P < 0.001$); $n = 3$. (After Bardgett, Whittaker & Frankland, 1993*b*.)

appear to be concurrent with a reduced total microbial biomass and activity. Changes in upland grassland management are also likely to result in shifts in the species composition of the fungal community. Shifts in the relative abundance of fungi and bacteria, and the abundance and composition of the soil biota in general, are likely to have a profound influence on processes of nutrient cycling and organic matter decomposition, and ultimately the composition and productivity of upland plant communities. Further work is urgently required to ascertain such relationships between soil microbial populations (particularly mycorrhizal fungi) and vascular plant communities, and the impact of management change on such soil–plant–microbe interactions.

Acknowledgements

Much of the research presented in this chapter was conducted at the Institute of Terrestrial Ecology, Merlewood Research Station and at Lancaster University under the guidance of Dr Juliet Frankland and Professor John Whittaker, respectively (Natural Environment Research Council C.A.S.E. Studentship). I am grateful to all who assisted at the above institutions, in particular Doreen Howard, Jan Poskitt, Phil Ineson, Peter Flint and Andy Proctor. For the more recent investigations conducted at the Institute of Grassland and Environmental Research, I am grateful for the assistance of David Leemans and Phil Hobbs. I am also grateful to Åsa Frostegård (University of Lund, Sweden) for her help with the phospholipid work and Mike Theodorou for his comments on the manuscript.

References

Alexander, M. (1977). *Introduction to Soil Microbiology*. 2nd Edition. New York: John Wiley.
Anderson, J.M. & Healey, I.N. (1972). Seasonal and interspecific variation in major gut contents of some woodland Collembola. *Journal of Animal Ecology* **41**, 359-68.
Anderson, J.M. & Ineson, P. (1982). A soil microcosm system and its application to measurement of respiration and nutrient leaching. *Soil Biology and Biochemistry* **14**, 415-16.
Askew, G.P, Payton, R.W. & Sheil, R.S. (1985). Upland soils and land clearance in Britain during the second millennium BC. In *Upland Settlement in Britain during the Second Millennium and After* ed. D. Spratt & C. Burgess, pp. 5-33. Oxford: BAR British Series.

Bååth, E., Berg, B., Lohm, U., Lundgren, B., Lundkvist, H., Rosswall, T., Söderström, B. & Wiren, A. (1980). Effects of experimental acidification and liming on soil organisms and decomposition in a Scots pine forest. *Pedobiologia* 20, 85-100.

Bååth, E, Frostegård, Å. & Fritze, H. (1993). Soil bacterial biomass, activity, phospholipid fatty acid pattern, and pH tolerance in an area polluted with alkaline dust deposition. *Applied and Environmental Microbiology* 58, 4026-31.

Bååth, E. & Söderström, B. (1979). Fungal biomass and fungal immobilization of plant nutrients in Swedish coniferous forest soils. *Revue d'Ecologie et de Biologie du Sol* 16, 477-89.

Bardgett, R.D. (1990). *The effects of changes in sheep management intensity on faunal/fungal interactions related to nutrient cycling in upland grasslands.* PhD Thesis, University of Lancaster, UK.

Bardgett, R.D. (1991). The use of the membrane filter technique for comparative measurements of hyphal lengths in different grassland sites. *Agriculture, Ecosystems and Environment* 34, 115-19.

Bardgett, R.D., Frankland, J.C. & Whittaker, J.B. (1993). The effects of agricultural practices on the soil biota of some upland grasslands. *Agriculture, Ecosystems and Environment*, 45, 25-45.

Bardgett, R.D. & Hobbs, P.J. (1994). Changes in soil microbial community structure following the cessation of fertiliser inputs, liming and grazing on upland pasture. *AFRC Institute of Grassland and Environmental Research, Annual Report*, 1993, Aberystwyth, UK.

Bardgett R.D. & Leemans, D.K. (1995). The effects of cessation of fertiliser application, liming and grazing on microbial biomass and activity in a reseeded upland pasture. *Biology and Fertility of Soils*, 19, 148-54.

Bardgett, R.D. & Marsden, J.H. (1992). *Heather Condition and Management in the Uplands of England and Wales.* Peterborough: English Nature.

Bardgett, R.D., Marsden, J.H. & Howard, D.C. (1995). The extent and condition of heather on moorland in the uplands of England and Wales. *Biological Conservation* 71, 155-61.

Bardgett, R.D., Whittaker, J.B. & Frankland, J.C. (1993a). The diet and food preferences of *Onychiurus procampatus* (Collembola) from upland grassland soils. *Biology and Fertility of Soils*, 16, 296-8.

Bardgett, R.D., Whittaker, J.B. & Frankland, J.C. (1993b). The effect of collembolan grazing on fungal activity in differently managed upland grassland soils. *Biology and Fertility of Soils*, 16, 255-62.

Floate, M.J.S. (1972). Plant nutrient cycling in hill land. *Proceedings of the North of England Soils Discussion Group*, 7, 1-27.

Floate, M.J.S., Eadie, J., Black, J.S. & Nicholson, I.A. (1973). The improvement of *Nardus* dominant hill pastures by grazing control and surface treatment and its economic assessment. *Colloquium Proceedings*, 3, 33-9.

Frostegård, Å, Bååth, E. & Tunlid, A (1993). Shifts in the structure of soil microbial communities in limed forests as revealed by phospholipid fatty acid analysis. *Soil Biology and Biochemistry*, 25, 723-30.

Frostegård, Å, Tunlid, A. & Bååth, E. (1993). Phospholipid fatty acid composition, biomass and activity of microbial communities from two soil types experimentally exposed to different heavy metals. *Applied and Environmental Microbiology*, 59, 3605-17.

Hanlon, R.D. (1981). Influence of grazing by Collembola on the activity of senescent fungal colonies grown on media of different nutrient concentration. *Oikos*, **36**, 363-7.

Harley, J.L. & Waid, J.S. (1955). A method of studying active mycelia on living roots and other surfaces in the soil. *Transactions of the British Mycological Society*, **38**, 104-18.

Healey, I.N. (1965). *Studies on the Production Biology of Soil Collembola, with Special Reference to a Species of Onychiurus.* PhD Thesis, University of Wales.

Hill M.O, Evans, D.F. & Bell, S.A. (1992). Long-term effects of excluding sheep from hill pastures in North Wales. *Journal of Ecology*, **80**, 1-13.

Keyworth, W.G. (1951). A Petri-dish moist chamber. *Transactions of the British Mycological Society*, **34**, 291-2.

Kjøller, A. & Struwe, S. (1982). Microfungi in ecosystems: fungal occurrence and activity in litter and soil. *Oikos*, **39**, 391-422.

Korner, J. & Laczkó, E. (1992). A new method for assessing soil microorganism diversity and evidence of vitamin deficiency in low diversity communities. *Biology and Fertility of Soils*, **13**, 58-60.

Latter, P.M., Cragg, J.B. & Heal, O.W. (1967). Comparative studies on the microbiology of four soils in the northern Pennines. *Journal of Ecology*, **55**, 445-64.

Marrs, R.H., Rizand, A. & Harrison, A.F. (1989). The effects of removing sheep grazing on soil chemistry, above-ground nutrient distribution, and selected aspects of soil fertility in long term experiments at Moor House National Nature Reserve. *Journal of Applied Ecology*, **26**, 647-61.

Milton, W.E.J. (1940). The effect of manuring, grazing and cutting on the yield, botanical and chemical composition of natural hill pastures. I. Yield and botanical section. *Journal of Ecology*, **28**, 326-56.

Newbould, P. (1985). Improvement of native grasslands in the uplands. *Soil Use and Management*, **1**, 43-9.

Newell, K. (1984). Interaction between two decomposer basidiomycetes and Collembola under Sitka spruce: distribution, abundance and selective grazing. *Soil Biology and Biochemistry*, **16**, 227-34.

Ratcliffe, D.A. & Thompson, D.B.A. (1988). The British uplands: their ecological character and international significance. In *Ecological Change in the Uplands*, ed. M.B. Usher & D.B.A. Thompson, pp. 9-36. Oxford: Blackwell.

Rawes, M. (1981). Further results of excluding sheep from high-level grasslands in the north Pennines. *Journal of Ecology*, **69**, 651-69.

Sydes, C. & Miller, G.R. (1988). Range management and nature conservation in the British uplands. In *Ecological Change in the Uplands*, ed. M.B. Usher & D.B.A. Thompson, pp. 323-337. Oxford: Blackwell.

Thornton, R.H. (1958). Biological studies on some tussock-grassland soils. *New Zealand Journal of Agricultural Research*, **1**, 922-37.

Tunlid, A. & White, D.C. (1992). Biochemical analysis of biomass, community structure, nutritional status, and metabolic activity of microbial communities in soil. In *Soil Biochemistry*, Vol 7, ed. G. Stotzky & J.M. Bollag. pp. 229-262. New York: Marcel Dekker.

Warcup, J.H. (1951). The ecology of soil fungi. *Transactions of the British Mycological Society*, **34**, 376-99.

Wardle, D.A. (1992). A comparative assessment of factors which influence microbial biomass carbon and nitrogen levels in soil. *Biological Reviews*, **67**, 321-58.

Webster, J. (1957). Succession of fungi on decaying cocksfoot culms. *Journal of Ecology*, **45**, 1-30.

Widden, P. (1987). Fungal communities in soils along an elevation gradient in northern England. *Mycologia*, **71**, 298-309.

Zelles, L., Bai, Q.Y., Beck, T. & Beese, F. (1992). Signature fatty acids in phospholipids and lipopolysaccharides as indicators of microbial biomass and community structure in agricultural soils. *Soil Biology and Biochemistry*, **24**, 317-23.

Zelles, L., Bai, Q.Y., Ma, R.X., Rackwitz, R., Winter, K. & Beese, F. (1994). Microbial biomass, metabolic activity and nutritional status determined from fatty acid patterns and poly-hydroxybutyrate in agriculturally-managed soils. *Soil Biology and Biochemistry*, **26**, 439-46.

12

Uptake and immobilization of caesium in UK grassland and forest soils by fungi, following the Chernobyl accident

J. DIGHTON AND G. M. TERRY

Introduction

Following the Chernobyl nuclear reactor accident in 1986, there has been much scientific effort to establish the extent of radiocaesium contamination of the terrestrial environment, and to determine the propensity for radiocaesium to transfer to plants and through the food chain to man. In upland ecosystems (grassland and forest) fungi play a major role in controlling cycling of nutrients in soil and, through mycorrhizal associations, the uptake of nutrients into plants. Little attention, however, has been directed at understanding the role of fungi in the movement and availability of Cs in soil.

It has been shown that radiocaesium may be accumulated in the basidiomes of basidiomycete fungi (Haselwandter, 1978; Eckl, Hofmann & Turk, 1986; Elstner et al., 1987; Byrne, 1988; Dighton & Horrill, 1988; Haselwandter, Berreck & Brunner, 1988; Oolberkkink & Kuyper, 1989; Watling et al., 1993). Post-Chernobyl levels of radiocaesium in fruiting structures range from background to 15 000 Bq kg^{-1} dry wt depending on author, fungal species and locality. Ectomycorrhizal basidiomycete species investigated by Byrne (1988) contained a range from background to 117 Bq kg^{-1} dry wt of total radiocaesium; Dighton and Horrill (1988) recorded a range of 3890–15 820 Bq kg^{-1} dry wt for the basidiomes of the ectomycorrhizal species *Lactarius rufus* and *Inocybe longicystis*. The more recent study of Watling et al. (1993) showed ^{137}Cs contents of fruiting structures from a range of fungal species from different habitats in the UK. The values ranged from background to 1479 Bq kg^{-1} dry wt and of the basidiomycetes, greater accumulation was found in the ectomycorrhizal species (particularly the Russulaceae) than in the saprotrophic groups. This may be related to habitat type and consequent availability of radiocaesium in soil water.

These reports have shown that the levels of radiocaesium in fungal fruiting structures can be very high and these fruiting structures can be an important food source for grazing animals and, thus, form a potential route to man. From the study of Dighton and Horrill (1988), using the fingerprint ratio of ^{137}Cs to ^{134}Cs from the Chernobyl emission, it was calculated that between 25 and 92% of the ^{137}Cs present in the basidiomes of the ectomycorrhizal fungi *L. rufus* and *I. longicystis* was of pre-Chernobyl origin (Table 12.1). Calculations made from the data of Byrne (1988) also showed that 13–69% of the ^{137}Cs contained in basidiomes of ectomycorrhizal fungi collected in Slovenia were from pre-Chernobyl sources. This suggests potential long-term accumulation and retention in fungal thalli enhancing the possibility of transfers through the food chain to man.

Records exist of radiocaesium content of fungal fruiting structures collected before Chernobyl. Eckl *et al.* (1986) reported a range of ^{137}Cs concentrations of between 0–21 000 Bq kg^{-1} dry wt of basidiomes of different fungal species, measured in 1981 and 1982 in Austria, 4–5

Table 12.1. *Radiocaesium content of basidiomes of two mycorrhizal basidiomycete fungi collected from different sites following the Chernobyl accident*

Site	^{134}Cs	^{137}Cs	% ^{137}Cs due to Chernobyl
Lactarius rufus			
MH86	155	3885	8
MH87	530	5510	19
St	448	3442	26
S2	1204	7297	33
SB	582	4243	27
B	669	1779	75
Inocybe longicystis			
SB	1760	14060	25
S4	735	8737	17

Note: Activities are expressed as Bq kg^{-1}, bulked samples. MH86 and MH87 are upland peat under *Pinus contorta*; S2, SB & S4 are upland peat under *Picea sitchensis*; St is a humic podsol under *Picea sitchensis* and B is a humic podsol under *Picea abies*.
Source: After Dighton & Horrill (1988).

years before the Chernobyl accident. These represent very high levels of radiocaesium. Differences in Cs concentration within the same species of fungus could be up to 2000 fold (*Amanita fulva*). They showed that the greatest concentration of ^{137}Cs in soil occurred in the upper 5 cm. By relating basidiomes to supporting mycelia in different substrata, they showed that *A. fulva, Leccinum scabrum, Russula emetica* and *Suillus variegatus* growing in peat soil had higher ^{137}Cs content (mean 12 000; range 5000–21 000 Bq kg^{-1} dry wt) than fungal species growing on other soil types (mean 1480; range 37–8600 Bq kg^{-1} dry wt).

Elstner *et al.* (1987) also reported pre-Chernobyl accumulation of ^{137}Cs, especially in *Xerocomus badius*, of 35–94 Bq kg^{-1} dry wt; their figures being very much in the lower range of those reported by Eckl *et al.* (1986).

These findings raised a number of questions, namely:

(i) Do a wide range of fungal species accumulate radiocaesium?

(ii) Is accumulation greater in mycorrhizal fungi than in saprotrophic fungi?

(iii) What proportion of the Cs taken into a fungal thallus remains in the hyphae?

(iv) Is there translocation of Cs within the thallus and directional transport in relation to source-sink strengths?

(v) Can accumulation into fungal thalli account for a significant proportion of the Cs present in soil?

(vi) Can mycorrhizal associations affect uptake of radiocaesium into host plants?

Of these questions, all but number (iv) were addressed in a study in the UK, the results of which are outlined here and have been published elsewhere (Clint, Dighton & Rees, 1991; Dighton, Clint & Poskitt, 1991; Clint & Dighton, 1992).

Radiocaesium accumulation in a range of mycorrhizal and saprotrophic fungi

The influx of radiocaesium into 18 fungal species was assessed to identify the rate of uptake of Cs and differences in uptake rate between mycorrhizal and saprotrophic fungi. The following fungal species (with abbreviated code) were used: *Amanita fulva* (AF), *Cenococcum geophilum* (CG), *Cystoderma amianthinum* (CA), *Elaphomyces muricatus* (EM),

Hebeloma crustuliniforme (HC), *H. sacchariolens* (HS), *Hymenoscyphus ericae* (HE), *Laccaria proxima* (LP), *Lactarius rufus* (LR), *Mycena polygramma* (MP), *M. sanguinolenta* (MS), *Paxillus involutus* (PI), *Rhizopogon roseolus* (RR), *Suillus bovinus* (SB), *S. luteus* (SL), *S. variegatus* (SV), *Suillus* sp. (BD), and *Trichoderma* sp. (TA).

Caesium influx was determined using hyphae grown in liquid culture on 6.5 cm diameter sterile nylon mesh circles in Pachlewska's medium (Clint *et al.*, 1991). Fungal mycelia grown in this way in the presence or absence of a 24 h CsCl pretreatment (at $5 \mu M$) were then transferred to modified plastic syringe barrels equipped with a stop tap. The mycelium was washed in buffered Pachlewska's medium (pH 6), followed by $5 \mu M$ CsCl labelled with ^{137}Cs at 1000 Bq ml^{-1} (200 Bq nmol^{-1} Cs). The labelling solution was drained off and replaced with fresh labelling solution for 15 min. Three separate washes of unlabelled medium were used following the uptake phase and all solutions were counted for radioactive content in a gamma counter.

The values for influx of radiocaesium were expressed per unit dry weight of the fungal mycelium and per unit hyphal surface area. Hyphal length was measured using a homogenized subsample of the mycelial mat, the homogenate diluted to 1 l, and 2 ml aliquots mixed with methylene blue, diluted to 15 ml and filtered through $1.2 \mu m$ Millipore filters (Sundman & Sivela, 1978). Hyphal length was measured under a microscope by counting hyphal intersections with a grid mounted in the eyepiece (Olson, 1950). The surface area of the hyphae was calculated from the hyphal length measure and the mean diameter calculated from 25 measurements for each fungal species, assuming the hyphae to be cylinders.

Caesium influx into the fungi is shown in Figs. 12.1 and 12.2 and shows a range of values from 85 to 276 nmol Cs g^{-1} dry wt (40 °C) h^{-1} or 0.01 to 2.81 nmol Cs m^{-2} h^{-1} when expressed on a surface area basis. Influx, based on unit mycelium weight, differed significantly between species, with saprotrophic species of *M. polygramma*, *M. sanguinolenta* and *C. amianthinum* having the highest values, whilst the non-basidiomycete mycorrhizal fungi *C. geophilum* and *H. ericae* had the lowest values. Small differences in influx rates between fungi pretreated or not pretreated for 24 h with $5 \mu M$ CsCl were found. These differences were only significant for *M. polygramma*, *S. luteus* and *P. involutus*, where pretreatment enhanced the subsequent influx of Cs.

When influx was expressed on a surface area basis, the ranking of species was different from that according to a weight basis, although

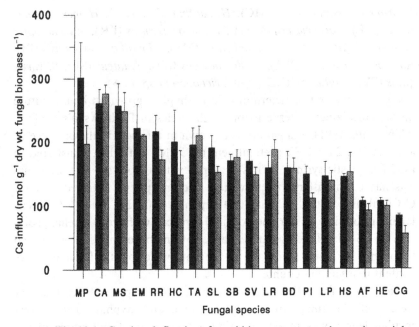

Fig. 12.1. Caesium influx into fungal biomass expressed on a dry weight basis (40 °C). Fungal codes; *Amanita fulva* (AF), *Cenococcum geophilum* (CG), *Cystoderma amianthinum* (CA), *Elaphomyces muricatus* (EM), *Hebeloma crustuliniforme* (HC), *H. saccharoliens* (HS), *Hymenoscyphus ericae* (HE), *Laccaria proxima* (LP), *Lactarius rufus* (LR), *Mycena polygramma* (MP), *M. sanguinolenta* (MS), *Paxillus involutus* (PI), *Rhizopogon roseolus* (RR), *Suillus bovinus* (SB), *S. luteus* (SL), *S. variegatus* (SV), *Suillus* sp. (BD), *Trichoderma* sp. (TA). see text. Solid bars indicate samples pre-treated for 24 h with $5\,\mu$M CsCl; hatched bars represent samples not pre-treated before influx measures. (After Clint, Dighton & Rees, 1991.) Values show means and standard error, $n = 6$.

the saprotroph *M. polygramma* still had the highest uptake rate. Again, there was little difference between pretreated and non-pretreated tissue.

In general, it can be seen that the influx of radiocaesium into fungal mycelia under laboratory conditions varies considerably between species, with uptake by saprotrophic fungi generally being greater than mycorrhizal fungi. This is contrary to the findings of studies of radiocaesium content of basidiomes (Eckl *et al.*, 1986; Elstner *et al.*, 1987; Horyna & Randa, 1988; Watling *et al.*, 1993) where accumulation in ectomycorrhizal species was greater than in saprotrophs. This may, however, be a difference in source-sink relationships within the fungal thallus at the time of fruiting; uptake here was measured in non-fruiting mycelium. The lack of difference in influx rates between pretreated and non-

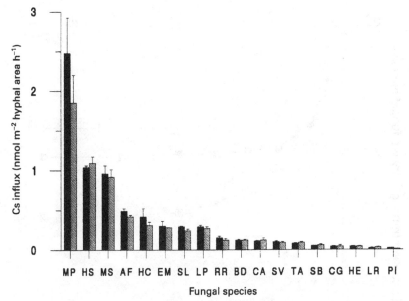

Fig. 12.2. Caesium influx into fungal biomass expressed on a hyphal surface area basis. Fungal codes as in Fig. 12.1. Solid bars indicate samples pre-treated for 24 h with $5\,\mu M$ CsCl; hatched bars represent samples not pre-treated before influx measures. (After Clint, Dighton & Rees, 1991.) Values show means and standard error, $n = 6$.

pretreated tissue suggests that the half-times for Cs accumulation are very long, or that the overall accumulation capacity is very large, or both.

Radiocaesium retention by fungal hyphae

Using a similar technique to that described above, influx of radiocaesium into three fungi (*Cladosporium cladosporioides*, *Trichoderma viride* and *Phoma* sp.), representative of saprotrophs of UK upland grasslands, was measured. The fungi were grown on Hagem's nutrient solution (see Dighton, *et al.*, 1991), presented with ^{137}Cs at 1000 Bq ml^{-1} in a $5\,\mu M$ CsCl solution in the Hagem's incubation medium, as above. Following Cs loading, Cs efflux was measured from successive washings with Hagem's incubation solution, with $5\,\mu M$ CsCl but without radioactive ^{137}Cs. This included initial washing with 4 separate 10 ml washes during the first 2 min, followed by successive 10 ml aliquots of the solution allowed to reside in contact with the mycelium for known durations over a 95 min experimental period. The radioactivity in each aliquot and the remainder in the fungal tissue were measured by gamma counting. A

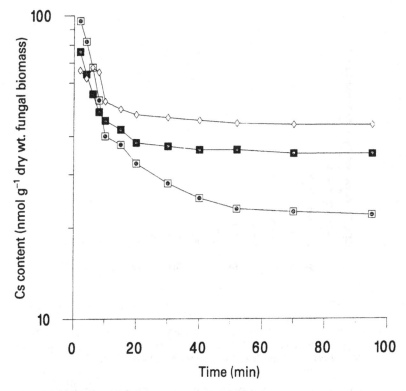

Fig. 12.3. Semi-log plot of tracer content against time for efflux of ^{137}Cs from three fungi. ◇ = *Phoma* sp., ■ = *Trichoderma viride*, □ = *Cladosporium cladosporioides*. (After Dighton, Clint & Poskitt, 1991.)

sum of the radioactivity in the eluting solution aliquots from t_{max} to t_0 provided a measure of efflux (Fig. 12.3 and Table 12.2) (Dighton *et al.*, 1991).

   The semi-log plot of Cs efflux (Fig. 12.3) shows that there were multiple exponential components visible in the efflux kinetics, suggesting that the Cs had been incorporated into a number of cellular compartments. At the end of the 95 min efflux period, between 17 and 53% of the initial radiocaesium content of the mycelium remained in the fungus, depending on the species. In addition, the time required for the original content to decline to 3% of the original was calculated at between 37 and 308 h (Table 12.2), suggesting that Cs is retained within the fungal tissue, making it less available for exchange with the outside medium.

Table 12.2. *Summary of efflux parameters over 95 min (t_{max}) for three grassland fungal species,* Cladosporium cladosporioides, Trichoderma viride *and* Phoma *sp.*

	C. cladosporioides	T. viride	Phoma sp.
Initial content (nmol g^{-1} dry wt)	125	110	80
Final content (nmol g^{-1} dry wt)	22	34	42
% original activity at t_{max}	17	32	53
Efflux at t_{max} (nmol g^{-1} dry wt h^{-1})	7	0.9	1.5
Half time at t_{max} (h)	12.4	77.0	36.5
Time required for content to reduce to 3% original (h)	37	308	146

Note: All dry weights at 40 °C.

Fungal immobilization of radiocaesium in soil

Using the figures for Cs influx for six species of grassland fungi, the three species named above and *Mortierella* sp, *Epicoccum nigrum* and *Fusarium* sp. (Dighton *et al.* 1991), an average figure of radiocaesium influx of 134 nmol Cs g^{-1} dry wt (40 °C) h^{-1} was calculated. Mean total hyphal length and biomass from two upland grassland sites (Black Combe and Corney Fell on the western side of the English Lake District, Cumbria) were determined from soil suspensions filtered through membrane filters. Hyphal length was measured by a grid line intersect method (Olson, 1950) and, using a mean hyphal diameter of 3 μm (an average of 30 measures of hyphal diameters) and an average hyphal density of 0.27 g dry wt cm^{-3} (Lodge, 1987), the fungal biomass (dry wt at 40 °C) was calculated from the volume of hyphal tissue in the soil on a unit area basis of 6 g dry wt m^{-2} for the Black Combe site and 2.6 g dry wt m^{-2} for the Corney Fell site. Assuming an influx rate of 134 nmol Cs g^{-1} h^{-1}, fungi at Black Combe had the capacity to accumulate 804 nmol Cs m^{-2} h^{-1} and 350 nmol Cs m^{-2} h^{-1} at Corney Fell. There are very few published figures of the Cs content of soil, but those of Oughton (1989) suggest that available Cs in soil solution is of the micromolar level. Thus, with the fungal accumulation rates calculated above, the fungi could immobilize a large proportion of radiocaesium input into the soil system, from Chernobyl deposition, over a relatively short period.

Effect of mycorrhizal associations on Cs uptake into host plants

In a series of experiments, the uptake of radiocaesium into mycorrhizal and non-mycorrhizal plants was investigated, using plants with differing types of mycorrhizal association.

Uptake into heather

Short term Cs influx into heather (*Calluna vulgaris*) was measured, as this plant shows high levels of radiocaesium activity compared to other plant species in upland ecosystems (Howard, Beresford & Nelson, 1988; Harrison et al., 1990). Since these areas are often grazed by sheep, this is potentially an important route for contamination and transfer to humans.

Shoot cuttings of *C. vulgaris* var. 'Darkness' were rooted in sterile sand. Plants were then grown on in sand containing 1 part of inoculum to 5 parts sand where the inoculum consisted of soil and root fragments from a local heather moor, to provide an ericaceous mycorrhizal inoculum for half the plants and microwaved (650 W for 2 min 500 g^{-1}; Ferriss, 1984) inoculum for the other half as a control. Cs influx into the plants was determined after 6 weeks' growth from plants whose washed roots were dipped into the following radiocaesium-spiked uptake solution.

Uptake was determined from the following four solutions containing $CaSO_4$, NaCl and 2-[*N*-Morpholino]ethanesulfonic acid (MES) at pH 4.0 (as described in detail by Clint & Dighton, 1992):

1. 5 μM CsCl, 5 μM KCl (low Cs, low K [LL])
2. 5 μM CsCl, 500 μM KCl (low Cs, high K [LH])
3. 500 μM CsCl, 5 μM KCl (high Cs, low K [HL])
4. 500 μM CsCl, 500 μM KCl (high Cs, high K [HH])

These uptake solutions were labelled with 1000 Bq ml^{-1} ^{137}Cs giving 200 and 2 Bq nmol^{-1} respectively for the 'low' and 'high' Cs level. Plant roots were immersed in vials containing 10 ml of the uptake solution. After known time intervals (15–60 min) plants were transferred into fresh vials of uptake solution. The ^{137}Cs remaining in each vial was determined by Cerenkov counting and, at the end of the uptake period, the plants were separated into roots and shoots, weighed, ashed and the ^{137}Cs content determined. Small portions of the roots were assessed for mycorrhizal infection after staining with acid fuchsin (Kormanic, Bryan & Schultz, 1980).

Mycorrhizal infection varied between 22 and 49% in inoculated plants and no infection was recorded in non-mycorrhizal plants. Fig. 12.4 shows the caesium content of whole plants over the influx period for all four Cs and K combinations. In all cases, uptake appeared to be biphasic with an initial fast uptake phase over the first hour followed by a slower rate of uptake over the remainder of the time course. In all cases, the caesium content of non-mycorrhizal plants was greater than that of mycorrhizal plants. Fig. 12.5 shows the shoot/root ratio of Cs accumulated in the plants at the end of the uptake period. It can be seen that the ratio is higher in mycorrhizal than non-mycorrhizal plants, indicating that, although there is a lower overall uptake in mycorrhizal plants, more Cs is translocated to the shoots than in the non-mycorrhizal plants.

Uptake by other plant species

Uptake of radiocaesium was also determined in the grass *Festuca ovina* and the clover *Trifolium repens* from labelled soil. The soil used was collected from Wasdale, Cumbria, sieved and mixed 1 to 3 with sharp sand (on a weight basis). The mixture was sterilized by microwave to eliminate mycorrhizal inocula. Non-sterilized soil was used as inoculum for half the pots. The soil was labelled with 100 Bq g^{-1} ^{137}Cs at a point source in each pot. Plants were harvested after 3 months' growth. Table 12.3 shows the Cs content in shoots and roots of plants in relation to the presence or absence of vesicular-arbuscular mycorrhizal infection.

It can be seen from Table 12.3 that, in *Festuca*, mycorrhizal infection did not stimulate any enhanced growth. Shoots showed a higher concentration of radiocaesium than roots and, similar to heather, translocation of Cs to shoots was increased in the presence of mycorrhizal infection. In *Trifolium*, however, mycorrhizal infection stimulated greater root and shoot growth. Mycorrhizal plants took up less radiocaesium than non-mycorrhizal plants and, in this species, there appeared to be no increased translocation of Cs to the shoot in mycorrhizally infected plants.

Discussion

There is now a considerable body of evidence that fungal basidiomes accumulated radiocaesium following the Chernobyl accident (Byrne, 1988; Dighton & Horrill, 1988; Haselwandter *et al.*, 1988; Oolberkkink & Kuyper, 1989; Yoshida, Muramatsu & Ogawa, 1994). Considerable variation in the Cs content of basidiomes of the same and different

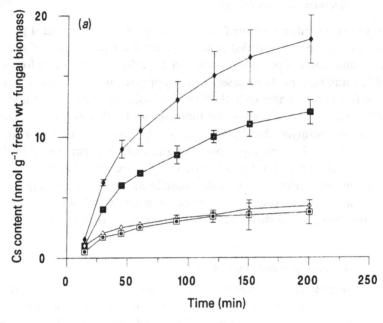

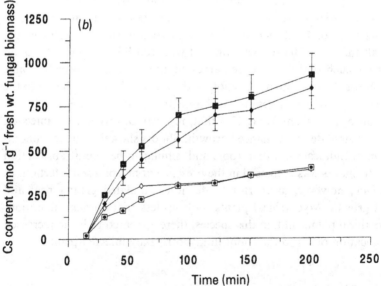

Fig. 12.4. Plot of Cs content against time for young heather plants in liquid medium at different ratios of Cs:K (see text) with no pre-treatment of CsCl. Closed symbols = non-mycorrhizal plants; open symbols = mycorrhizal plants; (a) ■, □ = 5 μM CsCl, 5 μM KCl externally; ◆, ◇ = 5 μM CsCl, 500 μM KCl externally; (b) ■, □ = 500 μM CsCl, 5 μM KCl externally; ◆, ◇ = 500 μM CsCl, 500 μM KCl externally. (After Clint & Dighton, 1992.) Values show means ± standard error, n = 6 (error bars not shown when smaller than plot symbols).

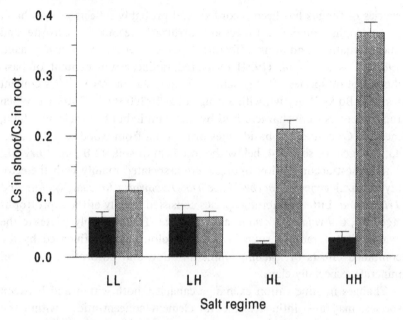

Fig. 12.5. Histogram of shoot : root ratio of accumulated Cs in heather plants after 240 min uptake from liquid medium with no pretreatment in CsCl. Solid bars = non-mycorrhizal, hatched bars = mycorrhizal plants; LL = low Cs, low K; LH = low Cs, high K; HL = high Cs, low K; HH = high Cs, high K as defined in the text. (After Clint & Dighton, 1992). Values show means and standard error, $n = 6$.

Table 12.3. *Growth and uptake of radiocaesium by* Festuca ovina *and* Trifolium repens *from soil*

Species		weight (g)		radiocaesium content (Bq g^{-1})		
		root	shoot	root	shoot	root/shoot
Festuca	NM	8.31	1.90	0.53	1.70	3.62
		±0.67	±0.08	±0.92	±0.40	±0.96
	M	8.95	2.15	0.41	2.31	5.24
		±1.16	±0.20	±0.12	±0.49	±0.24
Trifolium	NM	1.71	1.49	1.11	3.12	2.97
		±0.24	±0.28	±0.51	±0.12	±0.78
	M	3.63	2.15	0.76	2.89	2.95
		±0.61	±0.44	±0.19	±0.45	±0.65

Note: Values are means ± standard errors, $n = 5$; weights are dry weights at 40 °C. NM = non-mycorrhizal and M = mycorrhizal plants.

species of fungus has been recorded, and probably reflects growth habit (saprotrophs compared to ectomycorrhizas), resource, microsite and microclimatic conditions affecting uptake (Seeger & Schweinshaut, 1981). Yoshida *et al.* (1994) measured radiocaesium content of basidiomes of 69 species of 42 genera of fungi in Japan. Mean ^{137}Cs content was 483 Bq kg^{-1} dry wt, with a range of 2–16 300 Bq kg^{-1} dry wt. When fungal species were characterized by the main habitat of their mycelium, mean ^{137}Cs content of basidiomes originating from wood was 27; litter, 43; top 5 cm of soil, 464; below the top 5 cm of soil, 42 Bq kg^{-1} dry wt.. The highest accumulation figures were associated mainly with the ectomycorrhizal genera *Inocybe, Laccaria, Lactarius, Russula, Suillus* and *Tricholoma*. Litter-inhabiting species consisted mainly of the saprotrophs *Agaricus, Collybia, Marasmius* and *Ramaria*. This probably relates to the availability of radiocaesium in the soil soloution as influenced by the combined effects of leaching, mineralization and adsorption onto soil minerals, especially clays.

Changes in stipe cation exchange capacity, both within and between species, may also influence mineral element concentrations within the fruit body (McKnight, McKnight & Harper, 1990). Their study showed variation of CEC between 8.5 and 51.4 between fungal species. The interaction between CEC and radiocaesium may account for the differences between levels of radiocaesium content of basidiomes. Variation in CEC of different parts of the fungal thallus may also affect the location and distribution of radiocaesium in the thallus. They suggest that soil mineral and organic matter content does not control fungal tissue CEC values and that the CEC is significantly correlated to K$^+$, Ca^{2+} and Mg^{2+} composition of the tissue. This factor may also account for the high Cs accumulation in the flesh of basidiome pilei and low concentrations in the lamellae as observed by Seeger and Schweinshaut (1981).

The fate of radiocaesium in the fungal thallus has received less attention, so a number of questions concerning its fate were posed. From Cs influx and efflux studies on pure cultures of fungal mycelia in laboratory conditions, Clint *et al.* (1991) and Dighton *et al.* (1991) showed that there was considerable variation in accumulation of Cs between fungal species. In general, saprotrophic species had higher rates of accumulation than mycorrhizal fungal species. This is contrary to the findings of de Meijer, Aldenkamp and Jansen (1988) where mean ^{137}Cs content of mycorrhizal basidiomes was 2507 Bq kg^{-1} dry wt compared with 778 for saprotrophic species. From the efflux studies of Clint *et al.* (1991), it appeared that some 20–50% of the Cs taken into the fungal thallus can be retained; the chemistry of this retention is, as yet, unknown.

Using influx and accumulation data for upland grassland fungi in *in vitro* studies related to fungal mass in the field, the potential impact of fungi in immobilization of radiocaesium fallout from Chernobyl has been evaluated (Dighton, *et al.*, 1991). These figures are only approximate, as reliable data on the concentration and 'availability' of Cs in soil are scarce. The data, however, confirm the findings of Witkamp (1968) and of Witkamp and Barzansky (1968) who demonstrated the importance of microbial immobilization of ^{137}Cs in soil. Using *Trichoderma viride*, Witkamp (1968) demonstrated concentration factors of 38–42 for ^{137}Cs from mineralized solution derived from solid decomposing litter in controlled experiments. Significant increases in accumulation of ^{137}Cs from tagged soil by general microbial biomass growing on untagged, unsterile leaves and cellophane compared to sterilised were shown to occur by Witkamp and Barzansky (1968), indicating the potential importance of soil microflora in radiocaesium immobilization. Similarly, Bruckmann and Wolters (1991) demonstrated a microbially induced active transport of ^{137}Cs into the litter layer from surrounding substrata which was accelerated in the presence of mesofauna.The extent of perennial hyphal thalli in the soil system, particularly in organic forest soils, with the capacity to retranslocate elements within the thallus, implies that this component could be important in long-term immobilization of radionuclides.

Our results on the effects of mycorrhizal inoculation of either heather or grasses suggested a suppression of radiocaesium uptake by mycorrhizal plants but significant effects on partitioning within the plant (Clint & Dighton, 1992). In all cases, translocation of Cs to the shoot was increased by mycorrhizal association. There was, however, no enhanced translocation in clover. Data from shoots (only) of clover (*Melilotus officinalis*) and a grass (*Sorghum sudanese*), grown in previously contaminated soil, showed that mycorrhizal clover contained twice as much Cs as non-mycorrhizal, but there was no difference in uptake in grass (Rogers & Williams, 1986). The availability of radionuclides to plants, however, is very much dependent on the soil type and binding patterns, mineral matrices and competing ions. For example, Sikalidis, Misaelides and Alexiades (1988) showed that vermiculite was very effective in binding radiocaesium at low concentrations. Shalhevet (1973) demonstrated that uptake into *Sorghum vulgare* [*bicolor*] was almost 20 times greater in a kaolinitic than an illitic soil, indicating strong fixation of ^{137}Cs by illite. Dergunov, Abbazov and Mikulin (1980) and Paasikallio (1984) showed that the duration of growth of the plant in the soil influenced the amount of radiocaesium removed from soil. They showed a three–four fold decline in uptake by cotton by the end of the fifth year of growth. Similarly, after six years, the amount of radiocaesium removed from soil by ryegrass (*Lolium multiflorum*) was 75% of the amount removed in the first year (Paasikallio, 1984). These observations may relate to changes in the binding patterns in soil. In both cases, the mycorrhizal status of the plants was not assessed. The interactions and the influence of mycorrhizal fungi as potential immobilizers or mobilizers (due to enzymatic properties and pH modifications) are a subject for further study.

The investigations reported here have illustrated the potential for fungi (of all trophic/habitat groups) to take up and accumulate radiocaesium. The mechanisms of accumulation, directional translocation in relation to source/sink relationships within the thallus resulting in localized accumulation, however, needs further investigation. The role of mycorrhizas in plant uptake of caesium appears to contradict the general hypotheses of enhanced nutrient element uptake. This area and the possible influence of mycorrhizal hyphae in soil due to enzymatic properties also require further study.

Acknowledgements

Part of this work was funded under a NERC Special Topic Initiative on Environmental Radioactivity postdoctoral fellowship awarded to Gill Terry. We also wish to thank Anita Monster, Susan Rees and Jan Poskitt for their technical assistance.

References

Bruckmann, A. & Wolters, V. (1991). Caesium 137 als Tracer fur die Rolle der microbiellen Biomass im Nahrstoffkreislauf. *Berichte des Forschungszentrums Waldokosysteme/Waldsterben* (B), Bd. **22**, 394-9.
Byrne, A. R. (1988). Radioactivity in fungi in Slovenia, Yugoslavia, following the Chernobyl accident. *Journal of Environmental Radioactivity*, **6**, 177-83.
Clint, G. M. & Dighton, J. (1992). Uptake and accumulation of radiocaesium by mycorrhizal and non-mycorrhizal heather plants. *New Phytologist*, **121**, 555-61.
Clint, G. M., Dighton, J. & Rees, S. (1991). Influx of ^{137}Cs into hyphae of basidiomycete fungi. *Mycological Research*, **95**, 1047-51.

Dergunov, I. D., Abbazov, M. A. & Mikulin, R. G. (1980). Accumulation of ^{90}Sr and ^{137}Cs by plants in relation to their residence time in soil. *Agrokhimiya*, **8**, 125-9.

Dighton, J., Clint, G. M. & Poskitt, J. (1991). Uptake and accumulation of radiocaesium by upland grassland soil fungi: a potential pool of Cs immobilization. *Mycological Research*, **95**, 1052-6.

Dighton, J. & Horrill, A. D. (1988). Radiocaesium accumulation in the mycorrhizal fungi *Lactarius rufus* and *Inocybe longicystis*, in upland Britain, following the Chernobyl accident. *Transactions of the British Mycological Society*, **91**, 335-7.

Eckl, P., Hofmann, W. & Turk, R. (1986). Uptake of natural and man-made radionuclides by lichens and mushrooms. *Radiation and Environmental Biophysics*, **25**, 43-54.

Elstner, E.F., Fink, R., Holl, W., Lengfelder, E. & Ziegler, H. (1987). Natural and Chernobyl-caused radioactivity in mushrooms, mosses and soil samples of defined biotopes in S.W. Bavaria. *Oecologia*, **73**, 553-8.

Ferriss, R. S. (1984). Effects of microwave oven treatment on microorganisms in soil. *Phytopathology*, **74**, 121-6.

Harrison, A. F., Clint, G. M., Jones, H. E., Poskitt, J. M., Howard, B. J., Howard, D. M., Beresford, N. A. & Dighton, J. (1990). *Distribution and Recycling of Radiocaesium in Heather Dominated Ecosystems*. Report to Ministry of Agriculture, Fisheries and Food, London. Project N601.

Haselwandter, K. (1978). Accumulation of the radioactive nuclide ^{137}Cs in fruitbodies of basidiomycetes. *Health Physics*, **34**, 713-15.

Haselwandter, K., Berreck, M. & Brunner, P. (1988). Fungi as bioindicators of radiocaesium contamination: pre- and post-Chernobyl activities. *Transactions of the British Mycological Society*, **90**, 171-4.

Horyna, J. & Randa, Z. (1988). Uptake of radiocaesium and alkali metals by mushrooms. *Journal of Radioanalytical Nuclear Chemistry, Letters*, **127**, 107-20.

Howard, B. J., Beresford, N. A. & Nelson, W. A. (1988). *[134]Cs and [137]Cs Activity in Vegetation of North Yorkshire in January 1988*. Report to Ministry of Agriculture, Fisheries and Food, London. Project 494.

Kormanic, P. P., Bryan, W. C. & Schultz, R. C. (1980). Procedure and equipment for staining large numbers of plant roots for endomycorrhizal assay. *Canadian Journal of Microbiology*, **26**, 536-8.

Lodge, D. J. (1987). Nutrient concentrations, percentage moisture and density of field-collected fungal mycelia. *Soil Biology and Biochemistry*, **19**, 727- 33.

McKnight, K. B., McKnight, K.H. & Harper, K. T. (1990). Cation exchange capacities and mineral element concentrations of macrofungal stipe tissue. *Mycologia*, **82**, 91-8.

de Meijer, R. J., Aldenkamp, F. J & Jansen, A. E (1988). Resorption of caesium radionuclides by various fungi. *Oecologia*, **77**, 268-72.

Olson, F. C. W. (1950). Quantitative estimate of filamentous algae. *Transactions of the American Microscopical Society*, **59**, 171-4.

Oolberkkink, G. T. & Kuyper, T. W. (1989). Radioactive caesium from Chernobyl in fungi. *The Mycologist*, **3**, 3-6.

Oughton, D. H. (1989). *The Environmental Chemistry of Radiocaesium and Other Nuclides*. PhD Thesis, University of Manchester.

Paasikallio, A. (1984). The effect of time on the availability of ^{90}Sr and ^{137}Cs to plants from Finnish soils. *Annales Agriculturae Fenniae*, **23**, 109-20.

Rogers, R. D & Williams, S. E. (1986) Vesicular-arbuscular mycorrhiza: influence on plant uptake of caesium and cobalt. *Soil Biology and Biochemistry*, **18**, 371-6.

Seeger, R. & Schweinshaut, P. (1981). Vorkommen von Caesium in Hoheren Pilzen. *The Science of the Total Environment*, **19**, 253-76.

Shalhevet, J. (1973). Effect of mineral type and soil moisture content on plant uptake of ^{137}Cs. *Radiation Botany*, **13**, 165-71.

Sikalidis, C. A., Misaelides, P. & Alexiades, C. A. (1988). Caesium selectivity and fixation by vermiculite in the presence of various competing cations. *Environmental Pollution*, **52**, 67-79.

Sundman, V. & Sivela, S. (1978). A comment on the membrane filter technique for the estimation of length of fungal hyphae in soil. *Soil Biology and Biochemistry*, **10**, 399-401.

Watling, R, Laessoe, T, Whalley, A. J. S. & Lepp, N. W. (1993). Radioactive caesium in British mushrooms. *Botanical Journal of Scotland*, **46**, 487-7.

Witkamp, M. (1968). Accumulation of ^{137}Cs by *Trichoderma viride* relative to ^{137}Cs in soil organic matter and soil solution. *Soil Science*, **106**, 309-11.

Witkamp, M. & Barzansky, B. (1968). Microbial immobilization of ^{137}Cs in forest litter. *Oikos*, **19**, 392-5.

Yoshida, S, Muramatsu, Y & Ogawa, M. (1994). Radiocaesium concentrations in mushrooms collected in Japan. *Journal of Environmental Radioactivity*, **22**, 141-54.

13

Effects of pollutants on aquatic hyphomycetes colonizing leaf material in freshwaters

S. BERMINGHAM

Introduction

Leaf material is an important primary energy source in many northern temperate streams and rivers (Bärlocher & Kendrick, 1981). It is utilized by various macroinvertebrates which may be classified in terms of their feeding biology. Coarse particulate organic matter (CPOM; particle size > 1 mm) is utilized by a group described as the 'shredders'. These macroinvertebrates begin the incorporation of leaf material into the food web both by consumption and by producing fine particulate organic matter (FPOM), in the form of leaf fragments and faeces. FPOM (particle size < 1 mm) is utilized by a group of macroinvertebrates described as collectors. Both shredders and collectors are utilized by predators (Cummins, 1973; Anderson & Sedell, 1979).

The utilization of leaf material by shredders is therefore important to the community as a whole. Leaf material that has recently entered freshwaters is of low food quality for macroinvertebrates, as soluble compounds such as sugars and amino acids are rapidly leached into the surrounding waters. What remains after leaching are refractory compounds such as cellulose, pectins and lignin (Nykvist, 1962; Petersen & Cummins, 1974). When leaf material enters freshwater it is rapidly colonized by microorganisms and, especially in the early stages of decomposition, fungi dominate. Although some terrestrial fungi will be present on leaf material as it enters water, they are poorly adapted to conditions attained in temperate freshwaters and therefore do not persist (Bärlocher & Kendrick, 1974). In contrast, aquatic hyphomycetes are well adapted to the freshwater environment. They have low temperature requirements for growth and, in general, only sporulate when submerged. Furthermore, their conidia are morphologically adapted to impacting on substrata in flowing waters (Ingold, 1975; Webster, 1981). Aquatic hyphomycetes are important in improving the quality of leaf material as a food source for shredders.

Shredders appear to lack enzymes capable of degrading cellulose, pectin and lignin, etc, and it is unlikely that they absorb much energy directly from leaf material. They do, however, have enzymes which are capable of digesting microorganisms (Bjarnov, 1972; Monk, 1976; Martin et al., 1980). Several studies have indicated that shredders prefer material that has been colonized by fungi to sterile leaf material (Kaushik & Hynes, 1971; Bärlocher & Kendrick, 1976; Kostalos & Seymour, 1976). Furthermore, laboratory studies have indicated that both the rate of leaf decomposition and feeding preferences of shredders depend on the particular species of aquatic hyphomycete present (Bärlocher & Kendrick, 1981; Suberkropp & Arsuffi, 1984). Therefore, any perturbation of the fungal community could affect the incorporation of leaf material into the detrital food web, either by affecting the rate of decomposition or utilization by shredders. Such a perturbation could result from a pollutant entering the system.

The effects of pollutants on aquatic hyphomycetes have been investigated in laboratory and field studies (Bärlocher, 1993; Maltby, 1993), although, to our knowledge, no study has attempted to link observed effects of pollutants in the field to results of laboratory toxicity experiments and thus determine cause–effect relationships. This chapter first reviews previous studies investigating the influence of pollutants on aquatic hyphomycetes, and then reports on a case study where the effects of drainage waters from abandoned coal mines on the aquatic hyphomycete community were assessed using both field and laboratory-based experiments.

Field-based studies

A range of pollutants have been studied including acid mine drainage, low pH, sewage discharges and insecticides. Fewer species of aquatic hyphomycetes colonize leaf material in streams of low pH (pH < 5.5) and this lower diversity corresponds to a slower rate of leaf decomposition (Hall et al., 1980; Chamier, 1987). The drainage waters from abandoned coal mines, which are characteristically acidic and contaminated with metals (e.g. Olem, 1991), have also been shown to reduce the diversity of the fungal community colonizing leaf material, with the aquatic hyphomycete community being the most sensitive of the groups of fungi investigated (Maltby & Booth, 1991). Again, a reduction in diversity was related to a reduction in the rate of leaf processing.

Although pesticides are major inputs into freshwaters, only one pesticide has been studied. The insecticide methoxychlor did not affect the structure of the fungal community, based on measurements of conidia concentrations in the water column. In fact, significantly higher numbers of conidia were recorded in treated than in untreated streams, probably as a result of decreased shredder feeding rates (Suberkropp & Wallace, 1992). Similarly, sewage effluents had no effect on the assemblage of fungi colonizing leaf material (Suberkropp *et al.*, 1988). It therefore appears that the aquatic hyphomycete community is sensitive to low pH and metals, but few studies have been conducted on other types of pollutants.

Laboratory studies

There have been few studies on the effects of potential pollutants on aquatic hyphomycetes in the laboratory. In contrast to the field-based investigations, a range of pesticides have been studied. The herbicides 2,4-DB and paraquat inhibited the growth of *Flagellospora penicillioides*, *Lunulospora curvula* and *Phalangispora constricta*, at concentrations > 2.5 g 1^{-1}. The growth of all three species, however, was more sensitive to the fungicides Captafol and Mancozeb, with significant inhibition of growth recorded at concentrations > 0.1 g 1^{-1} (Chandrasheker & Kaveriappa, 1989). In contrast, the insecticide DDT stimulated biomass production by four species of aquatic hyphomycetes *Tetracladium setigerum, Varicosporium elodeae, Heliscus submersus* and *Clavariopsis aquatica*) at concentrations likely to be found in streams, i.e. 2 mg 1^{-1} (Dalton, Hodkinson & Smith, 1970).

The effects of cadmium and zinc on growth and sporulation of a number of aquatic hyphomycetes have been assessed (Abel & Bärlocher, 1984; M. A. Iles, unpublished observations). Both growth and sporulation of five species (*Alatospora acuminata, Tetracladium marchalianum, Clavariopsis aquatica, Heliscus lugdunensis, Flagellospora curvula*) were sensitive to cadmium (Abel & Bärlocher, 1984). Sporulation was more sensitive to cadmium than growth. The total number of conidia produced by leaf discs was significantly reduced at concentrations of cadmium in the range 0.1–100 mg Cd 1^{-1}. Interspecific differences in sensitivities were recorded, *A. acuminata* being the most sensitive and *T. marchalianum* being most tolerant. These experiments were conducted using field-incubated leaf material and therefore neither the distribution of fungi on leaf material nor the influence of leaf quality on sporulation could be standardized. Factors other than cadmium concentration could therefore explain the results from the sporulation experiments. Growth and sporulation have also been shown to be sensitive to zinc. Experiments using liquid media concluded that the biomass production of *Tetrachaetum elegans* was inhibited at $100 \, mg \, Zn \, l^{-1}$ and sporulation of *Tricladium gracile* was inhibited by concentrations of $1 \, mg \, Zn \, l^{-1}$ (M.A. Iles, unpublished observations).

In the laboratory, aquatic hyphomycetes have been shown to be sensitive to a range of pollutants, and, in the case of metals, sporulation appears to be more sensitive than growth. However, corresponding field studies have not been conducted to see if effects measured in the laboratory could result in changes in species assemblages colonising leaf material.

A case study: the effect of mine drainage

This study was conducted at a site on the River Don near Sheffield (National Grid Reference SE 206028). The effluent itself was acidic (pH~5.6) but did not result in a significant change in the pH of the receiving waters (pH~ 6.3). It did, however, result in significant elevations in the concentrations of iron, manganese, nickel and sulphate ions. The effects of this effluent on both the function (i.e. decomposition) and structure (i.e. assemblage) of the aquatic hyphomycete community immediately upstream and downstream of the discharge were assessed. Furthermore, laboratory toxicity experiments were conducted in an attempt to identify cause–effect relationships.

The effect of this discharge on the rate of alder (*Alnus glutinosa*) leaf decomposition was assessed over a three-month period using the leaf bag method (for review see Boulton & Booth, 1991). A mesh size of 9 mm was used to exclude macroinvertebrates. Consequently, the decomposition rates measured in this study were due to microbial processing and physical factors only. Weight loss from alder leaves followed a single phase negative exponential curve, and leaves incubated above the discharge decomposed at a faster rate than those below it (Fig. 13.1: ANCOVA $F > 4.51$, df = 3, $P < 0.05$). Statistical analysis for each time period indicated that between-station differences were not significant until after day 21 of incubation ($t = 5.60$, df = 2, P< 0.05).

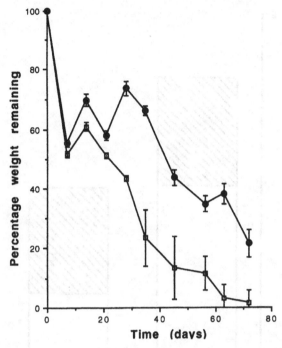

Fig. 13.1. Percentage weight remaining of alder leaves incubated at the upstream (□) and downstream (●) station, as a function of time of incubation. (Data presented as the mean of three replicates and bars denote one standard error.)

The difference in processing rates could have been the result of between-station differences in microbial processing and/or physical factors such as flow rates. Between-station differences in the activity of the microbial assemblages on leaf material were assessed by measuring oxygen uptake, antibiotics being used to separate bacterial and fungal activity. Measurements of respiration rates confirmed that there was more microbial activity on material incubated above the discharge than below it (Fig. 13.2: $t = 4.51$, df $= 46$, $P < 0.05$). Moreover, results from studies using the antibiotics penicillin and streptomycin indicated that the reduction in microbial activity was due to a decrease in fungal respiration, bacterial respiration being unaffected by the discharge. Any change in the activity of the fungal assemblage may result from either a change in total biomass and/or a change in the structure of that assemblage.

Scanning electron micrographs of the surface of alder leaves that had been incubated at the two stations indicated that, although a dense mat of mycelium was visible on material incubated at the upstream station,

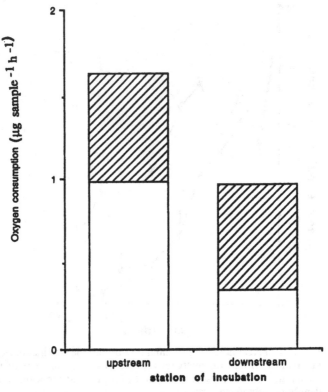

Fig. 13.2. Mean oxygen uptake of bacteria (hatched bars) and fungi (open bars) on alder leaves incubated at both the upstream and downstream stations. (Data presented as the mean of 40 replicates.)

little mycelium was detected on material incubated at the downstream station (Fig. 13.3(*a*),(*b*)). This difference in hyphal density at the two stations suggested that total fungal biomass was reduced at the downstream station. Problems were encountered when attempting to quantify fungal biomass present on leaf material. From the literature, ergosterol appeared to be the most promising biochemical marker of biomass (Gessner & Schwoerbel, 1991). However, *in vitro* investigations of the relationship between biomass and ergosterol content indicated that for six out of nine species of aquatic hyphomycetes investigated no simple relationship existed between biomass and ergosterol content (Bermingham, 1993). As there is no method that will accurately quantify the biomass of a mixed community, field studies of pollutant-induced changes in fungal biomass must be based on qualitative methods, such

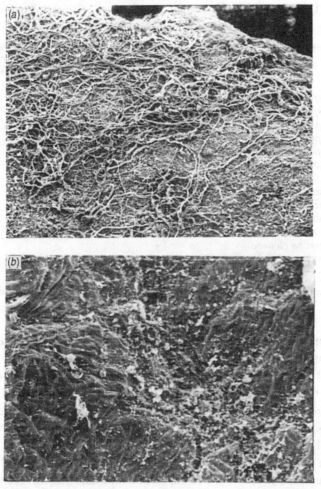

Fig. 13.3 SEM of the surface of alder leaves incubated at the upstream (*a*) and downstream (*b*) stations for 12 weeks.

as SEM. The observed difference in fungal biomass may reflect a change in the species assemblage colonizing leaf material.

The structure of the fungal assemblage was assessed using traditional methods, where the presence of species at a station was inferred from the presence of its conidia on leaf baits (Ingold, 1975). Whereas 14 species of aquatic hyphomycetes were recorded on material incubated at the upstream station, only seven were recorded on material incubated at the downstream station (Table 13.1). The lower number of species recorded on material incubated below the discharge resulted from the

Table 13.1. *The presence (+) and absence (−) of aquatic hyphomycetes on alder leaves at the two stations on the River Don*

Species	Upstream	Downstream
Alatospora acuminata	+	−
Anguillospora crassa	+	+
Anguillospora longissima	+	+
Clavariopsis aquatica	+	−
Clavatospora longibrachiata	+	+
Clavatospora stellata	+	−
Flagellospora curvula	+	−
Heliscus lugdunensis	+	+
Lemonniera aquatica	+	+
Tetrachaetum elegans	+	−
Tetracladium marchalianum	+	−
Tricladium attenuatum	+	+
Varicosporium elodeae	+	+
Varicosporium giganteum	+	−
Total	14	7

exclusion of a number of species, including *Alatospora acuminata*, *Flagellospora curvula*, *Tetrachaetum elegans* and *Tetracladium marchalianum*. Interestingly, no species of aquatic hyphomycetes were recorded on the leaf baits until day 21 of incubation. This corresponds to the time that significant differences were recorded between rates of decomposition at the two stations, suggesting that, prior to day 21, weight loss was the result of leaching and physical abrasion.

From the field study, it can be concluded that the effluent resulted in a decrease in the rate of leaf decomposition and that this was related to a reduction in fungal diversity and activity. As some species appeared to be excluded by the discharge, laboratory toxicity experiments were conducted to determine if their field distribution could be explained in terms of interspecific differences in susceptibilities to metals found in the discharge. Experiments were designed to assess the effects of iron and manganese, the principal components of the discharge, on all stages of the life cycle of three species of aquatic hyphomycete: *Articulospora tetracladia* (recorded at both stations,) and *Tetrachaetum elegans* and *Alatospora acuminata* (restricted to the upstream station). Metal concentrations used in these experiments spanned those recorded in the field (i.e. iron $0.2\,\mathrm{mg\,l^{-1}}$–$6.9\,\mathrm{mg\,l^{-1}}$; manganese $0.04\,\mathrm{mg\,l^{-1}}$–$1.48\,\mathrm{mg\,l^{-1}}$). The effects of metals on mycelial extension rates (MER) were assessed using solid media, whilst liquid media were used for bio-mass studies. Whereas the MER of *A. tetracladia* was stimulated by concentrations of iron above the control (Fig. 13.4(*a*): ANCOVA $F = 3.42$, df $= 4, 5, P < 0.05$), there was no significant effect of iron on the MER of either *T. elegans* or *A. acuminata* (ANCOVA F < 1.01, df $= 4, 50, P > 0.05$). Similarly, there was no significant effect of man-ganese on the MER of *T. elegans* or *A. acuminata* (Fig. 13.4(*b*): ANCOVA F< 3.19, df $= 4,5$ $P > 0.05$). In contrast to the MER studies, there was no significant effect of either metal on the biomass produced by the three species investigated (Fig. 13.5: ANOVA F< 2.01, df $= 4,9$, $P > 0.05$). One possible explanation for the lack of effect is that the metals in the media were not bioavailable and therefore were not taken up by the fungi. However, this can be discounted as chemical analysis of the exposed mycelium indicated that there was a significant effect of metal concentration in the media on that accumulated in the mycelium (Fig. 13.6: Fe: ANOVA F< 13.57, df $= 4,9$ $P < 0.001$; Mn: ANOVA F< 8.03, df $= 4,9$, $P < 0.01$). The fact that mycelium was accumulating metals indicated that at least a proportion of the metal added to the media was being taken up by the fungi and that these concentrations had no inhibitory effect on the growth of the three species investigated.

However, exposing mycelium to metals did influence subsequent spor-ulation (Fig. 13.7). Although, there was no significant effect of either iron or manganese on the number of conidia produced by *A. tetracladia* (H< 6.88, df $= 4, P > 0.05$), *T. elegans* did not produce conidia when exposed to iron concentrations greater than 1.2 mg Fe l^{-1} and produced significantly fewer conidia when exposed to manganese concentrations greater than 0.32 mg Mn l^{-1} (U> 21.0, $n_1 n_2 = 4, 4, P < 0.05$). Similarly, conidia production by *A. acuminata* was significantly reduced when exposed to iron concentrations greater than 1.2 mg Fe l^{-1} or manganese concentrations above 0.32 mg Mn l^{-1} (H> 12.85, df $= 4, P < 0.01$).

Therefore, although there is no evidence that iron and manganese inhibit growth, those species which were apparently restricted to the upstream station did not produce conidia when exposed to concentra-tions of metals recorded at the downstream station. It would therefore appear that metals in the mine discharge did have a direct toxic effect on aquatic hyphomycetes which in turn resulted in reduced rates of leaf decomposition. The identification of aquatic hyphomycetes depends upon the presence of conidia. Therefore, it is unclear whether those species absent from the downstream station are present as mycelium which is unable to sporulate and are thus not recorded, or whether

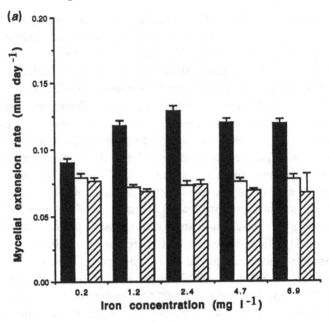

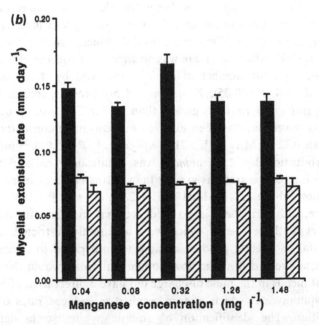

Fig. 13.4. Mycelial extension rates of *A. tetracladia* (solid bar), *T. ele-gans* (open bar) and *A. acuminata* (hatched bar) on media amended with differing concentrations of iron (*a*) and manganese (*b*). (Bars denote one standard error.)

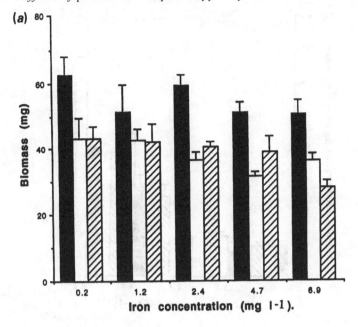

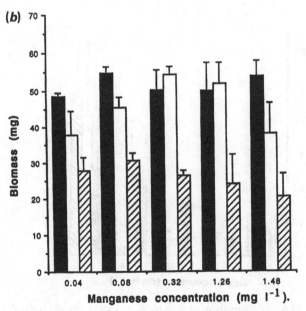

Fig. 13.5. Biomass production by *A. tetracladia* (solid bar), *T. elegans* (open bar) and *A. acuminata* (hatched bar) in liquid media amended with different concentrations of iron (*a*) and manganese (*b*). (Data presented as the mean of three replicates and bars denote one standard error.)

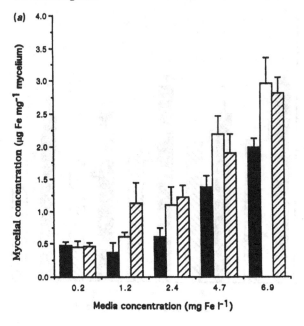

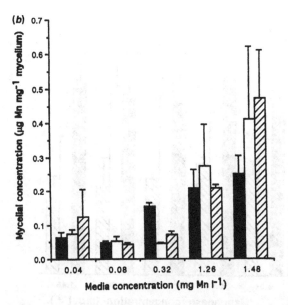

Fig. 13.6. Accumulation of iron by *A. tetracladia* (solid bar), *T. elegans* (open bar) and *A. acuminata* (hatched bar) from liquid media amended with different concentrations of iron (*a*) and manganese (*b*). (Data presented as the mean of three replicates and bars denote one standard error.)

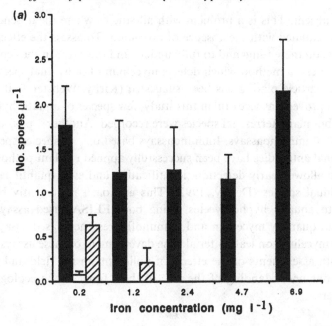

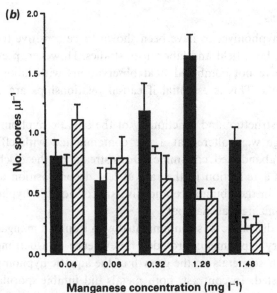

Fig. 13.7. Sporulation by mycelium of *A. tetracladia* (solid bar), *T. elegans* (open bar) and *A. acuminata* (hatched bar) grown on media of different concentrations of iron (*a*) and manganese (*b*). (Data presented as the mean of four replicates and bars denote one standard error.)

they are absent. This is a problem with all studies where the absence of conidia is equated with the absence of mycelium. To assess the effects of pollutants on these fungi and to fully understand their role in the ecology of freshwaters, a method which detects mycelium of individual species is required. Particle plating has been suggested (Kirby, Webster & Baker, 1990) but proved unsuccessful in this study; few species of aquatic hypho-mycetes but many terrestrial species were recorded. Another   possibility is the use of immunoassays. Immunoassays based on   species-specific monoclonal antibodies have been successfully applied to plant pathogens and have allowed early detection, identification and even quantification of individual species (Dewey, 1992). This approach is currently being applied to aquatic hyphomycetes, using both ELISA-based assays to detect and quantify mycelium and immunofluorescence-based assays to visualize mycelium on leaf material. The development of these assays will allow both assessments of the effects of pollutants in the field and also increase our understanding of the role of these fungi in the ecology of freshwaters.

Conclusions

1. Aquatic hyphomycetes have been shown to be sensitive to pol-lutants in both field and laboratory studies. However, previous studies have not combined field observations with laboratory experiments. This is essential if causal relationships are to be identified.

2. Both the structure and functioning of the aquatic hyphomycete assemblage were altered at a site contaminated with effluent from an abandoned coal-mine. Downstream of the discharge there was a reduction in the rate of leaf decomposition and a reduction in the abundance and diversity of aquatic hyphomy-cetes colonizing leaf material.

3. The mine discharge was contaminated with iron and manganese. Laboratory experiments provided no evidence of a detrimental effect of these metals on the growth of the aquatic hyphomycete species tested. In contrast, both metals did inhibit sporulation and those species apparently restricted to the upstream station did not sporulate at metal concentrations prevalent downstream of the discharge.

4. Results from the laboratory experiments suggest that iron and manganese toxicity may account for field distribution of selected

species. However, as identification of aquatic hyphomycetes is based on conidial morphology, it is unclear whether species apparently eradicated by the discharge were actually absent or present as non-sporulating mycelia. To assess the effects of pollutants on these fungi and to fully understand their role in the ecology of freshwater, a method which allows detection of mycelium of individual species is required.

References

Abel, T. H. & Bärlocher, F. (1984). Effects of cadmium on aquatic hyphomycetes. *Applied and Environmental Microbiology*, **48**, 245–51.

Andersen, N. H. & Sedell, J. R. (1979). Detritus processing by macroinvertebrates in stream ecosystems. *Annual Review of Entomology*, **24**, 351–7.

Bärlocher, F. (1993). Human interference. In *The Ecology of Aquatic Hyphomycetes*, ed. F. Bärlocher. pp. 173–182. Springer-Verlag: London.

Bärlocher, F. & Kendrick, B. (1974). Dynamics of fungal populations on leaves in streams. *Journal of Ecology*, **62**, 761–90.

Bärlocher, F. & Kendrick, B. (1976). Hyphomycetes as intermediaries in energy flow in streams. In *Recent Advances in Aquatic Mycology*. ed. E. B. G. Jones. pp. 325–344. Elek: London.

Bärlocher, F. & Kendrick, B. (1981). Role of aquatic hyphomycetes in the trophic structure of streams. In *The Fungal Community, Its Organisation and Role in the Ecosystem*, ed. D. T. Wicklow & G. C. Carroll. pp. 743–760. Marcel Dekker: New York.

Bermingham, S. (1993). Effects on coal mine effluent on leaf-processing in a freshwater stream. PhD Thesis. University of Sheffield.

Bjarnov, N. (1972). Carbohydrases in *Chironomus*, *Gammarus* and Trichoptera larvae. *Oikos*, **23**, 261–3.

Boulton, A. J. & Booth, P. I. (1991). A review of the methodology used to measure leaf litter decomposition in lotic environments: time to turn over a new leaf? *Australian Journal of Marine and Freshwater Research*, **42**, 1–43.

Chamier, A. C. (1987). Effect of pH on microbial degradation of leaf litter in seven streams of the English Lake District. *Oecologia*, **71**, 491–500.

Chandrashekar, K. R. & Kaveriappa, K. M. (1989). Effect of pesticides on the growth of aquatic hyphomycetes. *Toxicology Letters*, **48**, 311–15.

Cummins, K. W. (1973). Trophic relations of aquatic insects. *Annual Review of Entomology*, **18**, 183–206.

Dalton, S. A., Hodkinson, M. & Smith, K. A. (1970). Interactions between DDT and river fungi. I The effects of p–p'-DDT on the growth of aquatic hyphomycetes. *Applied Microbiology*, **20**, 662–6.

Dewey, F. M. (1992). Detection of plant-invading fungi by monoclonal antibodies. In *Techniques for the Rapid Detection of Plant Pathogens*, ed. J. M. Duncan & L. Torrance, pp. 47–62. Blackwell Scientific Publications: Oxford.

Gessner, M. O. & Schwoerbel, J. (1991). Fungal biomass associated with decaying leaf litter in a stream. *Oecologia*, **87**, 602–3.

Hall, R. J., Likens, G. E., Fiance, S. B. & Hendrey, G. R. (1980). Experimental acidification of a stream in the Hubbard Brook, Experimental Forests New Hampshire. *Ecology*, **61**, 976–89.

Ingold, C. T. (1975). *Guide to the Aquatic Hyphomycetes*. Scientific publication 30. Freshwater Biological Association: Windermere, UK.

Kaushik, N. K. & Hynes, H. B. N. (1971). The fate of dead leaves that fall into streams. *Archiv für Hyrobiologie*, **68**, 465–515.

Kirby, J. J., Webster, J. & Baker, J. H. (1990). A particle plating method for analysis of fungal community composition and structure. *Mycological Research*, **94**, 621–6.

Kostalos, M. & Seymour, R. L. (1976). Role of microbial enriched detritus in the nutrition of *Gammarus minus* (Amphipoda) *Oikos*, **27**, 512–16.

Maltby, L. (1993). Heterotrophic microbes. In *The Rivers Handbook*, ed. P. Calow & G. Petts, pp. 165–194. Blackwell Scientific Publications: Oxford.

Maltby, L. & Booth, R. (1991). The effect of coal mine effluent on fungal assemblages and leaf breakdown. *Water Research*, **25**, 247–50.

Martin, M. M., Martin, J. S., Kukar, J. J. & Merrit, R. W. (1980). The digestive enzymes of detritus-feeding stonefly nymphs (Plecoptera: Pteronarcyidae). *Canadian Journal of Zoology*, **59**, 1947–51.

Monk, D. C. (1976). The distribution of cellulase in freshwater invertebrates of different feeding habitats. *Freshwater Biology*, **6**, 471–5.

Nykvist, N. (1962). Leaching and composition of litter. V. Experiments on leaf litter of *Alnus glutinosa*, *Fagus silivatica* and *Quercus rubra*. *Oikos*, **13**, 232–48.

Olem, H. (1991). Minerals and mine drainage. *Journal of Water Pollution Control Federation*, **63**, 472–5.

Petersen, R. C. & Cummins, K. W. (1974). Leaf processing in a woodland stream. *Freshwater Biology*, **4**, 343–68.

Suberkropp, K. & Arsuffi, T. L. (1984). Degradation, growth and changes in palatability of leaves colonised by six aquatic hyphomycete species. *Mycologia*, **76**, 398–407.

Suberkropp, K., Michelis, A., Lorch, H-J. & Ottow, J. C. G. (1988). Effect of sewage treatment plant effluent on the distribution of aquatic hyphomycetes in the River Erms, Schwäbische Alb, F. R. G. *Aquatic Botany*, **32**, 141–53.

Suberkropp, K. & Wallace, J. B. (1992). Aquatic hyphomycetes in insecticide-treated and untreated streams. *Journal of North American Benthological Society*, **11**, 165–71.

Webster, J. (1981). Biology and ecology of aquatic hyphomycetes. In *The Fungal Community, its Organisation and Role in Ecosystems*, ed. D. T. Wicklow & G. C. Carroll, pp. 681–691. Marcel Dekker: New York.

14

Fungi and salt stress

L. ADLER

Introduction

Salinization of soils is a growing threat to the future of agriculture in many parts of the world. Cultivation of arid and semi-arid land seems to be inevitably linked to salt accumulation; evaporation of water used for irrigation concentrates the dissolved salt and cultivation may compact the soil and cause retention of salts in the upper layers (Epstein *et al.*, 1980). Hence, the understanding of the basic processes of salt tolerance and salt adaptation has important applied and environmental interests. An advantage of studying these processes in fungi is their experimental tractability, and the possibility of utilizing genetic approaches that are not available in other eukaryotic organisms.

Salt relations

Fungi occupy environments ranging from freshwater to cured food products and concentrated brines. Effects of increased salt concentration on fungal physiology are frequently explained in terms of the effects of a more general factor, the water potential (Ψ) of the environment. Often, this term has been used to describe the growth limits for a particular species below which growth does not occur. One should bear in mind though that the limiting value is not absolute, but depends on factors such as nutrition, temperature and the nature of the Ψ adjusting solute (Pitt & Hocking, 1977; Blomberg & Adler, 1992). The mycoflora of saline environments such as a salt marsh appears to differ little from that of more normal soils (Luard & Griffin, 1981). The majority of the soil fungi examined by Kouyeas (1964) ceased to grow at -10 MPa, corresponding to the water potential of a 2.1 M NaCl solution. At a still lower Ψ

217

the active mycoflora is dominated by species of *Aspergillus* and *Penicillium* (Griffin & Luard, 1979). These fungi tend to be solute indifferent and show similar growth responses irrespective of type of solute used to control Ψ. Extreme habitats generally sustain little species diversity and only a few fungi are able to grow in saturated NaCl solutions. Two species that are commonly encountered on salted dried fish, *Polypaecilum pisce* and *Basipetospora halophila*, grow vigorously in saline media and can be cultured in saturated conditions (Wheeler, Hocking & Pitt, 1988). Yeast species capable of growth in near saturated brines are found in several genera such as *Hansenula*, *Pichia*, *Zygosaccharomyces* and *Debaryomyces*. (Onishi, 1963; Tokuoka, 1993).

The physiology of adaptation of salt-tolerant yeasts and other fungi to concentrated environments has been recently reviewed (Blomberg & Adler, 1993; Clipson & Jennings, 1993; Tokuoka, 1993). For a discussion of the molecular biology of osmostress responses of the yeast *Saccharomyces cerevisiae*, the reader is referred to the review by Mager and Varela (1993). The present review calls attention to some recent advances in our understanding of fungal ion transport and osmoregulation that are relevant to fungal salt tolerance.

Ion transport

Fungal enzymes are generally rather sensitive to high concentrations of inorganic ions; under *in vitro* conditions NaCl is stimulatory up to 50–200 mM but inhibits enzyme activity as the concentration is further increased (Blomberg & Adler, 1993; Clipson & Jennings, 1993). Elimination of Na^+ from the cytosol is thus a key mechanism for averting toxic effects on biochemical processes in the cytosol. Exposed to high salinity, fungi like *Penicillium ochro-chloron* (Gadd et al., 1984) and *Aspergillus nidulans* (Beever & Laracy, 1986) appear to exclude Na^+ effectively, while in *Neurospora crassa* (Slayman & Tatum, 1964), *S. cerevisiae* (Camacho, Ramos & Rodriguéz-Navarro, 1981), and *Debaryomyces hansenii* (Norkrans & Kylin, 1969) internal Na^+ increases with the external NaCl load, so that at high salinities Na^+ becomes as prevalent as potassium, and the intracellular K^+/Na^+ ratio reaches about unity.

In *S. cerevisiae* influx of Na^+ occurs via the K^+ carrier and probably also in association with the high affinity uptake of phosphate (Borst-Pauwels, 1981). Na^+ co-transport with amino acids and sugars has been demonstrated in animals and bacteria (Stewart & Booth, 1983) and may exist also in fungi. Influx of Na^+ requires a counteracting exodus to maintain low internal levels. Progress in understanding the mechanisms behind sodium efflux in fungi has recently advanced by cloning and molecular analysis of sodium transporters in yeasts.

A putative electrogenic Na^+/H^+ antiporter working on the acid side was identified in *Schizosaccharomyces pombe* by selecting for increased lithium tolerance of cells transformed with a multycopy gene library (Jia *et al.*, 1992). The *sod2* gene cloned by this screen encodes a 52 kD protein that possesses 12 putative membrane spanning regions and demonstrates slight homology to human and bacterial Na^+/H^+ antiporters. Over-expression of *sod2* increased export Na^+ and influx of protons while a *sod2* disruptant showed no such transport activities. Cells carrying a *sod2* null mutation decrease their NaCl tolerance by approximately five-fold over a strain having this gene .

The electrochemical proton gradient that drives the energy-dependent uptake of ions and nutrients in fungi is generated by the plasma membrane H^+ ATPase (Serrano, 1991). This enzyme is a major membrane protein in fungi, estimated to consume 10–15% of the ATP produced by growing yeast cells. Direct evidence for its essential role came from disruption of the *PMA1* gene in *S. cerevisiae*, which generated lethality in haploid cells. The *PMA1* genes sequenced from *S. cerevisiae* (Serrano, 1991), *N. crassa* (Hager *et al.*, 1986), *S. pombe* (Ghislain, Schlesser & Goffeau, 1987) and *Zygosaccharomyces rouxii* (Watanabe, Shiramizu & Tamai, 1991) are highly conserved.

The NaCl tolerance of the salt-tolerant yeast *Z. rouxii* appears to be limited by the plasma membrane ATPase activity. The presence of vanadate, a potent inhibitor of this enzyme, causes growth inhibition at NaCl concentrations above 1 M, while cells continue to grow normally at sorbitol concentrations osmotically comparable to NaCl concentrations up to 3 M (Watanabe & Tamai, 1992). A similar but less pronounced growth inhibition was produced by the proton-ionophore CCCP (carbonylcyanide-*m*-chlorophenylhydrazone) at high concentrations of NaCl. These results suggest that salt tolerance contrasts with osmotolerance, in requiring an ATPase-driven proton gradient most likely needed to energise a putative Na^+/H^+ antiporter that becomes indispensable for growth at high external NaCl levels. In keeping with this notion, a rapid stimulation of the ATPase activity was observed on exposure of *Z. rouxii* to 1 M NaCl (Nishi & Yagi, 1992). Watanabe, Sanemitsu and Tamai (1993) subsequently demonstrated by Western and Northern analysis that the expression of the plasma membrane H^+ ATPase is also increased in cells cultured at elevated salinities.

220   L. Adler

In contrast with these findings, Vallejo and Serrano (1989) observed that *PMA1* promoter mutants of *S. cerevisiae*, exhibiting about 20% of the authentic ATPase activity, displayed growth that was similar to that of the wild-type even at 1.2 M NaCl (pH 6.0). The authors concluded that the ATPase is not rate-limiting for salt-tolerance in *S. cerevisiae*. Presumably, the maintained salt-tolerance despite the decreased H^+-ATPase activity is related to the recent identification of a novel type of ion motive ATPase in *S. cerevisiae*. Taking advantage of the fact that Li^+ is often transported by Na^+ carriers, Rodriguéz-Navarro and co-workers (Haro, Garciadeblas & Rodriguéz-Navarro, 1991) identified a gene involved in Na^+ efflux by complementing the Li^+ sensitivity of a low Li^+ efflux strain with a gene library. Sequencing revealed that this gene, called *ENA1* (exitus natrii), was identical to the previously identified *PMR2*, which on the basis of sequence homology was presumed to encode an ion motive plasma membrane ATPase (Rudolph *et al.*, 1989). Disruption of *ENA1/PMR2* resulted in increased sensitivity to Na^+ and Li^+ and a strongly defective efflux of these ions at alkaline pH. *ENA1* was suggested to encode an ion motive ATPase which is used to extrude Na^+ under alkaline conditions. In agreement with such a role, constructions in which *ENA1* was fused to a *lacZ* reporter gene demonstrated that expression of *ENA1* is induced by Na^+, Li^+ and alkaline pH (Garciadeblas *et al.*, 1992). Three homologues, *ENA2–ENA4*, are located in tandem array with *ENA1*; *ENA2* has a low level of expression at low as well as high extracellular salinity, while that of the other two isoforms remains to be examined. Na^+/H^+ antiporters which are suggested to play a major role in Na^+ and H^+ homeostasis in eukaryotes have been suggested to exist also in *S. cerevisiae* (Rodriguéz-Navarro & Ortega, 1982) but remain to be unequivocally demonstrated.

The striving for a cytoplasm rich in K^+ but low in Na^+ is a conserved feature of most cells. Why K^+ is the preferred intracellular cation is an unsolved question. An interesting contribution to resolve the K^+/Na^+ paradox has come from Wiggins (1990), who presented biophysical evidence that cells select in favour of K^+ in order to minimize the formation of extreme water structures in the cell. Rodriguéz-Navarro, Blatt and Slayman (1986) provided indirect evidence that *N. crassa* under K^+ deficiency expresses a H^+–K^+ symport, by demonstrating that the net inflow of positive charges was close to twice the K^+ uptake. This finding can be accommodated with the seemingly contradictory observation that potassium-starved fungi exhibit high rates of K^+ influx in exchange for H^+

(Ryan & Ryan, 1972), by proposing that the H^+ pump maintains electroneutrality during uptake of K^+ by mediating efflux of one H^+ per equivalent of charge transported by the symport (Sanders, 1988). In addition to the putative symporter, voltage-gated membrane channels selective for K^+ have been demonstrated in fungi, by the application of the patch-clamp technique on sphaeroplasts of *S. cerevisiae* (Gustin *et al.*, 1986). These channels which mediate high fluxes of K^+ along the electrochemical gradient, may bring about electrophoretic uptake of K^+ in energized, hyperpolarized cells and efflux of K^+ in depolarized cells (Serrano, 1991).

S. cerevisiae has dual affinities for uptake of K^+ (Rodriguéz-Navarro & Ramos, 1984). *TRK1*, the gene required for the high affinity K^+ transport, was isolated by complementation of the defective K^+ uptake in *trk1* cells (Gaber, Styles & Fink, 1988). The *TRK1* gene product is a membrane protein containing 12 putative membrane spanning regions. Deletion of *TRK1* is not lethal but leads to an about 50-fold increase in the K^+ requirement for normal growth. Ko and Gaber (1991) identified a second gene, *TRK2*, which encodes a protein with an overall structure similar to TRK1. Cells deleted for both *TRK1* and *TRK2* require a 10-fold higher concentration of K^+ than cells deleted for *TRK1* only. It was suggested that *TRK2* encodes the low affinity K^+ transporter (Ko, Buckley & Gaber, 1990), but a recent kinetic study indicates that TRK1 and TRK2 may be components of one and the same K^+ uptake system (Ramos *et al.*, 1994). It is not yet clear what kind of transport mechanism the *TRK*-encoded proteins mediate.

The K^+ uptake system mediates main Na^+ influx in *S. cerevisiae* and its ability to discriminate between K^+ and Na^+ depends on the growth conditions. In cells exposed to NaCl stress, K^+ uptake converts from its low affinity to its high affinity state (Ramos, Haro & Rodriguéz-Navarro, 1990). This transformation decreases the K_m ratio for K^+/Na^+ by about 20-fold. In keeping with this observation, a *trk1Δ* strain which is unable to switch to the high affinity mode is less effective in selecting for K^+ against Na^+ (Haro *et al.*, 1993), which results in about a 25% decrease in NaCl tolerance. The relatively slight decrease in salt tolerance resulting from loss of *TRK1* presumably is owing to a high intrinsic tolerance of *S. cerevisiae* to elevated internal Na^+ concentrations (Camacho *et al.*, 1981; Ölz *et al.*, 1993). However, in mutants lacking both *TRK1* and *ENA1*, there is an approximate 10-fold loss in NaCl tolerance (Haro *et al.*, 1993). This dramatic effect underlines the importance of internal cation regulation for cellular salt-tolerance.

Gaxiola and co-workers (1992) isolated the *HAL1* gene from *S. cerevisiae*, which upon increase in gene dosage improves growth under NaCl stress. The expression of the gene is increased in cells exposed to osmotic stress and over-expression improves intracellular accumulation of K^+, resulting in an increased internal K^+/Na^+ ratio. It was speculated that HAL1 interferes with K^+ uptake by changing the set point for the feedback inhibition exerted by the internal K^+ concentration. Interestingly, the *HAL1* gene is conserved in plants.

Role of the vacuole

Influx of Na^+ into a growing cell is counterbalanced not only by sodium efflux, but also by growth dilution. A preferential confinement of Na^+ in the vacuole will increase its contribution to the dilution effect, allowing for a reduced Na^+ concentration in the cytoplasm. Hence, compartmentation of ions into the vacuole, represents an option to reduce toxic effects in the cytosol. A vacuolar proton translocating H^+-ATPase is present on the fungal tonoplast which acidifies the vacuole and polarizes the vacuolar membrane (Klionsky, Herman & Emr, 1990). The proton gradient is proposed to drive uptake of ions by H^+ antiport. There is evidence for such transport of Zn^{2+} in *S. cerevisiae* (White & Gadd, 1987) as for Ca^{2+} in *S. cerevisiae* (Ohsumi & Anraku, 1983) and *N. crassa* (Cornelius & Nakashima, 1984). A membrane potential dependent cation channel with about equal selectivity for Na^+ and K^+ is also demonstrated in the vacuolar membrane (Klionsky *et al.*, 1990). Since membrane voltage is oriented with the cytosol negative, this channel is presumably outwardly directed in the polarized membrane. Little is known about selective sequestration of Na^+ and K^+ to the vacuole. Using X-ray microanalysis, Clipson, Hajibagheri and Jennings (1990) noted no preferential confinement of Na^+, K^+ and Cl⁻ to the vacuole in the marine fungus *Dendryphiella salina* cultured at 0.5 M NaCl. Differential extraction procedures have, on the other hand, indicated that K^+ can accumulate in the *S. cerevisiae* vacuole against its electrochemical gradient (Okorokov, Lichko & Kulaev, 1980; Perkins & Gadd, 1993), and the vacuolar accumulation of Li^+, which is often transported by Na^+ carriers, increases with increasing external Li^+ concentrations (Perkins & Gadd, 1993).

Indirect evidence for a function of the vacuole in NaCl tolerance also comes from studies of vacuolar mutants. A *vatc* mutant of *S. cerevisiae* lacking functional vacuolar H^+-ATPase is growth limited by a NaCl concentration less than half of that tolerated by the parental strain (Haro *et al.*, 1993). Among the vacuole protein sorting *vps* mutants isolated by Banta *et al.* (1988), all complementation groups (*vps*11, 16, 18, 33) that were devoid of any organelle resembling a normal vacuole were osmosensitive in the sense that no growth occurred in medium containing 1 M NaCl, 1 M KCl, or 2.5 M glycerol. Furthermore, among the 97 salt-sensitive *ssv* mutants of *S. cerevisiae* that were isolated by Latterich and Watson (1991), the majority were characterized by aberrant vacuole morphology and, strikingly, all the mutants mis-localized vacuolar proteins to the cell surface to various extents. Since the primary selection was not based on vacuolar defects, this finding points to an essential role for the vacuole in tolerance to high osmolarity. What can this role be? After shifting cells to high salinity, there is rapid water efflux, and an increased Na^+ influx into the cell (Norkrans & Kylin, 1969). A rapid sequestration of sodium into the vacuole might be essential to avert the toxic salt effects. Early stationary phase cells of the *ssv1-2* mutant, carrying a mutation in a gene essential for vacuole biogenesis, become unculturable within seconds on transfer to 1.5 M NaCl, while similarly treated parental cells maintain slow growth (Latterich & Watson, 1993). It was suggested that the vacuole mediates protection by allowing for rapid diffusion of water into the cytoplasm, thereby buffering the osmotic shock. However, as an alternative protective mechanism, a rapid mobilization of endogenous osmolytes might have occurred through hydrolysis of vacuolar polymers. Storage contents of the fungal vacuole include various cations, basic amino acids, and polyphosphate (Klionsky *et al.*, 1990). Due to the polyanionic nature of the polyphosphate it might reduce the osmotic contribution from divalent cations and basic amino acids. Hence, mobilization of vacuolar polyphosphate to cytoplasmic phosphate would result in a net increase in the level of free ions in both the cytosol and the vacuole. Such adjustments would serve well to accommodate small changes in turgor/cell volume and may perhaps also extend the limit for survival on harsh osmotic shock. Occurrence of mobilization of polyphosphate in response to osmotic shock was demonstrated by P-NMR analysis of growing *N. crassa*; an almost quantitative hydrolysis of vacuolar polyphosphate emerged concomitant with an increase in the concentration of inorganic phosphate in the cytosol (Yang, Bastos & Chen, 1993). Although these cells were subjected to hypo- rather than hyper-osmotic shock, the results indicate that the polyphosphate pool is responsive to changes in external osmolarity. Further experimentation will be needed to more directly address the osmoprotective effect of the vacuole.

Uptake of nutrients

Plasma membrane transport systems are exposed to the external environment and might, therefore, be sensitive targets for inhibition by high salinity. Consistent with this notion, Varela and co-workers (1992) observed that, at high concentrations of NaCl, *ura3* mutants of *S. cerevisiae* grow more slowly in a uracil-containing medium than the corresponding parental strain, implying impaired uptake of uracil under saline conditions. These authors also noted a significant reduction in the rate of uptake of methionine in cells shifted to high salinity. At 1.4 M NaCl, uptake was maintained below 10% of a control value, while at 0.7 M uptake recovered to control levels after about 2 h. Glucose is transported by facilitated diffusion in *S. cerevisiae* with dual affinities (Serrano, 1991). Using D-glucosamine as substrate, Lindman (1981) noted a three-fold increase of K_m for the high affinity uptake system (from 6 to 16 mM), as the salinity of the growth medium was increased from 0 to 0.7 M NaCl, while the V_{max} decreased by 30%. Hence, transport rates decrease at increased salinity and might become growth limiting. Gläser *et al.* (1993) isolated the *HAL2* gene which on increased dosage improves NaCl tolerance of *S. cerevisiae*. *HAL2* was found to be allelic to *MET22* and encodes a putative inositol phosphatase involved in methionine biosynthesis. Consistent with this observation, the authors demonstrated that fully prototrophic wild-type cells grew to a higher cell density in NaCl or LiCl media if methionine was added to 30 μg ml^{-1}. Since inositol phosphatases are sensitive to lithium salts it was suggested methionine biosynthesis becomes growth limiting by salt inhibition at high Li$^+$ or Na$^+$ concentrations.

Osmoregulation and osmotic signal transduction

The requirement for maintenance of a low and relatively constant ion concentration in the cytosol requires intracellular accumulation of alternative osmotica to sustain appropriate cell volume, metabolite concentration and turgor in saline environments. Fungi meet this demand by the production and accumulation of polyhydric alcohols (polyols) which, owing to their compatible nature, can be intracellularly accumulated to high concentrations without greatly disrupting enzyme functions (Brown, 1978). Polyols commonly accumulated by fungi include glycerol, erythritol, inositol, arabinitol, ribitol, xylitol and mannitol (Pfyffer, Pfyffer & Rast, 1986; Blomberg & Adler, 1992). In fungi subjected to salt stress, glycerol is normally the most responsive polyol, becoming increasingly predominant as the stress becomes more severe. A typical pattern of polyol accumulation is shown by the salt-tolerant yeast *D. hansenii*. The steady-state intracellular glycerol concentration in growing cells of this yeast is more or less linearly proportional to the salt concentration of the growth medium, and for cells cultured at 1.35 M NaCl, internal glycerol counteracts 50 to 60 % of the external osmotic potential (Larsson *et al.*, 1990). Arabinitol, the second polyol produced by *D. hansenii* is much less responsive to external salinity but becomes the predominant polyol in cells of ageing cultures, as glycerol decreases sharply in cultures entering the stationary phase (Adler & Gustafsson, 1980). The spectrum of higher polyols present in filamentous fungi is more varied than in yeasts (Pfyffer *et al.*, 1986; Meikle *et al.*, 1991) and the generalized response to salt-stress, as observed for *N. crassa* and species of *Aspergillus* and *Penicillium*, involves substantial accumulation of glycerol together with a less pronounced build up of a second polyol such as erythritol (Adler, Pedersen & Tunblad-Johansson, 1982; Beever & Laracy, 1986) or mannitol (Ellis, Grindle & Lewis, 1991). In filamentous fungi the accumulation of glycerol characteristically also occurs in growing cells, while in ageing cultures glycerol gradually disappears (Hocking, 1986).

Considering the diversity within the Fungal Kingdom, exceptions to the rule of glycerol as the primary compatible solute are to be expected, and recently demonstrated for yeast-like fungi belonging to the genera *Geotrichum* and *Endomyces* in which arabinitol or mannitol have taken the role as the main compatible solute (Luxo, Nobre & da Costa, 1993). In *S. cerevisiae*, on the other hand, glycerol serves as the exclusive compatible solute. Osmotic adaptation by means of glycerol accumulation is, however, a highly non-conservative process in *S. cerevisiae*. Salt-stress induces a strongly increased production, but most of the synthesised glycerol is lost to the surrounding medium (Brown, 1978). In contrast, osmotolerant yeasts like *Z. rouxii* and *D. hansenii* are capable of retaining a high proportion of the produced glycerol. In saline media, these yeasts may maintain a 1000 to 10 000-fold concentration gradient across the membrane. This efficient retention implies restricted leakage through the membrane or presence of a transport system that carries lost glycerol back into the cell. Hosono (1992) found that the overall unsaturation index of the acyl lipids in the plasma membrane of *Z. rouxii* decreased from 1.37 to 1.18 when the salinity of the growth medium was increased from 0 to 2.6 M NaCl, the most conspicuous changes being a decrease of linoleic acid (18:2) and a concomitant increase in oleic acid (18:1). Interestingly he also reported a three-fold increase in ergosterol in the membranes of salt-grown cells, resulting in an about five-fold increase in the sterol/phospholipid ratio. All these changes point to decreased fluidity and glycerol permeability of the membrane. More importantly, however, a transport system with specificity for glycerol has been identified in *D. hansenii* (Adler, Blomberg & Nilsson, 1985) and *Z. rouxii* (Van Zyl, Kilian & Prior, 1990). The energy required for the uptake of glycerol appears to be provided by the Na^+ gradient across the membrane which drives the glycerol uptake via a Na^+/glycerol symporter (Lucas, da Costa & van Uden, 1990; Van Zyl *et al.*, 1990). In *S. cerevisiae* the kinetics for glycerol uptake has been interpreted to indicate transport by passive diffusion (Blomberg & Adler, 1992). Recent recognition of the *FPS1* gene, encoding a *S. cerevisiae* homologue of the MIP family of channel-forming integral membrane proteins (Van Aelst *et al.*, 1991), may provide molecular insight into the glycerol movement across the membrane of this organism. The MIP family encompasses two bacterial glycerol facilitators, which thus opens the possibility that FPS1 may conduct channel-mediated glycerol transport. Exposure of *S. cerevisiae* grown at 0.7 M NaCl to hypotonic medium leads to rapid release of internally accumulated glycerol, while transfer of cells from basal medium to 0.7 M NaCl results in a decreased rate of glycerol efflux and an increased rate of intracellular glycerol accumulation (Blomberg & Adler, 1989). The presence of stretch-activated channels in the plasma membrane, which open by an outward osmotic pressure to selectively release glycerol, would explain this behaviour. Ion-conducting membrane channels gated by mechanical forces have been described in *S. cerevisiae* (Gustin *et al.*, 1988).

Glycerol is produced by diversion of the carbon flow from the glycolytic pathway to glycerol biosynthesis (Blomberg & Adler, 1992). The enzyme positioned at the branch point is NAD^+ dependent glycerol 3-phosphate dehydrogenase (GPD), which reduces dihydroxyacetone phosphate to glycerol 3-phosphate (G3P). The subsequent hydrolysis of G3P, to release glycerol and phosphate, is mediated by a phosphatase activity. In a screen for salt-sensitive mutants of *S. cerevisiae* with defective glycerol accumulation, isolates falling into four complementation groups (*OSG1-OSG4*) were isolated (Larsson *et al.*, 1993). By molecular complementation of the *osg1-1* mutant the *GPD1* gene, encoding cytoplasmic GPD was cloned. Cells containing a *gpd1* null allele, generated by deleting most of the *GPD1* open reading frame, exhibited a dramatic decrease of GPD activity and glycerol production resulting in an about 50% decrease in salt tolerance (R. Ansell, unpublished observations). Consistent with the existence of residual glycerol production in the *GPD1* null mutant, a second gene has been identified, designated *GPD2* (Eriksson *et al.*, 1995). Expression of reporter gene activity from the *GPD1* and *GPD2* promoters demonstrated that the expression of *GPD1*, but not *GPD2*, was induced by increased external osmolarity (Eriksson *et al.*, 1995).

Using strains essentially devoid of non-specific phosphatases, enzyme activity with high specificity for glycerol 3-phosphate (GPase), was demonstrated in *S. cerevisiae* (A-K. Påhlman *et al.*, unpublished observations). This GPase activity is always present in the cell but increases as the external osmolarity is increased. This osmotically induced increase is dependent on protein synthesis, since it is inhibited by cycloheximide. Hence, it is not unreasonable to assume that another osmo-induced gene(s) might be found behind this activity.

Taking the mutational approach to identify genes in the osmotically-induced glycerol response, Brewster and co-workers (1993) identified the *HOG1* and *HOG4* genes in a screen for mutants that could not grow on plates containing 0.9 M NaCl and failed to accumulate intracellular glycerol to wild-type levels. *HOG1* was shown to be a MAP kinase homologue, while *HOG4* proved identical to a previously identified gene, *PBS2* (Boguslawski, 1992), a putative MAP kinase activator. The MAP kinases comprise a family of protein kinases that are highly conserved in eukaryotic organisms, and link receptor activation to gene activation via a series of intracellular phosphorylation events (Neiman, 1993). *HOG1* and *PBS2* seem to function in the same genetic pathway, since null mutants for each gene or both exhibit the same phenotype: osmosensitivity and defective glycerol accumulation (Brewster *et al.*, 1993). Null mutants of *HOG1* or *PBS2* also suffered strong morphological aberrations at high salinity. It is unclear how this defect is linked to osmosensitivity, since a suppressor of the *pbs2* mutation repressed the morphology defects without having a large effect on salt sensitivity (Boguslawski, 1992). Using anti-phosphotyrosine antibodies, it was demonstrated that the *HOG1* gene product is tyrosine phosphorylated in a *PBS2*-dependent manner, only a minute after exposure of cells to 0.4 M NaCl (Brewster *et al.*, 1993). These findings strongly suggest the identification of an osmosensing signal transduction pathway that transmits signals in response to changes in external osmolarity. The phenotype of the signalling defective mutants indicates that the glycerol producing system is one of the targets for this pathway. In agreement with this proposition, null mutants of *HOG1* or *PBS2* exhibit strongly decreased expression of *GPD1*, as analysed by a *GPD1* reporter gene fusion, and blocked osmotic induction of the GPase activity (P. Eriksson & A-K. Påhlman, unpublished observations). As measured by tyrosine phosphorylation of HOG1, the transmission of the osmotic signal to downstream targets increased as cells were shifted to increased salinity to reach a maximum at 300 mM NaCl (Brewster *et al.*, 1993). This salinity corresponds to an osmotic potential (−1.4 MPa) similar to that of cells grown in basal medium (−1.5 MPa). Hence, the signal appears to increase as the osmotic gradient across the membrane decreases. What might be the primary osmosensory step in this signal translation? Considering the remarkable conservation of MAP cascades (Neiman, 1993), clues might come from the much better studied kinase cascade of the mating type response, by which peptide hormones elicit the switch from vegetative growth to the sexual cycle (Marsh, Neiman & Herskowitz, 1991). This signal is initiated by the binding of pheromone to its receptor. An attractive hypothesis advanced by Brewster and others (1993) suggests that a cell wall component may serve as a ligand for a plasma membrane receptor, triggering signal transmission as turgor pressure or membrane stretching modifies the interaction between ligand and receptor.

A third MAP kinase cascade identified in *S. cerevisiae* is mediated by protein kinase C (Errede & Levin, 1993). Mutations that block this pathway weaken the yeast cell wall, leading to bud lysis at low external osmolarity. There are no indications as yet, however, that this pathway is involved in cellular osmoregulation.

Conclusions

The response of a fungus to salt stress involves the integrated function of diverse capacities. The intracellular physiology is protected from the external salinity primarily by an effective exclusion of Na$^+$ and by a compensatory accumulation of polyols, mainly glycerol, to achieve an internal environment that is suitable for enzyme function and growth under salt stress. Mutational approaches have permitted significant advances in the understanding of the ion transport mechanism that regulates cation levels in the cytosol, and revealed part of an osmosensing signal transduction pathway in *S. cerevisiae* (Fig. 14.1). The increasing interest in genetic and physiological analysis of mechanisms involved in

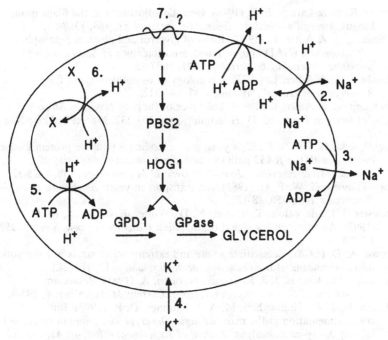

Fig. 14.1. Diagram of cellular functions involved in salt tolerance of fungi. (1) H^+-ATPase; (2) Na^+/H^+ antiporter; (3) Na^+-motive ATPase; (4) K^+ transporter; (5) vacuolar H^+-ATPase; (6) H^+ antiport; (7) osmosensing MAP kinase cascade controlling glycerol accumulation. See text for further details.

fungal salt- and osmotolerance promise rapidly-advancing insights into this exciting field.

References

Adler, L., Blomberg, A. & Nilsson, A. (1985). Glycerol metabolism and osmoregulation in the salt-tolerant yeast *Debaryomyces hansenii*. *Journal of Bacteriology*, **162**, 300-6.

Adler, L. & Gustafsson, L. (1980). Polyhydric alcohol production and intracellular amino acid pool in relation to halotolerance of the yeast *Debaryomyces hansenii*. *Archives of Microbiology*, **124**, 123-30.

Adler, L., Pedersen, A. & Tunblad-Johansson, I. (1982). Polyol accumulation by two filamentous fungi grown at different concentrations of NaCl. *Physiologia Plantarum*, **56**, 139.

Banta, L. M., Robinson, J. S., Klionsky, D. J. & Emr, S. D. (1988). Organelle assembly in yeast: characterization of yeast mutants defective in vacuolar biogenesis and protein sorting. *Journal of Cell Biology*, **107**, 1369-83.

Beever, R. E. & Laracy, E. P. (1986). Osmotic adjustment in the filamentous fungus *Aspergillus nidulans*. *Journal of Bacteriology*, **168**, 1358-65.

Blomberg, A. & Adler, L. (1989). Roles of glycerol and glycerol-3-phosphate dehydrogenase (NAD$^+$) in acquired osmotolerance of *Saccharomyces cerevisiae*. *Journal of Bacteriology*, **171**, 1087-92.

Blomberg, A. & Adler, L. (1992). Physiology of osmotolerance in fungi. *Advances in Microbial Physiology*, **33**, 145-212.

Blomberg, A. & Adler, L. (1993). Tolerance of fungi to NaCl, In *Stress Tolerance in Fungi*, ed. D. H. Jennings, pp. 209-232. Marcel Dekker: New York.

Boguslawski, G. (1992). *PBS2*, a yeast gene encoding a putative protein kinase, interacts with the *RAS2* pathway and affects osmotic sensitivity of *Saccharomyces cerevisiae*. *Journal of General Microbiology*, **138**, 2425-32.

Borst-Pauwels, G. W. F. H. (1981). Ion transport in yeast. *Biochimica et Biophysica Acta*, **650**, 88-127.

Brewster, J. L., de Valoir, T., Dwyer, N. D., Winter, E. & Gustin, M. C. (1993). An osmosensing signal transduction pathway in yeast. *Science*, **259**, 1760-3.

Brown, A. D. (1978). Compatible solute and extreme water stress in eukaryotic micro-organisms. *Advances in Microbial Physiology*, **17**, 181-242.

Camacho, M., Ramos, J. & Rodriguéz-Navarro, A. (1981). Potassium requirement of *Saccharomyces cerevisiae*. *Current Microbiology*, **6**, 295-9.

Clipson, N. J. W., Hajibagheri, M. A. & Jennings, D. H. (1990). Ion compartmentation in the marine fungus *Dendryphiella salina* in response to salinity X-ray-microanalysis. *Journal of Experimental Botany*, **41**, 190-202.

Clipson, N. J. W. & Jennings, D. H. (1993). *Dendryphiella salina* and *Debaryomyces hansenii*: models for ecophysiological adaptation to salinity by fungi that grow in the sea. *Canadian Journal of Botany*, **70**, 2097-105.

Cornelius, G. & Nakashima, H. (1984). Vacuoles play a decisive role in calcium homeostasis in *Neurospora crassa*. *Journal of General Microbiology*, **133**, 2341-7.

Ellis, S. W., Grindle, M., & Lewis, D. H. (1991). Effect of osmotic stress on yield and polyol content of dicarboximide-sensitive and -resistant strains of *Neurospora crassa*. *Mycological Research*, **95**, 457-64.

Epstein, E., Norlyn, J. D., Rush, D. W., Kingsbury, R. W., Kelley, D. B., Cunningham, G. A. & Wrona, A. F. (1980). Saline culture of crops: a genetic approach. *Science*, **210**, 399-404.

Eriksson, P., Andre, L., Ansell, R., Blomberg, A. & Adler, L. (1995). Molecular cloning of *GPD2*, a second gene encoding *sn*-glycerol 3-phosphate dehydrogenase (NAD$^+$) in *Saccharomyces cerevisiae*. *Molecular Microbiology* (in press).

Errede, B. & Levin, D. E. (1993). A conserved kinase cascade for MAP kinase activation in yeast. *Current Opinion in Cell Biology*, **5**, 254-60.

Gaber, R. F., Styles, C. A. & Fink, G. R. (1988). *TRK1* encodes a plasma membrane protein for high-affinity potassium transport in *Saccharomyces cerevisiae*. *Molecular and Cellular Biology*, **8**, 2848-59.

Gadd, G. M., Chudek, J. A., Foster, R. & Reed, R. H. (1984). The osmotic responses of *Penicillium ochro-chloron*: changes in internal solute levels in response to copper and salt stress. *Journal of General Microbiology*, **130**, 1969-75.

Garciadeblas, B., Rubio, F., Quintero, F. J., Banuelos, M. A., Haro, R. & Rodriguéz-Navarro, A. (1992). Differential expression of two genes encoding isoforms of the ATPase involved in sodium efflux in *Saccharomyces cerevisiae*. *Molecular and General Genetics*, **236**, 363-8.

Gaxiola, R., de Larrinoa, I. F., Villalba, J. M. & Serrano, R. (1992). A novel and conserved salt-induced protein is an important determinant of salt tolerance in yeast. *EMBO Journal*, **11**, 3157-64.

Ghislain, M., Schlesser, A. & Goffeau, A. (1987). Mutation of a conserved glycine residue modifies the vanadate sensitivity of the plasma membrane H^+-ATPase from *Schizosaccharomyces pombe*. *Journal of Biological Chemistry*, **262**, 17549-55.

Gläser, H.-U., Thomas, D., Gaxiola, R., Montrichard, F., Surdin-Kerjan, Y. & Serrano, R. (1993). Salt tolerance and methionine biosynthesis in *Saccharomyces cerevisiae* involve a putative phosphatase gene. *EMBO Journal*, **12**, 3105-10.

Griffin, D. M. & Luard, E. J. (1979). Water stress and microbial ecology. In *Strategies of Microbial Life in Extreme Environments*, ed. M. Shilo, pp. 49-63. Verlag Chemie:Weinheim.

Gustin, M. C., Martinac, B., Saimi, Y., Culbertson, M. R. & Kung, C. (1986). Ion channels in yeast. *Science*, **233**, 1195-97.

Gustin, M. C., Zhou, X.-L., Martinac, B. & Kung, C. (1988). A mechanosensitive ion channel in the yeast plasma membrane. *Science*, **242**, 762-5.

Hager, K. M., Mandala, S. M., Davenport, J. W., Speicher, D. W., Benz, E. J. J. & Slayman, C. W. (1986). Amino acid sequence of the plasma membrane ATPase of *Neurospora crassa*: deduction from genomic and cDNA sequences. *Proceedings of the National Academy of Sciences, USA*, **83**, 7693-7.

Haro, R., Banuelos, M. A., Quintero, F. J., Rubio, F. & Rodriguéz-Navarro, A. (1993). Genetic basis of sodium exclusion and sodium tolerance in yeast. A model for plants. *Physiologia Plantarum*, **89**, 868-74.

Haro, R., Garciadeblas, B. & Rodriguéz-Navarro, A. (1991). A novel P-type ATPase from yeast involved in sodium transport. *FEBS Letters*, **291**, 189-91.

Hocking, A. D. (1986). Effects of water activity and culture age on the glycerol accumulation patterns of five fungi. *Journal of General Microbiology*, **132**, 269-75.

Hosono, K. (1992). Effect of salt stress on lipid composition and membrane fluidity on the salt-tolerant yeast *Zygosaccharomyces rouxii*. *Journal of General Microbiology*, **138**, 91-6.

Jia, Z.-P., McCullough, N., Martel, R., Hemmingsen, S. & Young, P. G. (1992). Gene amplification at a locus encoding a putative Na^+/H^+ antiporter confers sodium and lithium tolerance in fission yeast. *EMBO Journal*, **11**, 1631-40.

Klionsky, D. J., Herman, P. K. & Emr, S. D. (1990). The fungal vacuole: composition, function, and biogenesis. *Microbiological Reviews*, **54**, 266-92.

Ko, C. H. & Gaber, R. F. (1991). *TRK1* and *TRK2* encode structurally related K^+ transporters in *Saccharomyces cerevisiae*. *Molecular and Cell Biology*, **11**, 4266-73.

Ko, C. H., Buckley, A. M. & Gaber, R. F. (1990). *TRK2* is required for low affinity K^+ transport in *Saccharomyces cerevisiae*. *Genetics*, **125**, 305-12.

Kouyeas, V. (1964). An approach to the study of moisture relations of soil fungi. *Plant and Soil*, **20**, 351-63.

Larsson, C., Morales, C., Gustafsson, L. & Adler, L. (1990). Osmoregulation of the salt-tolerant yeast *Debaryomyces hansenii* grown in a chemostat at different salinities. *Journal of Bacteriology*, **172**, 1769-74.

Larsson, K., Eriksson, P., Ansell, R. & Adler, L. (1993). A gene encoding *sn*-glycerol 3-phosphate dehydrogenase (NAD$^+$) complements an osmosensitive mutant of *Saccharomyces cerevisiae*. *Molecular Microbiology*, **10**, 1101-11.

Latterich, M. & Watson, M. D. (1991). Isolation and characerization of osmosensitive vacuolar mutants of *Saccharomyces cerevisiae*. *Molecular Microbiology*, **5**, 2417-26.

Latterich, M. & Watson, D. M. (1993). Evidence for a dual osmoregulatory mechanism in the yeast *Saccharomyces cerevisiae*. *Biochemical and Biophysical Research Communications*, **191**, 1111-17.

Lindman, B. (1981). On sugar uptake and halotolerance in the yeast *Debaryomyces hansenii*. PhD Thesis, University of Göteborg, Sweden.

Luard, E. & Griffin, D. M. (1981). Effect of water potential on fungal growth and turgor. *Transactions of the British Mycological Society*, **76**, 33-40.

Lucas, C., da Costa, M. & van Uden, N. (1990). Osmoregulatory active sodium- glycerol co-transport in the halotolerant yeast *Debaryomyces hansenii*. *Yeast*, **6**, 187-91.

Luxo, C., Nobre, M. F. & da Costa, M. S. (1993). Intracellular polyol accumulation by yeastlike fungi of the genera *Geotrichum* and *Endomyces* in response to water stress (NaCl). *Canadian Journal of Microbiology*, **39**, 868-73.

Mager, W. H. & Varela, J. C. S. (1993). Osmostress response of the yeast *Saccharomyces*. *Molecular Microbiology*, **10**, 253-8.

Marsh, L., Neiman, A. M. & Herskowitz, I. (1991). Signal transduction during pheromone response in yeast. *Annual Review of Cell Biology*, **7**, 699-728.

Meikle, A. J., Chudek, J. A., Reed, R. H. & Gadd, M. G. (1991). Natural abundance 13C-nuclear magnetic resonance spectroscopic analysis of acyclic polyol and trehalose accumulation by several yeast species in response to salt stress. *FEMS Microbiology Letters*, **82**, 163-8.

Neiman, A. M. (1993). Conservation and reiteration of a kinase cascade. *Trends in Genetics*, **9**, 390-5.

Nishi, T. & Yagi, T. (1992). A transient and rapid activation of plasma-membrane ATPase during the initial stages of osmoregulation in the salt tolerant yeast *Zygosaccharomyces rouxii*. *FEMS Microbiology Letters*, **99**, 95-100.

Norkrans, B. & Kylin, A. (1969). Regulation of the potassium to sodium ratio and of the osmotic potential in relation to salt tolerance in yeasts. *Journal of Bacteriology*, **100**, 836-45.

Ohsumi, Y. & Anraku, Y. (1983). Calcium transport driven by a proton motive force in vacuolar membrane vesicles of *Saccharomyces cerevisiae*. *Journal of Biological Chemistry*, **258**, 5614-17.

Okorokov, L. A., Lichko, L. P. & Kulaev, I. S. (1980). Vacuoles: main compartments of potassium, magnesium, and phosphate ions in *Saccharomyces carlsbergensis* cells. *Journal of Bacteriology*, **144**, 661-5.

Ölz, R., Larsson, K., Adler, L. & Gustafsson, L. (1993). Energy flow and osmoregulation of *Saccharomyces cerevisiae* grown in a chemostat under NaCl stress. *Journal of Bacteriology*, **175**, 2205-13.

Onishi, H. (1963). Osmophilic yeasts. *Advances in Food Research*, **12**, 53-90.

Perkins, J. & Gadd, G. M. (1993). Accumulation and intracellular compartmentation of lithium ions in *Saccharomyces cerevisiae*. *FEMS Microbiology Letters*, **107**, 255-60.

Pfyffer, G. E., Pfyffer, B. U. & Rast, D. M. (1986). The polyol pattern, chemotaxonomy, and phylogeny of the fungi. *Sydowia*, **39**, 160.

Pitt, J. I. & Hocking, A. D. (1977). Influence of solute and hydrogen ion concentration on the water relations of some xerophilic fungi. *Journal of General Microbiology*, **101**, 35-40.

Ramos, J., Alijo, R., Haro, R. & Rodriguéz-Navarro, A. (1994). *TRK2* is not a low-affinity potassium transporter in *Saccharomyces cerevisiae*. *Journal of Bacteriology*, **176**, 249-52.

Ramos, J., Haro, R. & Rodriguéz-Navarro, A. (1990). Regulation of potassium fluxes in *Saccharomyces cerevisiae*. *Biochimica et Biophysica Acta*, **1029**, 211-17.

Rodriguéz-Navarro, A., Blatt, M. R. & Slayman, C. L. (1986). A potassium-proton symport in *Neurospra crassa*. *Journal of General Physiology*, **87**, 649-74.

Rodriguéz-Navarro, A. & Ortega, D. (1982). The mechanism of sodium efflux in yeast. *FEBS Letters*, **138**, 205-8.

Rodriguéz-Navarro, A. & Ramos, J. (1984). Dual system for potassium transport in *Saccharomyces cerevisiae*. *Journal of Bacteriology*, **159**, 9440-5.

Rudolph, H. K., Antebi, A., Fink, G. R., Buckley, C. M., Dorman, T. E., LeVitre, J., Davidow, L. S., Mao, J. & Moir, D. T. (1989). The yeast secretory pathway is perturbed by mutations in *PMR1*, a member of a Ca^{2+} ATPase family. *Cell*, **58**, 133-45.

Ryan, J. P. & Ryan, H. (1972). The role of intracellular pH in the regulation of cation exchanges in yeast. *Biochemical Journal*, **128**, 139-46.

Sanders, D. (1988). Fungi. In *Solute Transport in Plant Cells and Tissues*, ed. D.A. Baker, & J.L Hall, pp. 106-165. Longman Scientific & Technical: Harlow, Essex, UK.

Serrano, R. (1991). Transport across yeast vacuolar and plasma membrane. In *The molecular and cellular biology of the yeast Saccharomyces: genome dynamics, protein synthesis, and energetics*, Vol. 1, ed. J. R. Broach, J. R. Pringle, & E. W. Jones, pp. 523-585. Cold Spring Harbor Laboratory Press: Cold Spring Harbor.

Slayman, C. W. & Tatum, E. L. (1964). Potassium transport in *Neurospora*. I. Intracellular sodium and potassium concentrations, and cation requirements for growth. *Biochimica et Biophysica Acta*, **88**, 578-92.

Stewart, L. M. D. & Booth, I. R. (1983). Na^+-involvement in proline transport in *Escherichia coli*. *FEMS Microbiology Letters*, **19**, 161-4.

Tokuoka, K. (1993). Sugar- and salt-tolerant yeasts. *Journal of Applied Bacteriology*, **74**, 101-10.

Vallejo, C. G. & Serrano, R. (1989). Physiology of mutants with reduced expression of plasma membrane H^+-ATPase. *Yeast*, **5**, 307-19.

Van Aelst, L., Hohmann, S., Zimmermann, K. F., Jans, A. W. H. & Thevelein, J. M. (1991). A yeast homologue of the bovine lens fibre MIP gene family complements the growth defect of a *Saccharomyces cerevisiae* mutant on fermentable sugars but not its defect in glucose-induced RAS-mediated cAMP signalling. *EMBO Journal*, **10**, 2095-104.

Van Zyl, P. J., Kilian, S. G. & Prior, B. A. (1990). The role of an active transport mechanism in glycerol accumulation during osmoregulation by

234   L. Adler

Zygosaccharomyces rouxii. Applied Microbiology and Biotechnology, **34**, 231-5.
Varela, J. C. S., van Beekvelt, C., Planta, R. J. & Mager, W. H. (1992). Osmostress-induced changes in yeast gene expression. *Molecular Microbiology*, **6**, 2183-90.
Watanabe, Y., Shiramizu, M. & Tamai, Y. (1991). Molecular cloning and sequencing of plasma membrane H + ATPase gene from the salt tolerant yeast *Zygosaccharomyces rouxii. Journal of Biochemistry*, **110**, 237-40.
Watanabe, Y. & Tamai, Y. (1992). Inhibition of cell growth in *Zygosaccharomyces rouxii* by proton-ionophore and plasma membrane ATPase inhibitor in the presence of high concentration of sodium chloride. *Bioscience, Biotechnology and Biochemistry*, **56**, 342-3.
Watanabe, Y., Sanemitsu, Y. & Tamai, Y. (1993). Expression of plasma membrane proton ATPase gene in salt-tolerant yeast *Zygosaccharomyces rouxii* is induced by sodium chloride. *FEMS Microbiology Letters*, **114**, 105-8.
Wheeler, K., Hocking, A. & Pitt, J. I. (1988). Influence of temperature on the water relations of *Polypaecilum pisce* and *Basipetospora halophila*, two halophilic fungi. *Journal of General Microbiology*, **134**, 2255-61.
White, C. & Gadd, G. M. (1987). The uptake and cellular distribution of zinc in *Saccharomyces cerevisiae. Journal of General Microbiology*, **133**, 727-37.
Wiggins, P. M. (1990). Role of water in some biological processes. *Microbiological Reviews*, **54**, 432.
Yang, Y., Bastos, M. & Chen, K. Y. (1993). Effect of osmotic stress and growth stage on cellular pH and polyphosphate metabolism in *Neurospora crassa* as studied by ^{31}P nuclear magnetic resonance spectroscopy. *Biochimica et Biophysica Acta*, **1179**, 141-7.

15

Fungal sequestration, mobilization and transformation of metals and metalloids

G. F. MORLEY, J. A. SAYER,
S. C. WILKINSON, M. M. GHARIEB
AND G. M. GADD

Introduction

The soil environment is complex and infinitely variable (Metting, 1992). The fate of metal contaminants is similarly diverse and dependent on many factors, such as mineral composition and organic content (Krosshavn, Steinnes & Varskog, 1993), and is mediated by physico-chemical processes (Zhu & Alva, 1993). Metal contaminants in the soil undergo complex interactions with both organic and inorganic components, and many studies have shown that the basic inorganic alumino-silicate clays, silts, sands and other mineral components are important metal sequestrants (Farrah & Pickering, 1976, 1977; Kuo & Baker, 1980; Harter, 1983; Krosshavn et al., 1993). It is important to recognize, however, that the soil organic component, which contains both living organisms and their decay or metabolic products, also exerts a strong influence on metal retention.

Fungi exist in soils primarily as saprotrophic degraders of organic matter, and also as pathogens of plants and in mycorrhizal associations with plant root systems. Fungi are ubiquitous soil microorganisms, predominant in acidic soils, often comprising the largest pool of biomass (including bacteria, microalgae, actinomycetes, protozoa, nematodes, earthworms and other invertebrates) and organic products under these conditions (Metting, 1992). This, combined with their high surface area to mass ratio, ensures that fungal–metal interactions are of primary importance in the organic soil environment (see Colpaert & Van Tichelin, Chapter 9). This is especially true of acidic soil conditions where metals are more likely to be speciated into soluble and more mobile forms (Hughes & Poole, 1991) and where metal ion/fungal/mineral interactions are more likely to occur due to the predominance of the fungal component of the biota in such soils. In the soil, fungi are important as both active sequestrants of metal ions, via cellular and physiological mechanisms and as passive sequestrants, by sorption to cellular components and by-products of both living and dead fungal biomass.

Many fungi are able to solubilize metals and metalloids from insoluble compounds such as ores, metal phosphates, sulphides and oxides (Burgstaller & Schinner, 1993). Fungi can also be important agents of corrosion, certain *Penicillium* spp. being able to remove iron from stainless iron–nickel–chromium alloys (Siegel, Galun & Siegel, 1990). The solubilization of essential metals by the soil microflora, including fungi and bacteria, is important for the nutrition of plants and other organisms, and is an important component of natural biogeochemical cycles. Most phosphate fertilizers are made insoluble in the soil by precipitation with metal cations of aluminium and iron in acidic soils, and calcium in calcareous soils (Jones *et al.*, 1991). These insoluble metal phosphates need to be solubilized to become biologically available. It follows that inessential and toxic metal compounds may be solubilized in the same way and become available to fungi and other soil organisms and plants. Fungal solubilization of metals may also have biotechnological potential for metal recovery from industrial by-products and low grade ores, and the detoxification of sewage sludge and other solid wastes. The two most important genera appear to be *Aspergillus* and *Penicillium*, of which *Aspergillus niger* and *Penicillium simplicissimum* have received most attention (Burgstaller & Schinner, 1993).

Several species of fungi, including both filamentous forms and unicellular yeasts, have been shown to carry out transformations of metals, metalloids and organometallic compounds. Such transformations include reduction, methylation and dealkylation (Table 15.1). These processes are of environmental importance in biogeochemical cycles, since transformation of a metal or metalloid may modify its mobility and toxicity. Methylation of various metal(loid)s such as selenium and mercury increases their volatility, resulting in atmospheric dispersal. Methyl derivatives of metals, for example, methylmercury, also exhibit increased lipid solubility which facilitates their passage across biological membranes. These compounds have attracted great interest because of their toxicity and biomagnification in food chains (Thayer, 1984, 1988). Methyl derivatives of selenium, however, are much less toxic than inorganic forms (Wilber, 1980). Reduction of oxyanions of metalloids, such as selenate or selenite to amorphous selenium, can result in their immobilization and detoxification (Ramadan *et al.*, 1988). Dealkylation of

Table 15.1. *Transformations of metals, metalloids and organometals by fungi*

Transformation	Fungal species	Reference
Reduction		
Se(IV)/Se(VI) → Se(0)	*Fusarium* sp.	Ramadan *et al.* (1988), Gharieb (1993), Gharieb *et al.* (1995)
Te(IV)/Te(VI) → Te(0)	*Schizosaccharomyces pombe*	Smith (1974)
Hg(II) → Hg(0)	*Candida albicans, Saccharomyces cerevisiae*	Yannai *et al.* (1991)
Ag(I) → Ag(0)	*Debaryomyces hansenii, C. albicans, S. cerevisiae, Rhodotorula rubra, Aureobasidium pullulans*	Kierans *et al.* (1991)
Cu(II) → Cu(I)	*D. hansenii*	Wakatsuki *et al.* (1988, 1991a,b)
Fe(III) → Fe(II)	*S. cerevisiae*	Lesuisse & Labbe (1989, 1992)
Methylation		
Se(IV)/Se(VI) → $(CH_3)_2$Se	*Alternaria alternata*	Thompson-Eagle *et al.* (1991)
Se(VI) → $(CH_3)_2$Se	*Acremonium falciforme*	Chasteen *et al.* (1990)
Se(VI) → $(CH_3)\widetilde{S}e_2$	*Penicillium citrinum*	
As(III) → $(CH_3)_3$As	*Scopulariopsis brevicaulis*	Challenger (1945)
$CH_3AsH_2O_3$ → $(CH_3)_3$As methylarsonic acid	*Penicillium* sp.	Huysmans & Frankenberger (1991)
Te(IV)/ Te(VI) → $(CH_3)_2$Te	*Penicillium* sp.	Huysmans & Frankenberger (1991)
Dealkylation		
(bis)tributyltin → di-, monobutylin, Sn(II)	Several	Barug (1981), Orsler & Holland (1992), Macaskie & Dean (1990), Gadd (1993b)

organotin compounds also results in a decrease in toxicity, a process relevant to fungal degradation of organometal(loid)-based fungicides (Blunden & Chapman, 1982).

Mechanisms of sequestration

Metal uptake by fungi can be described simply as a two-phase process. The first phase is metabolism independent and occurs whether the biomass is dead or alive, in an active or inactive metabolic state. The second,

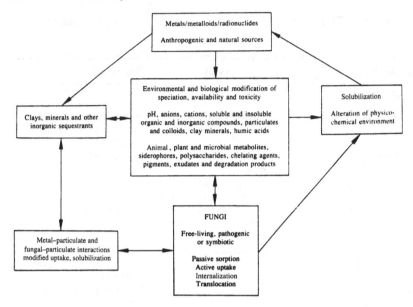

Fig. 15.1. Diagrammatic representation of the major mechanisms involved in fungal sequestration of metals and metalloids in the natural environment. (See Gadd, 1993a for a review.)

metabolism-dependent phase comprises processes which sequester, internalize and possibly transform the metal as well as including the production of extracellular metabolites which may interact with the metal species in the immediate environment.

Fungi can be regarded therefore as both simple passive sorbents (they and their components, extracellular products and decay detritus act as simple metal-sorbing organic surfaces) on to which metals, especially ionic metal species, sorb by passive physico-chemical processes. Living fungal biomass also interacts with the metal species by physiological mechanisms. Fig. 15.1 is a schematic representation of the physico-chemical and physiological interactions of fungi, and their metabolic products, with toxic metal species in the soil environment.

Physico-chemical mechanisms of sequestration of metals by fungi

Fungal cell walls are complex three-dimensional structures of organic macromolecules, predominantly chitins, chitosans and glucans but also containing proteins, lipids and other polysaccharides (Peberdy, 1990). This variety of structural components contains many different functional groups each with their own charge distribution and therefore able to bind metal ions to a greater or a lesser extent. Uptake of metals by fungal biomass corresponds to models with a multiplicity of non-equivalent, heterogeneous binding sites (Tobin, Cooper & Neufeld, 1990). Loci on fungal cell walls can act as precipitation nuclei, with precipitation of metals occurring in and around cell wall components. Studies have shown passive accumulation of metals, especially radionuclides, occurring in this manner with crystals of elements such as uranium and thorium precipitating around bound molecules of the metal resulting in structures which can be easily visualized by microscopy (Tsezos & Volesky, 1982*a*, *b*). Fungi are also responsible for the production of many extracellular products such as chelating and sequestering agents (for example, citric acid, siderophores), precipitating agents (for example, oxalic acid, which can precipitate metal oxalates, and hydrogen sulphide, which can precipitate metal sulphides), and pigments with metal binding abilities such as melanins (Gadd & White, 1989).

Adsorption of metals to fungal biomass has often been characterized according to various models of adsorbent/adsorbate interaction such as Langmuir, Freundlich and Brunauer–Emmett–Teller isotherm models and Scatchard plots (Langmuir, 1918; Freundlich, 1926; Brunauer, Emmett & Teller, 1938; Scatchard, 1949). Although the application of such models to biological systems is a simplistic approach, and often open to criticism (Harter & Baker, 1977), it is a useful way of differentiating and characterizing the types of interaction which may occur. It also allows comparisons with other organic and inorganic components of the microbial environment, and provides information on kinetics of binding including maximum metal-ion binding capacity. Adsorption of metals to fungal biomass has been shown to correspond to various models of metal–ligand interactions. The strength and extent of bonding is dependent on the electrostatic interactions between the metal ions and the ligand groupings, and these are themselves mediated by factors such as metal speciation, ionic radius, electronegativity, ionic charge and accessibility. Adsorptive preferences have frequently been shown to be related to the Irving-Williams scale of ion–ligand attraction (Irving & Williams, 1948) as well as to the Lewis theory of acid-base interactions (Hughes & Poole, 1991).

Passive uptake of metals is a pH dependent process, adsorption of cationic species often decreasing with external pH as protons compete more strongly with the metal ions for sorbing sites and as the charge distribution of the cell wall components is altered (Collins & Stotzky,

1992; Fourest & Roux, 1992). It has been shown, however, that the capacity of cell surfaces and components does not decrease with pH in the same manner and to the same extent as the metal-binding capacity of inorganic components of the soil environment (Collins & Stotzky, 1992; Hunter & James, 1992), probably due to the range and variety of metal binding sites present. Fungi are found in intimate association with soil organic and inorganic components including aluminosilicate clay minerals (Singleton, Wainwright & Edyvean, 1990). This may have a deleterious effect on metal binding as binding sites are masked, blocked or otherwise occluded by particles. The binding of clay particles to biomass may also have the effect of altering certain ligand charges which may contribute to both increased or decreased affinities for metals, dependent on clay type and the physico-chemical environment in which the interactions occur. This alteration of ligand charges has also been shown to occur when anions such as sulphate (SO_4^{2-}) bind, synergic effects increasing the affinity of toxic metals for binding sites (Hoins, Charlet & Sticher, 1993).

Physiological sequestration of metals by fungi

Many metals are essential for fungal growth and metabolism, and so for metals such as Na, Mg, K, Ca, Mn, Fe, Co, Ni, Cu and Zn mechanisms exist for their physiological sequestration. This is most often achieved by the incorporation of different ligand groupings, with differing specific and preferential binding capacities for the required metal ions, into the cell membrane. This system allows the cell to determine the type of metal species transported, and has a greater selectivity than the theoretical chemistry of such interactions would suggest (Hughes & Poole, 1989). Toxic metals, although generally of low abundance in the environment unless redistributed by anthropogenic activities, can often compete with physiologically essential ions for these ligand binding sites. Cadmium, for example, competes for both Mn^{2+} and Zn^{2+} transport systems and can substitute for the zinc ion in enzyme systems with deleterious results (Hughes & Poole, 1989).

Mechanisms of solubilization

Biochemical mechanisms for the fungal solubilization of metals and metalloids from insoluble compounds include the efflux of protons, the production of organic acids from which protons are available, phospha-

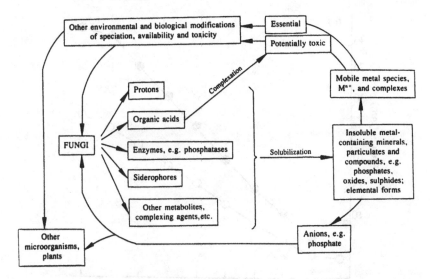

Fig. 15.2. Diagrammatic representation of the major mechanisms involved in fungal solubilization of insoluble metal compounds. (See Burgstaller & Schinner, 1993 for a review.)

tase enzymes which can release metal ions from insoluble phosphates, and chelating agents, such as iron (III)-specific siderophores (Fig. 15.2.).

The main source of extracellular protons is the H^+-translocating ATPase of the plasma membrane which functions in the maintenance of intracellular pH and the creation of electrochemical ionic gradients essential for the acquisition of many nutrients (Jones & Gadd, 1990; Sigler & Höfer, 1991).

Citrate and other organic anions may form complexes with metal ions, which can change their solubility and therefore bioavailability and mycotoxicity. Solubilization occurs by protonation of the anion, such as phosphate (Fig. 15.3a,b), which decreases its availability to the metal(loid) cation, and promotes dissolution of the solid. This is especially the case when the anion is capable of forming a weak acid such as phosphoric acid (Hughes & Poole, 1991). Conditions required for maximal citric acid production by *A. niger* include optimal concentrations of sugar, phosphate and nitrate, adequate oxygen availability, a suitable initial pH value and a deficiency of manganese ($< 0.2\,mM$) (Schreferl, Kubicek & Röhr, 1986). This manganese deficiency has been shown to lead to a change in the morphology of the hyphae which results in the fungus forming hard compact pellets (Schreferl, Kubicek & Röhr, 1986). The medium composition can be manipulated for the over-production of

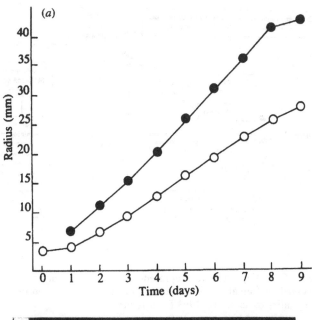

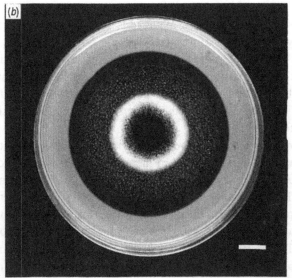

Fig. 15.3. Solubilization of insoluble metal compounds in agar medium by *Aspergillus niger*. (*a*) Growth of *A. niger* (○) and rate of clearing (●) of 0.3% (w/v) ZnO when incorporated in malt extract agar (MEA) and incubated at 25 °C. Average values shown from 3 replicate determinations: errors < 5%. (*b*) Photograph of *A. niger* grown at 25 °C for 6 days on 0.5% (w/v) $Co_3(PO)_4$ in MEA showing the clear zone of solubilization around the colony (see Sayer, Raggett & Gadd, 1995). Bar marker = 1 cm.

different organic acids. For example, the pH of the medium in which *A. niger* is grown influences which organic acid is produced; when the pH of the medium increases, citric, gluconic and oxalic acids are produced, in that order (Burgstaller & Schinner, 1993).

Franz, Burgstaller and Schinner (1991) reported that the production of citric acid by *P. simplicissimum* was induced by the adsorption of zinc oxide by the mycelium, with 1 mol citric acid being produced for every 1.5 mol zinc oxide added to the medium. In contrast to *A. niger*, which does not need the presence of an insoluble metal salt to induce citric acid production, citric acid is produced by *P. simplicissimum* only in the presence of the zinc oxide. The solubilized zinc forms a complex with the organic anion, limiting its availability and toxicity to the fungus. Burgstaller, Zanella and Schinner (1993) have demonstrated that citric acid production by *P. simplicissimum* is coupled with proton extrusion. Zinc oxide appears to act as a buffer when added to the growth medium, consuming 2 mol H^+ per mol ZnO, and preventing them from flowing back into the cell; excreted citrate anions balance this efflux of H^+.

Soluble metal cations may be mobilized enzymatically from insoluble metal phosphates by phosphatase enzymes. In yeast cells, there are two main groups of phosphatases. Alkaline phosphatases operate under phosphate-limited conditions, with acid phosphatases having a wider range of functions including protection from acidic conditions (Galabova, Tuleva & Balasheva, 1993). Zyla (1990) showed that waste *A. niger* mycelium from citric acid production contains many enzymes, including an acid phosphatase. In fact, this waste mycelium was considered to be a unique source of intracellular acid phosphatase which was most stable between pH 4.5 and 5.

A further mechanism of metal solubilization is the production of low molecular weight iron-chelating organic compounds, such as siderophores, which solubilize iron(III). The concentrations of iron(II) and iron(III) in the soil decrease by factors of one hundred and one thousand times respectively for every unit increase in pH. Siderophores are the most common means of acquisition of iron by bacteria and fungi, and are effective in a wide range of soils, including calcareous soils. The most common fungal siderophore is ferrichrome (Crichton, 1991).

Mechanisms of transformation

Reduction

There are several reports of the reduction of metals and metalloids by fungi, although this area has not received as much attention as analogous processes in bacteria. Reduction of Ag(I) to Ag(0) at concentrations up to $5\,mmol\,l^{-1}$ by several yeasts has been reported by Kierans et al. (1991) with growth on $AgNO_3$-containing solid media resulting in blackened colonies. Scanning electron microscopy (SEM) and X-ray microprobe analysis (XRMA) showed that metallic Ag(0) was accumulated in and around cell walls. Cu(II) ions ($200\,\mu mol\,l^{-1}$) were reduced to Cu(II) by cell wall materials prepared from the yeast Debaryomyces hansenii (Wakatsuki, Iba & Imahara, 1988). Cu(II) uptake by whole yeast cells was found to be stimulated by reducing agents and inhibited by oxidizing agents, while cell wall material stimulated Cu(II) uptake by sphaeroplasts. Both enzymic and non-enzymic Cu(II)-reducing systems were subsequently purified from D. hansenii cell walls (Wakatsuki et al., 1991a). The enzymic system was able to utilize both NADH and NADPH as an electron donor for Cu(II) reduction and required flavine mononucleotide (FMN) as a co-enzyme (Wakatsuki et al., 1991b). The authors suggested that the enzyme may be involved in the control of Cu(II) uptake. The reduction of Hg(II) to Hg(0) by fungi has been demonstrated (Yannai, Berdicevsky & Duek, 1991), but there have been few detailed studies.

The ability of fungi to reduce metalloids has also been demonstrated. A Fusarium sp., isolated from Sinai soil, was able to reduce selenite (SeO_3^{2-}) to Se(0) at concentrations up to $2\,mmol\ l^{-1}$ Na_2SeO_3 which resulted in a red colouration of colonies (Ramadan et al., 1988). SEM of the Fusarium sp. incubated in media containing $50\,mmol\,l^{-1}$ Na_2SeO_3 revealed the presence of crystals on the surfaces of hyphae and spores (Fig. 15.4); XRMA showed these crystals to be Se(0) (Gharieb, 1993; Gharieb et al., 1995). Colonies of several Mortierella species appeared pink or orange when grown on SeO_3^{2-}-containing media, also indicating the deposition of Se(0) (Zieve et al., 1985). Fungi have also been shown to reduce tellurite (TeO_3^{2-}) to Te(0), giving a black or dark grey colouration to colonies (Smith, 1974).

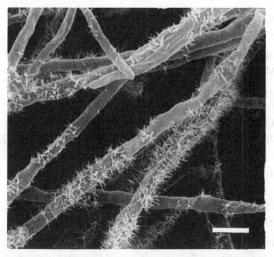

Fig. 15.4. Scanning electron microscopy of *Fusarium* sp. grown for 10 days at 25 °C on Czapex–Dox agar containing 50 mmol l^{-1} sodium selenite showing crystals of amorphous elemental selenium on the hyphae. Control cultures incubated in the absence of sodium selenite did not exhibit such crystals. Bar marker = 10 μm. (Gharieb, 1993; Gharieb *et al.*, 1995.)

Methylation

The biological methylation (= biomethylation) of several metal(loid)s has been demonstrated in filamentous fungi and yeasts. Methylation of some metals and metalloids results in their volatilization and, in certain cases, the appearance of a garlic-like odour (see Gadd, 1993*b* for a review). Biomethylation of arsenic and selenium has received most attention in the literature. The production of trimethylarsine (TMA) from arsenite by several fungi is well known and a biochemical pathway for this process was first suggested by Challenger (1945). Subsequent studies showed that several other fungi, such as *Gliocladium roseum*, *Candida humicola* and *Penicillium* sp., were able to convert monomethylarsonic acid (MMA) to TMA (see Tamaki & Frankenberger, 1992, for a review). Huysmans and Frankenburger (1991) demonstrated the conversion of MMA to TMA in a *Penicillium* sp. isolated from contaminated evaporation pond water, but no detectable TMA production was measured when arsenate or arsenite were used as the sole source of arsenic in culture media. The pathway for arsenic methylation involves the transfer of methyl groups as carbonium (CH_3^+) ions by S-adenosylmethionine (see Gadd, 1993*b*). The amino acids phenylalanine, isoleucine and glutamine were found to stimulate TMA production in the *Penicillium* sp.; optimal conditions for the process were medium pH 5–6, an incubation temperature of 20 °C and a range of phosphate concentrations from 0.1 to 50 mmol l^{-1}. MMA was the best source of arsenic for TMA production (Huysmans & Frankenberger, 1991).

Numerous fungi have been shown to convert both selenite (SeO$_3^{2-}$) and selenate (SeO$_4^{2-}$) to methyl derivatives of selenium such as dimethyl-selenide ((CH$_3$)$_2$Se, DMSe) and dimethyldiselenide ((CH$_3$)$_2$Se$_2$, DMDSe) (Fig. 15.5) (see Thompson-Eagle, Frankenberger & Longley, 1991, for a review). Se-methylating organisms present in evaporation pond water were studied by Thompson-Eagle, Frankenberger and Karlson (1989). *Alternaria alternata* was identified as being capable of volatilizing selenium, the volatile product being DMSe. Inorganic forms of Se (SeO$_3^{2-}$, SeO$_4^{2-}$) were methylated more rapidly than organic forms such as Se-containing amino acids and purines/pyrimidines. The mechanism for Se-methylation appears to be similar to that for arsenic (Gadd, 1993*b*). A *Penicillium* sp. capable of arsenic volatilization also produced several

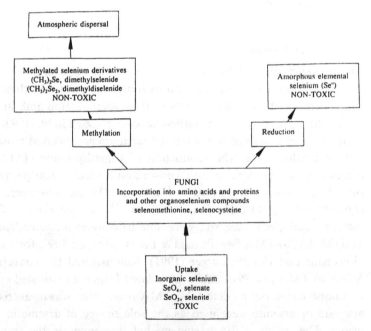

Fig. 15.5. Diagrammatic representation of the major mechanisms involved in fungal selenium transformations. (See Thompson-Eagle & Frankenberger, 1992 and Gadd, 1993*b* for reviews.)

volatile Se compounds, including DMSe and DMDSe from SeO_4^{2-} (Huysmans & Frankenberger, 1991).

There have been few detailed studies of fungal biomethylation of other metals and metalloids. Mercury biomethylation by fungal species has been reported (Yannai *et al.*, 1991; Gadd, 1992*a*, *b*), while there is evidence of dimethyltelluride $((CH_3)_2Te)$ and dimethylditelluride $((CH_3)_2Te_2)$ production from TeO_3^{2-} and TeO_4^{2-} by a *Penicillium* sp. (Huysmans & Frankenberger, 1991).

Dealkylation

Degradation of organometallic compounds can be carried out by fungi, either by biotic action (enzymes) or by facilitating abiotic degradation, for instance by alteration of pH and excretion of metabolites. However, as with other aspects of metal transformations, less information is available on the dealkylation of organometallic compounds by fungi than by bacteria (see Gadd, 1993*b*). Organotin compounds, such as tributyltin oxide and tributyltin naphthenate, may be degraded to mono- and dibutyltins by fungal action, inorganic Sn(II) being the ultimate degradation product (Barug, 1981; Orsler & Holland, 1992). Organomercury compounds may be detoxified by conversion to Hg(II) by fungal organo-mercury lyase, the Hg(II) being subsequently reduced to Hg(0) by mercuric reductase, a system analogous to that found in bacteria (Tezuka & Takasaki, 1988). Trimethyllead degradation has been demonstrated in an alkyllead-tolerant yeast (Macaskie & Dean, 1987) and in the wood-decay fungus *Phaeolus schweinitzii* (Macaskie & Dean, 1990). Few studies have been carried out on the demethylation of methylarsenic compounds by fungi.

Environmental significance of metal sequestration and solubilization

Table 15.2 shows the metal sorbing capacities of some soil fungi compared to other inorganic soil components. It can be seen that the fungi are efficient sorbents of metal ions over a wide range of pH values. It can also be seen that, although the fungi sorb less metal per unit dry weight, they are more efficient sorbents per unit surface area. It is also noticeable that the capacity of the fungi varies less with changing pH than the clay minerals. This is probably due to the wide variety and heterogeneity of the ligand-binding sites in the biomass matrix, which will all react

Table 15.2. Metal sorbing capacities of some clay minerals and soil fungi at different pH values

Sorbent	pH	Cd^{2+}	Cu^{2+}	Zn^{2+}
montmorillonite	4	48 (0.1)	80 (0.1)	45 (0.1)
	5	220 (0.3)	245 (0.3)	130 (0.2)
	6.5	435 (0.5)	–	240 (0.3)
kaolinite	4	13 (0.3)	15 (0.3)	11 (0.2)
	5	67 (1.3)	75 (1.5)	57 (1.1)
	6.5	195 (3.1)	–	105 (2.1)
Rhizopus arrhizus	4	14 (7.6)	16 (8.7)	14 (7.6)
	5	27 (14.6)	24 (13.0)	28 (15.1)
	6.5	49 (26.5)	–	34 (18.4)
Trichoderma viride	4	16 (4.4)	27 (7.3)	15 (4.1)
	5	42 (11.4)	35 (9.5)	21 (5.7)
	6.5	78 (21.2)	–	47 (12.8)

Note: All values were obtained using 1 g dry wt sorbent l^{-1} which was allowed to reach equilibrium in buffered (5 mM 2-[N-morpholino]-ethanesulphonic acid, pH as stated), stirred batch experiments in the presence of 1 mmol l^{-1} metal sulphates. Uptakes are expressed as μmol metal removed from solution (g^{-1} dry weight). Numbers in brackets are μmol metal removed from solution (m^{-2} surface area). Data for removal of Cu^{2+} at pH 6.5 are not shown as the metal precipitated out of solution, probably as the hydroxide, at pH > 6. (Morley & Gadd, 1995).

differently to charge alteration as the pH decreases, ensuring that a considerable binding capacity for the whole system extends across the whole pH range. The cation-exchange sites on the alumino-silicate clay-minerals are more homogeneously distributed and more consistent in their properties, being mainly flaws formed in the crystalline matrix during the clay morphogenesis, caused by the isomorphous substitution of the original Al(III) or Si(II) by elements of lower valency (Metting, 1992). These sites react more uniformly to the decrease in pH than the more heterogeneous fungal surfaces, and binding is affected to a greater extent as the H^+ concentration increases. Thus, although clay minerals are the major metal-sorbing components in the soil environment, biological systems such as fungi must have an important and, so far, unappreciated role.

The mycorrhizal relationship that many fungi have with plants may mediate metal toxicity to these plants, allowing them to survive in soil conditions with potentially toxic amounts of metal species present (see Gadd, 1993a). The sequestering ability of the fungi might effectively reduce the apparent metal content of the soil adjacent to the plant root systems. This relationship, though, might not always be beneficial, as it is possible that fungal accumulation of toxic metals increases the apparent root metal concentration if soil physico-chemical conditions alter or when the fungi die, degrade or otherwise release the accumulated metals.

The physiology of fungal systems allows interconnection of hyphal compartments and continuity of cytoplasm. This allows fungi to cross nutrient-poor regions by supplying the hyphae in these regions with nutrients drawn from that part of the mycelium in a nutritionally adequate locality. This ability of fungi to translocate nutrients (and indeed organelles, oxygen, wastes and structural components) is also a mechanism that is important in the translocation of metal species. Metal species when internalized by fungi are often sequestered and contained by organelles (such as vacuoles); continuity of cytoplasmic contents means that the cytosol as well as metal-laden organelles may be free to move, allowing their translocation and concentration in specific regions of the fungal thallus; elevated metal concentrations in basidiomes of basidiomycetes are well known (Gadd, 1993a).

Soil fungi, including mycorrhizal species, may increase inorganic nutrient availability to plants and other microorganisms by increasing the mobility of essential metal cations and anions. It is widely accepted that mycorrhizas increase phosphate uptake by the host plant. Mycorrhizal fungi obtain the majority of their nutrients from organic matter in the soil, mostly using protease and phosphatase enzymes.

Root 'infection' is favoured in nutrient-deficient soils, and this promotes increased growth of the host plant. However, mycorrhizas are a considerable drain on the carbon reserves of the plant and mycorrhizal infection is less common in nutrient-rich soils, and is decreased by the addition of nitrogen and phosphate sources. Read (1991) has proposed that the role of mycorrhizas is to provide host plants with growth-limiting resources at the stages in their growth when they are most essential. Solubilization may therefore be an important means for the release of essential metal ions.

In neutral soils, essential and potentially toxic metals are usually precipitated or complexed but may be available in exchangeable forms. When cations are leached from a soil, they are exchanged for protons. If the soil has a low cation exchange capacity there is low cation retention, and the removal of cations and replacement by H^+ can reduce the pH of the soil; many toxic metals have maximum solubility and bioavailability in cationic form. In acidic soils, fungi become the dominant component of the soil microflora, and certain organic acids may accumulate in their undissociated form which can affect the speciation and availability of toxic metal cations (Read, 1991).

Other environmental considerations include the dumping of metal-containing wastes in land-fill sites. There may be species of fungi at the site which are able to solubilize the particular metal concerned and mobilize it into the environment. Alternatively, fungi may be used to remove the metal from the waste prior to dumping (see later).

Biotechnological significance of fungal–metal interactions

Fungi and their by-products have been studied as possible sorbent materials in clean-up procedures for metal-contaminated wastes and effluents. Their case as suitable material is helped by the ease with which they are grown and their availability as waste products from industry, for example, *Aspergillus niger* is used in commercial citric acid production while *Saccharomyces cerevisiae* is used in brewing. Many studies have shown their effectiveness in sorbing metal contaminants either as living or dead sorbents in pellet, whole-cell or disassembled forms, and as mobile or immobilized sorbents in batch and continuous processes. They are also amenable to removal of the loaded metals by acids, alkalis or chelating agents, retaining a high percentage of their original sorption capacity after repeated regenerative stages. This is relevant to the development of strategies for limiting the spread of contamination in the soil environment, as well as other roles in biotechnology and bioremediation (Gadd, 1993*a*).

Most biotechnological processes for the winning or removal of metals and metalloids from ores by solubilizing or leaching processes involve chemoautotrophic bacteria of the genus *Thiobacillus*. However, the pH of growth media may be increased by many industrial metal-containing wastes, for example, filter dust from metal processing industries, and most *Thiobacillus* spp. cannot solubilize metals effectively above pH 5.5, with an optimum of pH 2.4 (Burgstaller & Schinner, 1993).

Although fungi need a source of carbon and constant aeration, they can solubilize metals at higher pH values and so could perhaps become more important where leaching with bacteria is not possible. However, fungi may not readily be used *in situ*, and specific bioreactors would be required. Leaching of metals with fungi can be very effective although a high level of organic acid production may need to be maintained. Burgstaller *et al.* (1992) used *P. simplicissimum* for the leaching of filter dust from a copper smelting plant, which contained 50% zinc, 20% lead and 6% tin, as high levels of citric acid production could not be maintained with *A. niger* which was inhibited by the toxic concentrations of metals present. With *A. niger*, acid production decreased with increasing filter dust levels, but with *P. simplicissimum* the filter dust served to stimulate production of citric acid.

Another possible application of fungal metal solubilization is in the leaching of metals from low grade ores, and the removal of unwanted contaminants such as phosphates (Burgstaller & Schinner, 1993). Furthermore, it is possible that leaching technologies may combine with those relating to biosorption (Gadd, 1993*a*).

The ability of fungi to transform metals and metalloids has been utilized in the bioremediation of contaminated land and water. Fungal methylation of selenium results in its volatilization, a process which has been used to remove selenium from the San Joaquin Valley and Kesterson Reservoir, California, using evaporation pond management and primary pond operation (Thompson-Eagle & Frankenberger, 1992). Incoming Se-contaminated drainage water was evaporated to dryness and the process repeated until the sediment Se concentration approached $100\,mg\,kg^{-1}$ on a dry weight basis. The volatilization process was then optimized until the selenium concentration in the sediment fell to an acceptable level. The carbon source, moisture, temperature and aeration were found to be important in optimizing selenium volatilization. This process produced selenium levels in air which were within acceptable limits, and in the San Joaquin Valley, local environmental conditions were thought to result in selenium being blown towards Se-deficient areas on the east side of the valley (Thompson-Eagle & Frankenberger, 1992). The economics of implementing the selenium volatilization process have been reported; calculated annual costs in 1990/91 ranged between \$227 and \$372 ha^{-1} (\$92 and \$151 acre^{-1}, respectively), depending on land and water quality (Thompson-Eagle et al., 1991). Fungal metal transformation technology could also be exploited for the removal of alkylleads from water (Macaskie & Dean, 1987, 1990).

Acknowledgements

G. M. G. gratefully acknowledges receipt of a Scottish Office Education Department/Royal Society of Edinburgh Support Research Fellowship and financial support from the AFRC/NERC Special Topic Programme 'Pollutant Transport in Soils and Rocks' and BBSRC (GR/J48214, 94/SPC02812) for some of the work described. J. A. S. and G. F. M. gratefully acknowledge receipt of a NERC and an AFRC/NERC postgraduate research studentship, respectively. M. M. G. gratefully acknowledges support from the Egyptian Government via the UK/Egyptian Government Channel Scheme.

References

Barug, G. (1981). Microbial degradation of bis(tributyltin) oxide. *Chemosphere*, **10**, 1145–54.

Blunden, S. J. & Chapman, A. H. (1982). The environmental degradation of organotin compounds – a review. *Environmental Technology Letters*, **3**, 267–72.

Brunauer, S., Emmett, P. H. & Teller, E. (1938). Adsorption of gases in multimolecular layers. *Journal of the American Chemical Society*, **60**, 309–19.

Burgstaller, W. & Schinner, F. (1993). Leaching of metals with fungi. *Journal of Biotechnology*, **27**, 91–116.

Burgstaller, W., Strasser, F., Wöbking, H. & Schinner, F. (1992). Solubilization of zinc oxide from filter dust with *Penicillium simplicissimum*: bioreactor leaching and stoichiometry. *Environmental Science and Technology*, **26**, 340–6.

Burgstaller, W., Zanella, A. & Schinner, F. (1993). Buffer stimulated citrate efflux in *Penicillium simplicissimum*: an alternative charge balancing ion flow in the case of reduced proton backflow? *Archives of Microbiology*, **161**, 75–81.

Challenger, F. (1945). Biological methylation. *Chemical Reviews*, **36**, 15–61.

Chasteen, T. G., Silver, G. M., Birks, J. W. & Fall, R. (1990). Fluorine-induced chemiluminescence detection of biologically mediated tellurium, selenium and sulfur compounds. *Chromatographia*, **30**, 181–5.

Collins, Y. E. & Stotzky, G. (1992). Heavy metals alter the electrokinetic properties of bacteria, yeasts and clay minerals. *Applied and Environmental Microbiology*, **58**, 1592–600.

Crichton, R. R. (1991). *Inorganic Biochemistry of Iron Metabolism*. Chichester: Ellis Horwood.

Farrah, H. & Pickering, W. F. (1976). The sorption of zinc species by clay minerals. *Australian Journal of Chemistry*, **29**, 1649–56.

Farrah, H. & Pickering, W. F. (1977). The sorption of lead and cadmium species by clay minerals. *Australian Journal of Chemistry*, **30**, 1417–22.

Fourest, E. & Roux, J-C. (1992). Heavy metal biosorption by fungal mycelial by-products, mechanisms and influence of pH. *Applied Microbiology and Biotechnology*, **37**, 399–403.

Franz, A., Burgstaller, W. & Schinner, F. (1991). Leaching with *Penicillium simplicissimum*. Influence of metals and buffers on proton extrusion and citric acid production. *Applied and Environmental Microbiology*, **57**, 769–74.

Freundlich, H. (1926). *Colloid and Capillary Chemistry*. London: Methuen.

Gadd, G. M. (1992*a*). Microbial control of heavy metal pollution. In *Microbial Control of Environmental Pollution*, ed. J. C. Fry, G. M. Gadd, R. A. Herbert, C. W. Jones & I. Watson-Craik, pp. 59–88. Cambridge: Cambridge University Press.

Gadd, G. M. (1992*b*). Molecular biology and biotechnology of microbial interactions with organic and inorganic heavy metal compounds. In *Molecular Biology and Biotechnology of Extremophiles*, ed. R. A. Herbert & R. J. Sharp, pp. 225–57. Glasgow: Blackie & Sons.

Gadd, G. M. (1993*a*). Interactions of fungi with toxic metals. *New Phytologist*, **124**, 25–60.

Gadd, G. M. (1993*b*). Microbial formation and transformation of organometallic and organometalloid compounds. *FEMS Microbiology Reviews*, **11**, 297–316.

Gadd, G. M. & White, C. (1989). Heavy metal and radionuclide accumulation and toxicity in fungi and yeasts. In *Metal–Microbe Interations*, ed. R. K. Poole & G. M. Gadd, pp. 19–38. Oxford: IRL Press.

Galabova, D., Tuleva, B. & Balasheva, M. (1993). Phosphatase activity during growth of *Yarrowica lipolytica*. *FEMS Microbiology Letters*, **109**, 45–8.

Gharieb, M. M. (1993). *Selenium Toxicity, Accumulation and Metabolism by Fungi and Influence of the Fungicide Dithane*. PhD thesis, University of Dundee.

Gharieb, M. M., Wilkinson, S. C. & Gadd, G. M. (1995). Reduction of selenium oxyanions by unicellular, polymorphic and filamentous fungi: cellular location of reduced selenium and implications for tolerance. *Journal of Industrial Microbiology*, **14**, 300–11.

Harter, R. D. (1983). Effect of soil pH on adsorption of lead, copper, zinc and nickel. *Soil Science Society of America Journal*, **47**, 47–51.

Harter, R. D. & Baker, D. E. (1977). Applications and misapplications of the Langmuir equation to soil adsorption phenomena. *Soil Science Society of America Journal*, **41**, 1077–80.

254   G. F. Morley et al.

Hoins, U., Charlet, L. & Sticher, H. (1993). Ligand effect on the adsorption of heavy metals: the sulfate–cadmium–goethite case. *Water, Air and Soil Pollution*, **68**, 241–55.

Hughes, M. N. & Poole, R. K. (1989). Metal mimicry and metal limitation in studies of metal-microbe interactions. In *Metal–Microbe Interactions*, ed. R. K. Poole & G. M. Gadd, pp. 1–18. Oxford: IRL Press.

Hughes, M. N. & Poole, R. K. (1991). Metal speciation and microbial growth – the hard (and soft) facts. *Journal of General Microbiology*, **137**, 725–34.

Hunter, R. J. & James, M. (1992). Charge reversal of kaolinite by hydrolyzable metal ions: an electroacoustic study. *Clays and Clay Minerals*, **40**, 644–9.

Huysmans, K. D. & Frankenberger, W. T. (1991). Evolution of trimethylarsine by a *Penicillium* sp. isolated from agricultural evaporation pond water. *The Science of the Total Environment*, **105**, 13–28.

Irving, H. & Williams, R. J. P. (1948). Order of stability of metal complexes. *Nature*, **162**, 747–7.

Jones, R. P. & Gadd, G. M. (1990). Ionic nutrition of yeast – physiological mechanisms involved and implications for biotechnology. *Enzyme and Microbial Technology*, **12**, 402–18.

Jones, D., Smith, B. F. L., Wilson, M. J. & Goodman, B. A. (1991). Phosphate solubilizing fungi in a Scottish upland soil. *Mycological Research*, **95**, 1090–3.

Kierans, M., Staines, A. M., Bennett, H. & Gadd, G. M. (1991). Silver tolerance and accumulation in yeasts. *Biology of Metals*, **4**, 100–6.

Kisser, M., Kubicek, C. P. & Röhr, M. (1980). Influence of manganese on morphology and cell wall composition of *Aspergillus niger* during citric acid fermentation. *Archives of Microbiology*, **128**, 26–33.

Krosshavn, M., Steinnes, E. & Varskog, P. (1993). Binding of Cd, Cu, Pb and Zn in soil organic matter with different vegetational background. *Water, Air and Soil Pollution*, **71**, 185–93.

Kuo, S. & Baker, A. S. (1980). Sorption of copper, zinc and cadmium by some acid soils. *Soil Science Society of America Journal*, **44**, 969–74.

Langmuir, I. (1918). The adsorption of gases on plane surfaces of glass, mica and platinum. *Journal of the American Chemical Society*, **40**, 1361–403.

Lesuisse, E. & Labbe, P. (1989). Reductive and non-reductive mechanisms of iron assimilation by the yeast *Saccharomyces cerevisiae*. *Journal of General Microbiology*, **135**, 257–63.

Lesuisse, E. & Labbe, P. (1992). Iron reduction and trans-membrane electron transfer in the yeast *Saccharomyces cerevisiae*. *Plant Physiology*, **100**, 769–77.

Macaskie, L. E. & Dean, A. C. R. (1987). Trimethyllead degradation by an alkyllead-tolerant yeast. *Environmental Technology Letters*, **8**, 635–40.

Macaskie, L. E. & Dean, A. C. R. (1990). Trimethyllead degradation by free and immobilized cells of an *Arthrobacter* sp. and by the wood decay fungus *Phaeolus schweinitzii*. *Applied Microbiology and Biotechnology*, **38**, 81–7.

Metting, F. B. (1992). Structure and physical ecology of soil microbial communities. In *Soil Microbial Ecology, Applications in Agricultural and Environmental Management*, ed. F. B. Metting, pp. 3–25. New York: Marcel Dekker.

Morley, G. F. & Gadd, G. M. (1995). Sorption of toxic metals by fungi and clay minerals. *Mycological Research* (in press).

Orsler, R. J. & Holland, G. E. (1992). Degradation of tributyltin oxide by fungal culture filtrates. *International Biodeterioration Bulletin*, **18**, 95–8.

Peberdy, J. F. (1990). Fungal cell walls – a review. In *Biochemistry of Cell Walls and Membranes in Fungi*, ed. P. J. Kuhn, A. P. J. Trinci, M. J. Jung, M. W. Goosey & L. E. Copping, pp. 5–30. Berlin: Springer-Verlag.

Ramadan, S. E., Razak, A. A., Yousseff, Y. A. & Sedky, N. M. (1988). Selenium metabolism in a strain of *Fusarium*. *Biological Trace Element Research*, **18**, 161–70.

Read, D. J. (1991). Mycorrhizas in ecosystems. *Experientia*, **47**, 376–91.

Sayer, J. A., Raggett, S. L. & Gadd, G. M. (1995). Solubilization of insoluble metal compounds by soil fungi: development of a screening method for solubilizing ability and metal tolerance. *Mycological Research*, **99**, 987–93.

Scatchard, G. (1949). The attraction of proteins for small molecules and ions. *Annals of the New York Academy of Science*, **51**, 600–72.

Schreferl, G., Kubicek, C. P. & Röhr, M. (1986). Inhibition of citric acid accumulation by manganese ions in *Aspergillus niger* mutants with reduced citrate control of phosphofructokinase. *Journal of Bacteriology*, **165**, 1019–22.

Siegel, S. M., Galun, M. & Siegel, B. Z. (1990). Filamentous fungi as biosorbents: a review. *Water, Air and Soil Pollution*, **53**, 335–44.

Sigler, K. & Höfer, M. (1991). Mechanisms of acid extrusion in yeast. *Biochimica et Biophysica Acta*, **1071**, 375–91.

Singleton, I., Wainwright, M. & Edyvean, R. G. J. (1990). Some factors influencing the adsorption of particulates by fungal mycelium. *Biorecovery*, **1**, 271–89.

Smith, D. G. (1974). Tellurite reduction in *Schizosaccharomyces pombe*. *Journal of General Microbiology*, **83**, 389–92.

Tamaki, S. & Frankenberger, W. T. (1992). Environmental biochemistry of arsenic. *Reviews of Environmental Contamination and Toxicology*, **124**, 79–110.

Tezuka, T. & Takasaki, Y. (1988). Biodegradation of phenylmercuric acetate by organomercury-resistant *Penicillium* sp. MR-2. *Agricultural and Biological Chemistry*, **52**, 3183–5.

Thayer, J. S. (1984). *Organometallic Compounds and Living Organisms*, New York: Academic Press.

Thayer, J. S. (1988). *Organometallic Chemistry, An Overview*, Weinheim: VCH Verlagsgesellschaft.

Thompson-Eagle, E. T. & Frankenberger, W. T. (1992). Bioremediation of soils contaminated with selenium. In *Advances in Soil Science, Vol. 17*, ed. R. Lal & B. A. Stewart, pp. 261–309. New York: Springer.

Thompson-Eagle, E. T., Frankenberger, W. T. & Karlson, U. (1989). Volatilization of selenium by *Alternaria alternata*. *Applied and Environmental Microbiology*, **55**, 1406–13.

Thompson-Eagle, E. T., Frankenberger, W. T. & Longley, K. E. (1991). Removal of selenium from agricultural drainage water through soil microbial transformations. In *The Economics and Management of Water and Drainage in Agriculture*, ed. A. Dinar & D. Zilberman, pp. 169–86. New York: Kluwer.

Tobin, J. M., Cooper, D. G. & Neufeld, R. J. (1990). Investigation of the mechanism of metal uptake by denatured *Rhizopus arrhizus* biomass. *Enzyme and Microbial Technology*, **12**, 591–5.

Tsezos, M. & Volesky, B. (1982a). The mechanism of uranium biosorption by *Rhizopus arrhizus*. *Biotechnology and Bioengineering*, **24**, 385–401.

Tsezos, M. & Volesky, B. (1982b). The mechanism of thorium biosorption by *Rhizopus arrhizus*. *Biotechnology and Bioengineering*, **24**, 955–69.

Wakatsuki, T., Hayakawa, S., Hatayama, T., Kitamura, T. & Imahara, H. (1991a). Solubilization and properties of copper reducing enzyme systems from the yeast cell surface in *Debaryomyces hansenii*. *Journal of Fermentation and Bioengineering*, **72**, 79–86.

Wakatsuki, T., Hayakawa, S., Hatayama, T., Kitamura, T. & Imahara, H. (1991b). Purification and some properties of copper reductase from cell surface of *Debaryomyces hansenii*. *Journal of Fermentation and Bioengineering*, **72**, 156–61.

Wakatsuki, T., Iba, A. & Imahara, H. (1988). Copper reduction by yeast cell wall materials and its role in copper uptake by *Debaryomyces hansenii*. *Journal of Fermentation Technology*, **66**, 257–65.

Wilber, C. G. (1980). Toxicology of selenium: a review. *Clinical Toxicology*, **17**, 171–230.

Yannai, S., Berdicevsky, I. & Duek, L. (1991). Transformations of inorganic mercury by *Candida albicans* and *Saccharomyces cerevisiae*. *Applied and Environmental Microbiology*, **57**, 245–7.

Zhu, B. & Alva, A. K. (1993). Differential adsorption of trace metals by soils as influenced by exchangeable cations and ionic strength. *Soil Science*, **155**, 61–6.

Zieve, R., Ansell, P. J., Young, T. W. K. & Peterson, P. J. (1985). Selenium volatilization by *Mortierella* species. *Transactions of the British Mycological Society*, **84**, 177–9.

Zyla, K. (1990). Acid phosphatases purified from industrial waste mycelium of *Aspergillus niger* used to produce citric acid. *Acta Biotechnologica*, **10**, 319–27.

16

Urban, industrial and agricultural effects on lichens

D. H. BROWN

Introduction

Lichens have acquired a reputation as valuable monitors of environmental pollution. Pollution gradients have been related either to differential species sensitivity, creating characteristic floristic changes, or to the capacity of lichens to acquire and retain specific chemicals. These properties may be partly related to the nature of the lichen symbiosis and the structure of the resulting thallus. A brief review of lichen biology is provided to explain how the symbionts interact, what structural and physiological flexibility may exist and how far lichens react to the temporal and spatial heterogeneity of the natural, uncontaminated environment. Reference has been made to a number of valuable reviews that quote a wider literature.

Lichen biology

The lichen symbiosis consists of a fungal component, acting as the interface with the environment, and a photosynthetic component that is to a greater or lesser extent surrounded by the fungal tissues. Ascomycotina represent the main fungal group comprising the lichens, although Basidiomycotina and Deuteromycotina also occur frequently (Hawksworth, 1988a). More than 40% of the Ascomycotina are lichenized and some genera have both free-living and lichenized members. Whereas the photosynthetic component is usually either cyanobacterial or chlorophycean, thalli with both types of photobionts in different regions of the same thallus also occur (Hawksworth, 1988b; Tschermak-Woess, 1988). There is still doubt as to whether certain commonly lichenized chlorophycean algae, particularly the genus *Trebouxia*, exist in the free-living condition (Ahmadjian, 1993).

The fundamentals of the lichenized state consist of the juxtaposition of a fungal hypha and a photobiont cell with the transfer of chemicals between these symbionts (Honegger, 1991). From this basic structural requirement a variety of thallus forms has developed (Jahns, 1988). One form can be a simple mass of mycobiont and photobiont cells (leprose condition), resting on the substratum or, apart from the aperture of the reproductive structures, embedded within rock (endolithic) or bark (endophloedal). More often thalli possess a layered arrangement whereby a dense, purely fungal, upper, mostly living cortex covers a defined layer of fungal and photobiont cells (photobiont layer) that rests on a fungal medulla possessing many air-gaps between the hyphae. This form is described as crustose when the medulla is adpressed to the substratum or is referred to as squamulose when raised above the substratum. Those possessing a lower fungal cortex like that of the upper layer, with or without fungal strands (rhizinae) attaching the thallus to its substratum, represent the foliose condition, whereas a more erect, often radially symmetrical form constitutes the fruticose state.

It is not known whether the above rather 'sandwich-like' organization represents the equivalent of a mycorrhizal form, whereby the fungal tissue channels scarce and not readily available chemicals to the photobiont, in exchange for autotrophic products formed by the latter, or a situation resembling slavery where the fungus deprives the photobiont of not only its own products but also access to environmental chemicals (Crittenden, 1989). In either case the photobiont fails to retain its full complement of autotrophic products and may, therefore, be relatively sensitized to additional applied stresses. Most studies on the responses of lichens to environmental conditions have tended to emphasize photosynthetic responses or attribute the overall sensitivity of the thallus to the photobiont partner. More research is needed to establish whether the mycobiont may be the 'weak partner' in the symbiosis.

When lichenized, the photobiont is stimulated by an as yet unknown mechanism to release specific photosynthetic products (Hill, 1976). The released products are genus-specific sugar alcohols with four to six-carbon atoms in chlorophycean systems and a glucan, degraded to glucose before fungal assimilation, in cyanobacterial systems. In most instances the mycobiont converts the incoming carbohydrate into mannitol, which can accumulate (Lines *et al.*, 1989), although transfer between the symbionts is decreased at lower thallus water contents (Tysiaczny &

Kershaw, 1979). Although the mycobiont dominates the bulk of most lichen thalli, it is not certain how much of the respiratory activity of the thallus derives from the mycobiont. Pearson and Brammer (1978) reported that the photobiont layer, with fungi closest to the source of respirable material but also containing a respiring photobiont, showed the highest activity and the purely fungal medulla the least activity per unit surface area.

The earliest experiments identifying the nature of mobile carbohydrates were performed with immersed thalli incubated with $^{14}CO_2$ in the presence of the presumptive mobile chemical (Hill, 1976). Some of the studies that identified ammonium ions as the mobile nitrogenous compound released by cyanobacterial cells in lichen thalli also included immersion in the presence of inhibitors or detergents, causing damage to the eukaryotic fungal cells (Rai, 1988). Like many immersion experiments, these overcame the natural barriers and controls of the intact thalli and undoubtedly provided a quantitatively inaccurate measure of fluxes of chemicals. Electron microscopy studies now tend to emphasize an intimate association, even partial ensheathing of photobiont cells by water-repellent layers derived from mycobiont cell walls (Honegger, 1991). There is no evidence of free liquid water occurring between cells in normally functional thalli, the mycobiont cell wall being the anticipated conduit for water delivery to photobiont cells (Jahns, 1984).

Unlike chlorophycean species, which can attain most of their photosynthetic capacity with water acquired from atmospheric humidity, dry cyanobacterial lichens require liquid water to reactivate photosynthesis (Lange, Killian & Ziegler, 1986). However, excessive water contents depress photosynthesis in all lichens (Kershaw, 1985; Kappen, 1988; Rundel, 1988). Although not directly demonstrated, it is anticipated that this is due to water filling the air gaps in cortical and/or medullary tissues, thereby impeding CO_2 diffusion to the photosynthetic cells. No comparable depression of nitrogen fixation occurs at high water contents, due to its greater solubility in water. The fungus may exert some degree of control on gaseous diffusion by the nature of the cracks, pores and superficial deposits of the cortical layers. (Jahns, 1984, 1988).

Adaptability of lichens

The relative amounts of the different tissues have been shown to alter in ways that appear to relate to differing environmental conditions. Snelgar and Green (1981) showed that thicker medullary and rhizinal layers occurred in *Pseudocyphellaria* thalli from sunnier environments. This was interpreted as an adaptation for water retention rather than irradiation protection. Multilayered thalli (Jahns, 1984), differing proportions of tissue layers (Sancho & Kappen, 1989), or altered fenestration of thalli (Nash *et al.*, 1990) have also been related to water retention, which in turn has been related to environmental conditions. However, although the water content of *Ramalina menziesii* thalli has been correlated with both proximity to the coast and degree of fenestration, this is related more to the surface deposition of sea-water salts than any physiological response due to morphological adaptations (Larson, Matthes-Sears & Nash, 1986).

Seaward (1976) reported a thinner photobiont layer and thicker medulla in thalli of *Lecanora muralis* growing under urban conditions. He also reported a higher growth rate with increasing distance from urban centres. In the absence of any clear illumination gradient, such differences appear to be related to some kind of pollution stress. There is some evidence to show that lichen growth rate can be increased by supplying fertilizing nutrients, but most growth enhancement is related to more frequent applications of water (Armstrong, 1988). Goyal and Seaward (1982*a*) related changes in tissue dimensions in species of *Peltigera* from uncontaminated and disused mine sites to heavy-metal stress. Unfortunately, the authors did not provide evidence to exclude the possibility that such changes could have been induced by the relative water availability of the different habitats.

Pigmentation of fungal hyphae to produce dark-coloured thalli has been shown to elevate the temperature of the thallus during drying events, which may have ecological relevance for thalli growing associated with snow cover (Kappen, 1988). For many years there has been speculation as to whether the deposition of specific lichen substances in the upper thallus layers provides protection for the photobiont against excessive irradiation or whether greater deposition reflects enhanced photosynthetic waste-product synthesis. Emphasis is now being placed on UV-B protection of both bionts by such chemicals despite the occurrence of thalli in well-lit situations that neither accumulate or produce such chemicals. All these changes might be ways in which the mycobiont could extend the environmental range of the photobiont. Lange *et al.* (1990), however, found no differences in the photosynthetic response to available water by algae within or released from thalli.

Photosynthetic rates are usually greatest in the youngest tissues at the margins of foliose or the tips of fruticose thalli (Nash, Moser & Link

1980), although umbilicate lichens have been shown to possess a mosaic of photosynthetic but not respiratory capacity (Larson, 1983). Field observations have shown, for example, that the aspect of the substratum can considerably influence the diurnal extent and duration of metabolic activity (Kappen, 1988). Many laboratory studies have reported on the response of lichen photosynthesis, nitrogen fixation and respiration to water content, light intensity and temperature (Kershaw, 1985; Kappen, 1988; Rundel, 1988). Such studies have shown that lichens have a considerable capacity to adapt to even relatively short-term alterations in environmental conditions, in both space and time (Kershaw, 1985; Green & Lange, 1991). This acclimation capacity emphasizes the flexibility of lichen thalli to respond to environmental conditions but complicates experiments involving exposure to potentially damaging pollutants, especially in defining suitable control or reference material. Most changes occur without, apparently, the necessity for new growth.

The nature of pollution stresses experienced by lichens

Other chapters in this volume discuss the form and extent of pollution that may derive from urban, industrial and agricultural sources. Only those aspects that are of major significance to lichens are mentioned here. The following reviews should be consulted for specific examples related to lichen studies: Gilbert (1973), Nash and Wirth (1988), Nash and Gries (1991) and Richardson (1988, 1992).

In the past, urban conditions were equated with high levels of particulate matter and sulphur dioxide emitted from low-elevation, coal-burning domestic fires. More recent emphasis might be placed on the decline in both kinds of emissions as a result of Clean Air Acts and alterations in energy sources for heating houses (but see Chapter 6). Urban pollution studies tend now to emphasize vehicular transport emissions, especially lead (and other heavy metals) and oxides of nitrogen.

Industrial installations with taller chimneys burning fossil fuel result in the dispersion of sulphur dioxide and other gases to much greater distances, with the potential for chemical modifications before encountering vegetation. Industrial sources can achieve a multiplicity of other chemical emissions that may be gaseous, for example fluorine from aluminium smelters (Perkins, 1982), or a mixture of soluble and particulate metallic chemicals. The long-distance dispersion of gaseous substances and their interactions mean that sulphur dioxide can impact on a lichen as the gas, in a soluble form where chemical speciation will depend on solution pH, dry deposition (frequently after association with cations such as ammonium ions) or in combination with oxides of nitrogen to produce 'acid rain' (Bates & Farmer, 1992). As photochemical oxidants are generated by a series of chemical reactions, their main damage effects are usually at a distance from the source of their precursor chemical emissions (Sigal & Nash, 1983; Nash & Gries, 1991).

Agricultural emissions may be related to either potentially extreme levels of nutrient emissions from animal waste or fertilizer usage, mostly investigated as nitrogenous chemicals (Brown, 1992; van Dobben, 1993), or to deliberate or accidental pesticide (herbicide, fungicide or insecticide) applications (Brown, 1992; Brown et al., 1995).

Most lichenological studies relate to relatively unique sets of pollution conditions that rarely consist of single chemical emissions (Herzig et al., 1989). The forms in which pollutants impinge on the lichen may not be identical to those measured by chemical means. In addition, pollutant quantities, and plant responsiveness, are in a dynamic state through normal diurnal and annual cycles (Boonpragob, Nash & Fox, 1989; Boonpragob & Nash, 1990) as well as showing long-term alterations as a result of changes in legislation or industrial activity. Despite the academic uncertainty that derives from the multiplicity and subtlety of the interactions that occur, a number of generalizations are possible.

Responses of lichens to gaseous pollutants

Sulphur dioxide

For a number of years it has been apparent that a comparison of species distributions from contemporary occurrences and herbarium records showed certain species regressing or advancing (Hawksworth, Coppins & Rose, 1974; Seaward & Hitch, 1982). In the 1960s Evernia prunastri, for example, was not reported from many of the Midland regions of the UK and its occurrence correlated with patterns of elevated winter SO_2 emission levels (Fig. 16.1). Records also showed that fertile specimens were considerably rarer and confined to the cleaner parts of the UK mainland. Whether this was a direct effect of this pollutant on the reproductive capacity of the fungus or an indirect response reflecting an overall lack of vigour is not known. Parmeliopsis ambigua, by contrast, showed an extension of its former preference for acid-barked trees to more southerly deciduous trees, following an increase in acidic precipitation, stem-flow and bark acidity. The progressive west- and northwards spread of

Lecanora conizaeoides, a species known only since the mid-nineteenth century, has been related to a decline in the competitive lichen flora and a tolerance to acidification by SO_2 products (Fig. 16.2(*a*)) (Seaward & Hitch, 1982).

Some new lichen distribution patterns have been related to non-pollutant-induced environmental changes. The decline of *Teloschistes flavicans* (Fig. 16.2(*b*)) partly reflects over-collecting and the destruction of suitable habitats by management practices, for example, ancient wooded parklands (Seaward & Hitch, 1982). Rose developed a list of species from which he derived an Index of Ecological Continuity for such old woodlands; high indices indicate a sustained but slow cycle of woodland usage (see Rose, 1992). Some species are treated as relict and currently appear to require assistance in order to colonise potentially suitable trees. Many of these species are also sensitive to gaseous pollutants, especially 'acid rain' (Farmer, Bates & Bell, 1992). The near extinction of other species, such as *Caloplaca luteoalba*, resulted from the destruction of elm trees due to *Ceratocystis ulmi*.

Pollution-related floristic changes have also been defined with distance from pollution sources, such as urban areas (Fig. 16.3). Many reports have been written showing sequences of increasing species number and biomass with increasing distance from an arbitrary urban centre. Superficially, the progression from adpressed crustose, through foliose, to erect or pendulous fruticose thalli observed under these conditions represents increasingly more surface tissue exposed to the aerial environment and a reduction in the degree of association with the substratum. However, differential species sensitivities exist with the same thallus morphology, so other factors contribute to the final sensitivity observed. Artificial fumigation experiments showed that SO_2 uptake was reduced with decreasing thallus water content (Grace, Gillespie & Puckett, 1985*a*) and membrane damage was related to SO_2 uptake but not directly to the SO_2 concentration supplied (Grace, Gillespie & Puckett, 1985*b*). Huebert, L'Hirondelle and Addison (1985) related photosynthetic damage to SO_2 concentration rather than time but agreed with Grace *et al.* (1985*b*) that SO_2 uptake was more important than dosage (concentration × time), thereby emphasizing the importance of peak rather than average concentrations for prediction of damage. Whilst physiological recovery after SO_2 fumigation is related to SO_2 uptake, Coxson (1988) showed dry deposition of SO_2 could enhance photosynthesis on rehydration. Although measurements of metabolic parameters following immersion in sulphite solutions can reflect biochemical differences between species,

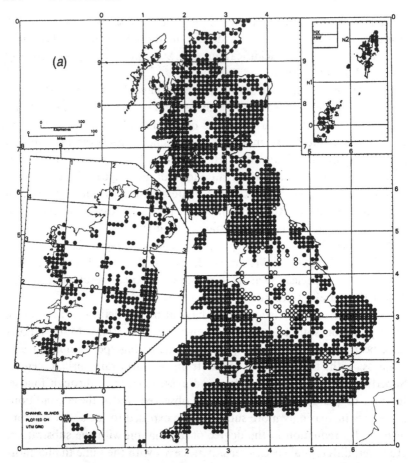

Fig. 16.1. Distribution of *Evernia prunastri* in the British Isles in (*a*) 1982 and (*b*) May 1994 showing recent range extensions. Records pre-1960 (○), post-1960 (●). Based on (*a*) Seaward and Hitch (1982) and (*b*) computer generated map from the British Lichen Society's Mapping Database, University of Bradford.

such treatments may overcome the subtleties of thallus form involved in the trapping and presentation of the pollutant to sensitive cells (Nash, 1988).

A combination of the above types of floristic observations have been employed in the construction of bioindicative schemes (Hawksworth & Rose, 1970). In these, associations of species and their position on the trunks of isolated trees have been correlated with specific winter SO_2 concentrations. Although relatively subjective, such schemes have subsequently been used by other researchers to estimate environmental levels

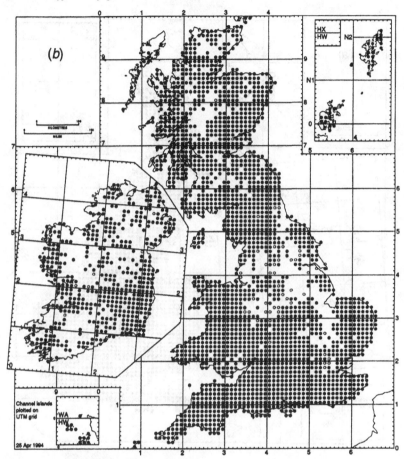

Fig. 16.1. (*b*) *Evernia prunastri*

of SO$_2$. Preliminary scales have been introduced for other pollutants such as secondary oxidants (e.g. O$_3$, PANs and NO$_x$) (Sigal & Nash, 1983) and ammonia-rich farming emissions (van Herk, 1991). Several more objective methods of describing floristic damage induced by pollutants have also been devised. Some use an Index of Atmospheric Purity, which includes measures of frequency/coverage for each species and the number of associated species (Showman, 1988). Such indices take into account the influence of both pollution and other environmental factors and hence great care must be taken in the selection of uniform trees from which to derive data. Computer-assisted analyses now permit the use of more randomly generated floristic data from which to derive species assemblages that can be linked to physical, chemical and other habitat

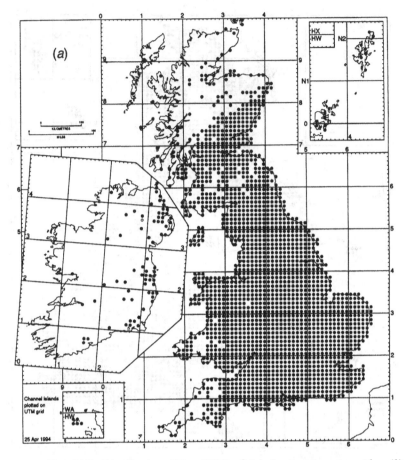

Fig. 16.2. Distribution (May 1994) of (a) *Lecanora conizaeoides*, (b) *Teloschistes flavicans* and (c) *Lobaria scrobiculata* in the British Isles. Records pre-1960 (○), post-1960 (●). Computer generated maps from the British Lichen Society's Mapping Database, University of Bradford.

data. These can produce potentially biologically meaningful correlations between environmental conditions and the occurrence of particular lichens (Nimis, Castello & Perotti, 1990).

More recent studies using many of the above indices failed to find the anticipated lichen assemblages at sites with known SO_2 concentrations (Gilbert, 1992). Such deviations have been attributed to differences in the chemical measurements employed (Richardson, 1988) and, with increasing confidence, the failure of species to immediately recolonize habitats where SO_2-pollution levels are falling. In addition, it is possible that the weaker correlations reflect declining SO_2 concentrations combined with

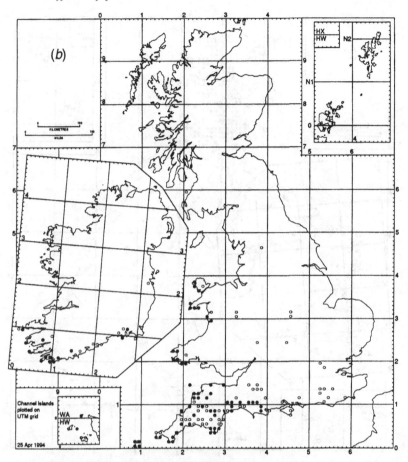

Fig. 16.2. (*b*) *Teloschistes flavicans*

rising levels of acidifying oxides of nitrogen (Fig. 16.4). Sulphur dioxide in solution develops a particular pH and, depending on the solution pH, a range of chemical forms. Most workers consider low pH forms are the more toxic. Acidification by other chemicals may enhance the toxicity of SO_2 in solution beyond the level anticipated from SO_2 alone.

Reinvasion of SO_2-depleted environments by lichens (Fig. 16.3) results in the return of only certain species in a progression described as 'zone-skipping' or 'zone-dawdling' (Gilbert, 1992). The differences may reflect subtle habitat distinctions related to the age of the colonized bark, its natural buffering capacity and, more especially, how far it is hydrated by percolating water or atmospheric humidity. Species with a preference for the well-leached regions of a tree trunk appear to colonize first.

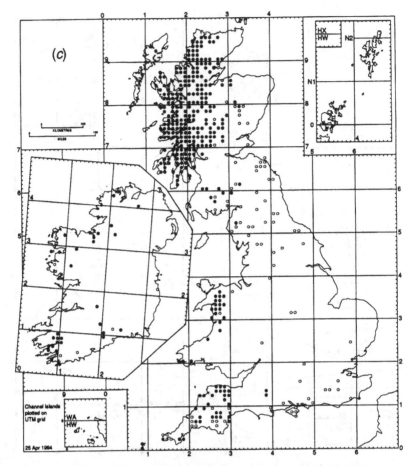

Fig. 16.2. (c) *Lobaria scrobiculata*

Acid rain

The pH of rain, throughfall and stem-flow solutions as well as bark pH have profound effects on certain species apparently sensitive to 'acid rain'. Amongst these are a number of cyanobacterial lichens (Fig. 16.2(c)) that appear to be especially sensitive (Gilbert, 1992). Transplant studies, linked to careful analysis of the chemistry of the above solutions in the original and polluted sites appear to support the suggestion that damage is related to pH rather than the balance of sulphur and nitrogen oxides (Farmer *et al.*, 1992). However, it is known that the proportions of these two chemicals does have some effect, for example, low pH solutions dominated by HNO_3 may act as a fertilizer to promote growth whereas, with solutions dominated by H_2SO_4, damage is more apparent (Scott & Hutchinson, 1989). Most studies still emphasize the damage caused to photosynthesis and nitrogen fixation by spraying or immersion of lichens in solutions below *c.* pH 3 (Nash & Gries, 1991). Although there are some photosynthetic effects attributable to nitrogen oxide damage (Nash, 1976), these often reflect environmentally unrealistic concentrations. Nitrate salts have been used as substitutes for NO_x, resulting in enhancement of chlorophyll concentrations when supplied alone but causing apparently synergistic damage in combination with SO_2 (Balaguer & Manrique, 1991). Field observations suggest gaseous oxidants may induce floristic modifications (Sigal & Nash, 1983), but laboratory fumigations using ozone did not show the photosynthetic sensitivity of different species to be as predicted (Ross & Nash, 1983). McCune (1988) found floristic patterns related more to mean, rather than peak, SO_2 concentrations and not to O_3 emissions.

Agricultural effects

Agricultural emissions are also characterized by elevated nitrogenous chemicals, but these are based more on ammonia gas or ammonium ions (Brown, 1992). Studies in the Netherlands have shown a shift from an acidophytic flora of species previously associated with high SO_2 concentrations to a flora enriched with species that are generally referred to as 'nitrophytic' (van Herk, 1991; van Dobben, 1993). There has been a clear decline in SO_2 emissions in the Netherlands, which has exposed a high background level of ammoniacal pollution, derived predominantly from farm animals, with a concomitant increase in substratum pH (Fig. 16.4). The correlation between a decline in SO_2 levels and expansion of such species in the Netherlands might suggest considerable sensitivity of nitrophytic species to this gas. However, as mentioned above, declining SO_2 levels do not invariably result in the species shift seen in the Netherlands, especially in the absence of farming activity. It is not yet clear whether nitrophytic species of genera such as *Physcia* and *Xanthoria* are currently successful because they require ammonia, ammonium cations, or a high pH, although computer-assisted analyses show correlations with such factors (van Dobben, 1993). Comparable communities are found in the absence of agricultural activity in Mediterranean regions, or on trees subjected to dust impregnation, especially from calcareous sources (Gilbert, 1976).

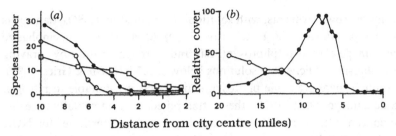

Fig. 16.3. Effect of distance from the centre of Newcastle-upon-Tyne on species distribution. (a) Change in species number on asbestos (□), sandstone (●) and ash trees (○). (b) Change in relative cover, showing replacement of crustose *Lecanora conizaeoides* (●) by fruticose *Evernia prunastri* (○) (Redrawn from Gilbert, 1973.)

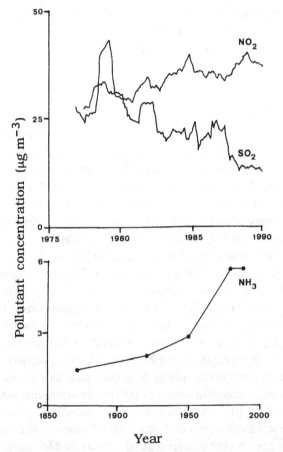

Fig. 16.4. Changes in aerial concentrations of sulphur dioxide, nitrogen dioxide (measured) and ammonia (estimated from livestock numbers) in The Netherlands in recent years. (Redrawn from van Dobben, 1993.)

The above examples emphasize that a polluted environment is rarely dominated by a single type of emission. In an environment of mixed pollutants, although tempting, it is difficult to establish whether one chemical is solely responsible for observed changes. Measurements should be made of all potentially contributory materials (Herzig *et al.*, 1989). For example, variations in the supply of physiologically essential or potentially toxic elements can modify pollution patterns ascribed to gaseous pollutants. Some, for example, strontium, nickel and zinc, can alleviate damage while others, for example, copper and lead, exacerbate the problem of SO_2 damage (Nieboer *et al.*, 1979; Richardson *et al.*, 1979).

Uptake of chemicals by lichens

For a pollutant to cause damage it must in some way interact with the organism involved. Chemical analysis of lichens has shown many situations where correlations exist between floristic or physiological damage and recovery of the anticipated pollutant, mainly sulphur, in the lichen thallus (Grace *et al.* 1985*b*; Folkeson & Andersson-Bringmark, 1987; Richardson, 1992). Fluorine, emitted into the otherwise relatively unpolluted environment around a rural aluminium smelter, showed damage related to the exposure aspect of the thallus (Perkins, 1982). Analyses of fluorine content showed that damage was proportional to the fluorine concentration recovered in the thallus, irrespective of exposure aspect.

There are many reports whereby concentrations of heavy metals recovered from lichens have been used to create dispersion maps related to distance from emission sources (factories, smelters, traffic on roads, etc), wind directions and topography (Puckett, 1988; Walther *et al.*, 1990; Gailey & Lloyd, 1993). In situations where emissions of metal-rich particulate matter dominate, an exponential decline in lichen metal content (Fig. 16.5) is frequently found with transplanted or *in situ* material (Nieboer *et al.*, 1982; Déruelle, 1984). Metal contents can be related to atmospheric metal burdens in particulate trappers and rain gauges (e.g. Pilegaard, 1979; Kansanen & Venetvaara, 1991), and correlations between metals often coincide with their joint occurrence in particles (Nieboer *et al.*, 1982).

Metal-rich particles have been observed within lichen thalli (Olmez, Cetin Gulovali & Gordon, 1985), but superficial contamination by dry deposition will also occur (Boonpragob & Nash, 1990). Reports of the accumulation of metals with increasing thallus age can be an accurate

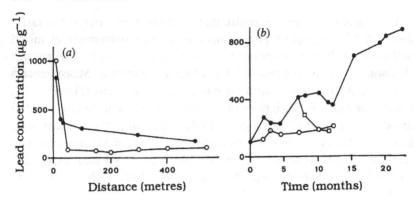

Fig. 16.5. Recovery of lead from lichens: (*a*) sampled at different distances from a road *Parmelia caperata* (○) *Cladonia portentosa* (●) and (*b*) *Parmelia caperata* transplanted to 8 m (●) and 100 m (○) from a road and reverse transplanted after 7 months (□). (Redrawn from Déruelle, 1984.)

measure of a sequential accumulation of particles but, if soluble positively charged elements are involved, also reflects the increasing cation exchange capacity with maturation and/or senescence of tissues (Brown, 1987). Accumulation of soil particles at the base of terricolous fruticose species, for example, *Cladonia* spp., may reflect splash contamination. Rhizinae may behave as filter systems for the trapping of particles (Goyal & Seaward, 1981; Brown, 1991).

In addition to reporting dispersion patterns, biomonitoring of metal emissions should be working towards being quantitatively capable of predicting the nature and concentration of recent environmental contaminants. It may take days or weeks to achieve equilibrium with the prevailing conditions (Pilegaard, 1979), following the introduction of thalli to contaminated areas, although changes resulting from a single period of rain have been reported (Brown & Brown, 1991). Losses of contaminant metals have been shown when plants were returned to uncontaminated environments (Fig. 16.5) (Déruelle, 1984) or when emission rates decline (Walther *et al.*, 1990). The rate of change in concentration is dependent on the chemical involved and its form and availability. Superficial deposits can be washed off with prolonged rainfall whereas entrapped particles or biogenically formed extracellular inclusions, for example, chelates with oxalates or phenolic lichen substances, tend to be retained (Boonpragob & Nash, 1990; Brown, 1991; Brown & Brown, 1991). Soluble cations equilibrate with extracellular exchange sites, including the cell wall and exterior of the plasma membrane.

Binding to these sites conforms to physico-chemically controlled processes, dominated by the affinity of elements for the chemically different exchange sites and the relative proportions of introduced and pre-bound elements. Equilibration can be rapid and under field conditions the exchangeable fraction is liable to be the most dynamic and unstable fraction. When lichens are stored in a moist condition, ammonium ions can be lost from extracellular exchange sites within hours of collection or artificial addition (Brown *et al.*, 1994). Elements initially bound at this site can be released for subsequent entry into the cell; this cellular location represents a dynamic reservoir rather than a uni-directional repository of elements (Brown & Beckett, 1985). For elements entering the cell interior, passage across the plasma membrane requires some kind of biologically controlled carrier system (see Brown & Brown, 1991). Entry may again involve inter-element competition and be influenced by factors such as pH and the energetic capacity of the cell.

Heavy metal tolerance

Biomonitoring studies have often reported lichens containing apparently excessive concentrations of potentially toxic elements (Brown, 1991). This has led to comments implying lichens have considerable tolerance to such elements. The basis for such apparent tolerance frequently rests on the metals being contained in insoluble particles trapped within the thallus. Studies in which soluble elements have been supplied to lichens in the laboratory (see Brown & Beckett, 1985; Brown & Brown, 1991 for references) have shown that, when taken from unpolluted sites, species containing cyanobacterial photobionts are considerably more sensitive to added heavy metals than are green-algae-containing species (often within the same genus). However, cyanobacterial members of the genus *Peltigera* are often encountered on abandoned heavy-metal-contaminated mine sites. Such material is enriched with heavy metals and shows less depression of photosynthesis when supplied with otherwise toxic levels of the same metals encountered in the field. This inferred tolerance has been coupled to a demonstration of reduced intracellular uptake of cadmium by material taken from polluted sites (Fig. 16.6). This implied tolerance could be the result of an exclusion mechanism associated with modified membrane properties. Subsequently, it has been shown, initially with bryophytes, that stripping physiological elements from the exterior exchange sites with high levels of potassium eliminates any differences in intracellular uptake rates (Wells, Brown & Beckett,

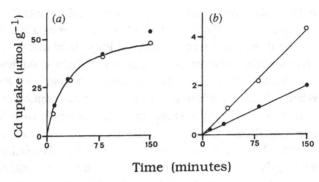

Fig. 16.6. Uptake of cadmium by *Peltigera membranacea* to (*a*) extra-cellular and (*b*) intracellular sites with material from an unpolluted (○) and a heavy metal contaminated (●) site. (Redrawn from Brown & Beckett, 1985.)

1995). This was the result of removing cations that would otherwise compete with cadmium for trans-membrane carrier systems. Such work emphasises the need to interpret carefully the basis for apparent tolerance mechanisms because many of these may be more fortuitous than genuine tolerance.

Biochemical changes related to pollution stress

The value of laboratory studies designed to investigate changes that may be induced by exposure to polluted environments must be treated with some degree of caution. It is tempting to apply acute stresses and imply that chronic conditions would result in a less dramatic response of the same kind. Chronic stress requires more notice to be taken of interactions with other environmental conditions. It is clear that the capacity of lichens to withstand (perhaps require) alternating diurnal wetting and drying phases, and to show adaptations to shifts in environmental conditions further complicates the multiplicity of factors required to be considered when assessing pollution stress. Transplantation into polluted environments has become a popular approach for assessing the extent of pollution damage.

Besides the physiological processes of photosynthesis, respiration and nitrogen fixation, a number of biochemical parameters have been tested for sensitivity to pollutant stress. These may not be specific to particular stresses but are perhaps more sensitive to chronic pollution incidents. Chlorophyll damage is not easy to assess because many of the organic solvents used can also extract, from certain species, acidic lichen substances that can cause phaeophytinization during extraction. The use of dimethyl sulphoxide as a solvent, with or without prior extraction of lichen substances or the addition of a neutralizing salt (combined with the measurement of wavelengths particular to chlorophyll and phaeophytin), has become a method of generalized damage assessment (Barnes *et al.*, 1992). Phaeophytin formation in the lichen may be a natural symptom of stress, especially in conditions involving acidification within the cell (Kardish *et al.*, 1987; Garty *et al.*, 1988). Chlorophyll fluorescence has not received much attention in relation to pollution damage.

M. Galun and co-workers have employed ATP concentrations, ratios of ATP:ADP:AMP, or energy charge for general stress monitoring (Kardish *et al.*, 1987; Garty *et al.*, 1988). Free radical damage has recently become fashionable and measurements of glutathione concentration, the proportions of oxidized to total thiol content, or ascorbate content have also been used as measures of pollutant damage (Kranner *et al.*, 1992; M. Galun, personal communications). There are also preliminary reports proposing measurement of enzymic changes associated with free-radical-scavenging mechanisms, for example, peroxidase and superoxide dismutase, for stress monitoring in laboratory and field (transplanted) systems (Köck, Schlee & Metzger, 1985; M. Galun, personal communications). These have much potential but so far have only been of value for detecting generalized damage responses. More work needs to be done on nitrate- and ammonium-assimilating enzymes under normal conditions (Rai, 1988; Brown & Tomlinson, 1993). It is very probable that these inducible enzymes will respond specifically to nitrogenous pollution.

Contribution of the fungus

The above examples have mostly concentrated on photobiont sensitivity. How far the fungus takes part in such responses to environmental change has been hard to establish. More attention could be paid to the changes in carbohydrate metabolism under stress, especially noting changes in transfer of carbohydrates or synthesis of exclusively fungal products (Fields & St Clair, 1984). Metal uptake studies have indicated high levels of heavy metals recovered from purely fungal tissues such as rhizinae or the medulla, following surgical excision (Goyal & Seaward, 1981. Because of the variability of the results and the use of exposure techniques that overcome the normal barriers for solution penetration (Goyal

276   D. H. Brown

& Seaward, 1982*b*), Brown (1991) cast doubt on the validity of these observations. They are, however, all that is currently available. Perhaps future studies should attempt to achieve differential damage to the symbionts. Digitonin has been used to selectively damage eukaryotic fungal cells in studies on the release of carbon and nitrogen compounds from the photobiont of cyanobacterial lichens (Rai, 1988). Applications of selective pesticides have already shown physiological damage to lichens (Brown, 1992). Careful selection of fungicides or herbicides may act as a probe for such studies.

References

Ahmadjian, V. (1993). *The Lichen Symbiosis*. New York: John Wiley.
Armstrong, R. A. (1988). Substrate colonization, growth, and competition. In *CRC Handbook of Lichenology, Vol. II*. ed. M. Galun, pp. 3–16. Boca Raton: CRC Press.
Balaguer, L. & Manrique, E. (1991). Interaction between sulfur dioxide and nitrate in some lichens. *Environmental and Experimental Botany*, **31**, 223–7.
Barnes, J. D., Balaguer, L., Manrique, E., Elvira, S. & Davison, A. W. (1992). A reappraisal of the use of DMSO for the extraction and determination of chlorophyll *a* and *b* in lichens and higher plants. *Environmental and Experimental Botany*, **32**, 85–100.
Bates, J. W. & Farmer, A. M. (1992). *Bryophytes and Lichens in a Changing Environment*. Oxford: Clarendon Press.
Boonpragob, K., Nash III, T. H. & Fox, C. A. (1989). Seasonal deposition patterns of acidic ions and ammonium to the lichen *Ramalina menziesii* Tayl. in southern California. *Environmental and Experimental Botany*, **29**, 187–97.
Boonpragob, K. & Nash, III, T. H. (1990). Seasonal variation of elemental status in the lichen *Ramalina menziesii* Tayl. from two sites in southern California: evidence for dry deposition accumulation. *Environmental and Experimental Botany*, **30**, 415–28.
Brown, D. H. (1987). The location of mineral elements in lichens; implications for metabolism. *Bibliotheca Lichenologica*, **25**, 361–75.
Brown, D. H. (1991). Lichen mineral studies – currently clarified or confused? *Symbiosis*, **11**, 207–23.
Brown, D. H. (1992). Impact of agriculture on bryophytes and lichens. In *Bryophytes and Lichens in a Changing Environment*, ed. J. W. Bates & A. M. Farmer, pp. 259–83. Oxford: Clarendon Press.
Brown, D. H. & Beckett, R. P. (1985). Uptake and effects of cations on lichen metabolism. *Lichenologist*, **16**, 173–88.
Brown, D. H. & Beckett, R. P. (1985). The role of the cell wall in the intracellular uptake of cations by lichens. In *Lichen Physiology and Cell Biology*, ed. D. H. Brown, pp. 247–58. London: Plenum Press.
Brown, D. H. & Brown, R. M. (1991). Mineral cycling and lichens: the physiological basis. *Lichenologist*, **23**, 293–307.
Brown, D. H., Standell, C. J. & Miller, J. E. (1995). Effects of agricultural chemicals on lichens. *Cryptogamic Botany*, **5**, (In press).

Brown, D. H. & Tomlinson, H. (1993). Effects of nitrogen salts on lichen physiology. *Bibliotheca Lichenologica*, **53**, 27–34.

Coxson, D. S. (1988). Recovery of net photosynthesis and dark respiration on rehydration of the lichen *Cladina mitis*, and the influence of prior exposure to sulphur dioxide while desiccated. *New Phytologist*, **108**, 483–7.

Crittenden, P. D. (1989). Nitrogen relations of mat-forming lichens. In *Nitrogen, Phosphorus and Sulphur Utilization by Fungi*, ed. L. Boddy, R. Marchant & D. J. Read, pp. 243–68. Cambridge: Cambridge University Press.

Déruelle, S. (1984). L'utilisation des lichens pour la détection et l'estimation de la pollution par le plomb. *Bulletin Ecologique*, **15**, 1–6.

Farmer, A. M., Bates, J. W. & Bell, J. N. B. (1992). Ecophysiological effects of acid rain on bryophytes and lichens. In *Bryophytes and Lichens in a Changing Environment*, ed. J. W. Bates & A. M. Farmer, pp. 284–313. Oxford: Clarendon Press.

Fields, R. C. & St. Clair, L. L. (1984). The effects of SO_2 on photosynthesis and carbohydrate transfer in the two lichens: *Collema polycarpon* and *Parmelia chlorochroa*. *American Journal of Botany*, **71**, 986–98.

Folkeson, L. & Andersson-Bringmark, E. (1987). Impoverishment of vegetation in a coniferous forest polluted by copper and zinc. *Canadian Journal of Botany*, **66**, 417–28.

Gailey, F. A. Y. & Lloyd, O. L. (1993). Spatial and temporal patterns of airborne metal pollution: the value of low technology sampling to an environmental epidemiology study. *Science of the Total Environment*, **133**, 201–19.

Garty, J., Kardish, N., Hagemeyer, J. & Ronen, R. (1988). Correlations between the concentration of adenosine triphosphate, chlorophyll degradation and the amounts of airborne heavy metals and sulphur in a transplanted lichen. *Archives of Environmental Contamination and Toxicology*, **17**, 601–11.

Gilbert, O. L. (1973). Lichens and air pollution. In *The Lichens*, ed. V. Ahmadjian & M. E. Hale, pp. 443–472. New York: Academic Press.

Gilbert, O. L. (1976). An alkaline dust effect on epiphytic lichens. *Lichenologist*, **8**, 173–8.

Gilbert, O. L. (1992). Lichen reinvasion with declining air pollution. In *Bryophytes and Lichens in a Changing Environment*, ed. J. W. Bates & A. M. Farmer, pp. 159–77. Oxford: Clarendon Press.

Goyal, R. & Seaward, M. R. D. (1981). Metal uptake in terricolous lichens. I. Metal localization within the thallus. *New Phytologist*, **89**, 631–45.

Goyal, R. & Seaward, M. R. D. (1982a). Metal uptake in terricolous lichens. II. Effects on the morphology of *Peltigera canina* and *P. rufescens*. *New Phytologist*, **90**, 73–84.

Goyal, R. & Seaward, M. R. D. (1982b). Metal uptake in terricolous lichens. III. Translocation in the thallus of *Peltigera canina*. *New Phytologist*, **90**, 85–98.

Grace, B., Gillespie, T. J. & Puckett, K. J. (1985a). Sulphur dioxide threshold concentration values for *Cladina rangiferina* in the Mackenzie Valley, N. W. T. *Canadian Journal of Botany*, **63**, 806–12.

Grace, B., Gillespie, T. J. & Puckett, K. J. (1985b). Uptake of gaseous sulphur dioxide by the lichen *Cladina rangiferina*. *Canadian Journal of Botany*, **63**, 797–805.

Green, T. G. A. & Lange, O. L. (1991). Ecophysiological adaptations of the lichen genus *Pseudocyphellaria* and *Sticta* to south temperate rainforests. *Lichenologist*, **23**, 267–82.

Hawksworth, D. L. (1988a). The fungal partner. In *CRC Handbook of Lichenology, Vol. I.* ed. M. Galun, pp. 35–8. Boca Raton: CRC Press.

Hawksworth, D. L. (1988b). The variety of fungal–algal symbioses, their evolutionary significance, and the nature of lichens. *Botanical Journal of The Linnean Society*, **96**, 3–20.

Hawksworth, D. L., Coppins, B. J. & Rose, F. (1974). Changes in the British lichen flora. In *The Changing Flora and Fauna of Britain*, ed. D. L. Hawksworth, pp. 47–78. London: Academic Press.

Hawksworth, D. L. & Rose, F. (1970). Qualitative scale for estimating sulphur dioxide air pollution in England and Wales using epiphytic lichens. *Nature*, **227**, 145–8.

Herzig, R., Liebendörfer, L., Urech, M., Ammann, K., Buecheva, M. & Landolt, W. (1989). Passive biomonitoring with lichens as a part of an integrated biological measuring system for monitoring air pollution in Switzerland. *International Journal of Environmental Analytical Chemistry*, **35**, 43–57.

Hill, D. J. (1976). The physiology of lichen symbiosis. In *Lichenology: Progress and Problems*, ed. D. H. Brown, D. L. Hawksworth & R. H. Bailey, pp. 457–96. London: Academic Press.

Honegger, R. (1991). Functional aspects of the lichen symbiosis. *Annual Review of Plant Physiology and Plant Molecular Biology*, **42**, 553–78.

Huebert, D. B., L'Hirondelle, S. J. & Addison, P. A. (1985). The effects of sulphur dioxide on net CO_2 assimilation in the lichen *Evernia mesomorpha*. *New Phytologist*, **100**, 643–51.

Jahns, H. M. (1984). Morphology, reproduction and water relations – a system of morphological interactions in *Parmelia saxatilis*. *Beiheft zur Nova Hedwigia*, **79**, 716–32.

Jahns, H. M. (1988). The lichen thallus. In *CRC Handbook of Lichenology, Vol. I*, ed. M. Galun, pp. 95–143. Boca Raton: CRC Press.

Kansanen, P. H. & Venetvaara, J. (1991). Comparison of biological collectors of airborne heavy metals near ferrochrome and steel works. *Water, Air and Soil Pollution*, **60**, 337–59.

Kappen, L. (1988). Ecophysiological relationships in different climatic regions. In *CRC Handbook of Lichenology, Vol. II*. ed. M. Galun, pp. 37–100. Boca Raton: CRC Press.

Kardish, N., Ronen, R., Bubrick, P. & Garty, J. (1987). The influence of air pollution on the concentration of ATP and on chlorophyll degradation in the lichen *Ramalina duriaei* (De Not.) Bagl. *New Phytologist*, **106**, 697–706.

Kershaw, K. A. (1985). *Physiological Ecology of Lichens*. Cambridge: Cambridge University Press.

Köck, M., Schlee, D. & Metzger, U. (1985). Sulfite-induced changes of oxygen metabolism and the action of superoxide dismutase in *Euglena gracilis* and *Trebouxia* sp. *Biochemie und Physiologie der Pflanzen*, **180**, 213–24.

Kranner, I., Guttenberger, H., Grill, D., Delefant, M. & Türk, R. (1992). Investigations of thiols in lichens. *Phyton (Horn)*, **32**, 69–73.

Lange, O. L., Kilian, E. & Ziegler, H. (1986). Water vapor uptake and photosynthesis of lichens: performance differences in species with green and blue-green algae as phycobionts. *Oecologia*, **71**, 104–10.

Lange, O. L., Pfanz, H., Kilian, E. & Meyer, A. (1990). Effect of low water potential on photosynthesis in intact lichens and their liberated algal components. *Planta*, **182**, 467–72.

Larson, D. W. (1983). The pattern of production within individual *Umbilicaria* lichen thalli. *New Phytologist*, **94**, 409–19.

Larson, D. W., Matthes-Sears, U. & Nash, III, T. H. (1986). The ecology of *Ramalina menziesii*. II. Variation in water relations and tensile strength across an inland gradient. *Canadian Journal of Botany*, **64**, 6–10.

Lines, C. E. M., Ratcliffe, R. G., Rees, T. A. V. & Southon, T. E (1989). A ^{13}C NMR study of photosynthate transport and metabolism in the lichen *Xanthoria calcicola* Oxner. *New Phytologist*, **111**, 447–56.

McCune, B. (1988). Lichen communities along O_3 and SO_2 gradients in Indianapolis. *Bryologist*, **91**, 223–8.

Nash III, T. H. (1976). Sensitivity of lichens to nitrogen dioxide fumigation. *Bryologist*, **79**, 183–6.

Nash III, T. H. (1988). Correlating fumigation studies with field effects. *Bibliotheca Lichenologica*, **30**, 201–16.

Nash III, T. H., Boucher, V. L., Gebauer, R. & Larson, D. W. (1990). Morphological and physiological plasticity in *Ramalina menziesii*: studies with reciprocal transplants between a coastal and inland site. *Bibliotheca Lichenologica*, **38**, 357–65.

Nash III, T. H. & Gries, C. (1991). Lichens as indicators of air pollution. In *The Handbook of Environmental Chemistry*, ed. O. Hutzinger, pp. 1–29. Berlin: Springer-Verlag.

Nash III, T. H., Moser, T. J. & Link, S. O. (1980). Nonrandom variation of gas exchange within arctic lichens. *Canadian Journal of Botany*, **58**, 1181–6.

Nash III, T. H. & Wirth, V. (1988). *Lichens, Bryophytes and Air Quality*. *Bibliotheca Lichenologica*, **30**.

Nieboer, E., Richardson, D. H. S., Lavoie, P. & Padovan, D. (1979). The role of metal-ion binding in modifying the toxic effects of sulphur dioxide on the lichen *Umbilicaria muhlenbergii*. I. Potassium efflux studies. *New Phytologist*, **82**, 621–32.

Nieboer, E., Richardson, D. H. S., Boileau, L. J. R., Beckett, P. J., Lavoie, P. & Padovan, D. (1982). Lichens and mosses as monitors of industrial activity associated with uranium mining in Northern Ontario, Canada – Part 3: Accumulations of iron and titanium and their mutual dependence. *Environmental Pollution (Series B)*, **4**, 181–92.

Nimis, P. L., Castello, M. & Perotti, M. (1990). Lichens as biomonitors of sulphur dioxide pollution in La Spezia (northern Italy). *Lichenologist*, **22**, 333–44.

Olmez, I., Cetin Gulovali, M. & Gordon, G. E. (1985). Trace element concentrations in lichens near a coal-fired power plant. *Atmospheric Environment*, **19**, 1663–9.

Pearson, L. C. & Brammer, E. (1978). Rate of photosynthesis and respiration in different lichen tissues by the Cartesian Diver technique. *American Journal of Botany*, **65**, 276–81.

Perkins, D. F. (1992). Relationship between fluoride contents and loss of lichens near an aluminium works. *Water, Air, and Soil Pollution*, **64**, 503–10.

Pilegaard, K. (1979). Heavy metals in bulk precipitation and transplanted *Hypogymnia physodes* and *Dicranoweisia cirrata* in the vicinity of a Danish steelworks. *Water, Air and Soil Pollution*, **11**, 77–91.

Puckett, K. J. (1988). Bryophytes and lichens as monitors of metal deposition. *Bibliotheca Lichenologica*, **30**, 231–67.

Rai, A. N. (1988). Nitrogen metabolism. In *CRC Handbook of Lichenology, Vol. I*. ed. M. Galun, pp. 201–37. Boca Raton: CRC Press.

Richardson, D. H. S. (1988). Understanding the pollution sensitivity of lichens. *Botanical Journal of the Linnean Society*, **96**, 31–43.

Richardson, D. H. S. (1992). *Pollution Monitoring with Lichens*. Naturalists' Handbooks, **19**. Slough: Richmond Publishing Co. Ltd.

Richardson, D. H. S., Nieboer, E., Lavoie, P. & Padovan, D. (1979). The role of metal-ion binding in modifying the toxic effects of sulphur dioxide on the lichen *Umbilicaria muhlenbergii*. II. ^{14}C-fixation studies. *New Phytologist*, **82**, 633–43.

Rose, F. (1992). Temperate forest management: its effects on bryophyte and lichen floras and habitats. In *Bryophytes and Lichens in a Changing Environment*, ed. J. W. Bates & A. M. Farmer, pp. 211–33. Oxford: Clarendon Press.

Ross, L. J. & Nash III, T. H. (1983). Effect of ozone on gross phyotosynthesis of lichens. *Environmental and Experimental Botany*, **23**, 71–7.

Rundel, P. W. (1988). Water relations. In *CRC Handbook of Lichenology, Vol. II*, ed. M. Galun, pp. 17–36. Boca Raton: CRC Press.

Sancho, L. G. & Kappen, L. (1989), Photosynthesis and water relations and the role of anatomy in Umbilicariaceae (lichenes) from Central Spain. *Oecologia*, **81**, 473–80.

Scott, M. G. & Hutchinson, T. C. (1989), A comparison of the effects on Canadian boreal forest lichens of nitric and sulfuric acids as sources of rain acidity. *New Phytologist*, **111**, 663–71.

Seaward, M. R. D. (1976). Performance of *Lecanora muralis* in an urban environment. In *Lichenology: Progress and Problems*, ed. D. H. Brown, D. L. hawksworth & R. H. Bailey, pp. 323–57. London: Academic Press.

Seaward, M. R. D. & Hitch, C. J. B. (1982). *Atlas of the Lichens of the British Isles. Vol. 1*. Cambridge: Institute of Terrestrial Ecology.

Showman, R. E. (1988). Mapping air quality with lichens, the North American experience. *Bibliotheca Lichenologica*, **30**, 67–89.

Sigal, L. L. & Nash III, T. H. (1983). Lichen communities on conifers in Southern California mountains: an ecological survey relative to oxidant air pollution. *Ecology*, **64**, 1343–54.

Snelgar, W. P. & Green, T. G. A. (1981). Ecologically-linked variation in morphology, acetylene reduction, and water relations in *Pseudocyphellaria dissimilis*. *New Phytologist*, **87**, 403–11.

Tschermak-Woess, E. (1988). The algal partner. In *CRC Handbook of Lichenology, Vol. I*, ed. M. Galun, pp. 39–92. Boca Raton: CRC Press.

Tysiaczny, M. J. & Kershaw, K. A. (1979). Physiological-environmental interactions in lichens. VII. The environmental control of glucose movement from alga to fungus in *Peltigera canina* var. *praetextata* Hue. *New Phytologist*, **83**, 137–46.

van Dobben, H. F. (1993). Vegetation as a monitor for deposition of nitrogen and acidity. PhD thesis, University of Utrecht.

van Herk, C. M. (1991). *Korstmossen als indicator voor zure depositie. Basisrapport; Samenvattend Rapport*. Arnhem: Provincie Gelderland, dienst MW en dienst RGW.

Walther, D. A., Ramelow, G. J., Beck, J. N., Young, J. C., Callahan, J. D. & Marcon, M. F. (1990). Temporal changes in metal levels of the lichens

Parmotrema praesorediosum and *Ramalina stenospora*, southwest Louisiana. *Water, Air and Soil Pollution*, **53**, 189–200.

Wells, J. M., Brown, D. H. & Beckett, R. P. (1995). Kinetic analysis of Cd uptake in Cd-tolerant and intolerant populations of the moss *Rhytidiadelphus squarrosus* (Hedw.) Warnst. and the lichen *Peltigera membranacea* (Ach.) Nyl. *New Phytologist* **129**, 477–86.

17

Fungal interactions with metals and radionuclides for environmental bioremediation

I. SINGLETON AND J. M. TOBIN

Introduction

Microorganisms, including fungi, are known to accumulate metals from their external environment and the possibility of using fungi as a means of treating metal/radionuclide-containing effluents is well recognized (Siegel, Galun & Siegel, 1990; Gadd, 1993). However, to date, there are no commercial systems in operation which specifically use fungi as a basis for a metal treatment system. This is despite the fact that certain fungal species, under optimal conditions, are as effective as ion exchange resins in the removal of metals from solution (Tsezos & Volesky, 1981). As yet, the development of this potential from scientific curiosity to commercial fact remains to be demonstrated.

The mechanisms of microbial metal uptake may be either independent of, or dependent on, cell metabolism (Huang, Huang & Morehart, 1990; Avery & Tobin, 1992). Metal uptake which is independent of cell metabolism will be referred to as biosorption in this work. It is generally regarded that biosorptive metal uptake mechanisms would be more appropriate for use in a metal treatment system (Kuyucak, 1990). This is because environmental conditions in most effluents may be too toxic for microbial growth. Biosorption, in many cases, accounts for most of the metal accumulated by the cell, and can represent 10–20% or more of the cell dry weight (Luef, Prey & Kubicek, 1991; Gadd, 1993). The process occurs by either physical or chemical means and usually involves surface interactions of metals with microbial cell walls or excreted cell products. This contrasts with energy dependent metal uptake where the mechanisms involved result in intracellular metal accumulation.

A review of the literature reveals that a wide variety of methods have been used to assess microbial metal-binding capacities. The differing methodologies and varying uptake mechanisms involved make the interpretation and comparison of results difficult. Further complications arise as many other factors, including cell age (Junghans & Straube, 1991; Volesky, May & Holan, 1993) and the composition of the medium used to grow the organisms (Treen-Sears, Martin & Volesky, 1984), can affect metal uptake characteristics. A more standardized approach would clearly facilitate the interpretation of experimental results and contribute to more rapid developments in the field. In addition, the mechanisms and factors involved in fungal metal uptake are not fully understood and require further elucidation in order that actual process optimization may ultimately be achieved.

This discussion will concentrate mainly on the processes involved in metal biosorption by fungi and will attempt to assess the work in a critical fashion. Some of the more recent advances being made in the area of metal/fungal interactions for environmental bioremediation will also be considered.

Methods used to assess fungal metal uptake

The initial assessment of the affinity of fungal (also bacterial and algal) biomass is commonly made using a stirred batch contacting system. Essentially the method involves the incubation of a known amount of biomass with different concentrations of metal. Incubation is typically carried out in a fixed volume of solution at a single pH value for a period of time that allows the metal concentration in solution to reach equilibrium. Results may be plotted as adsorption isotherms in the form of amount of adsorbed metal (mmol metal g^{-1} biomass, q_e) versus final or equilibrium metal concentration (c_f). Various methods are available to analyse the results and these are described in the following sections.

The metal concentrations used typically range from values at which not all the binding sites available on the biomass are occupied to values corresponding to biomass saturation, that is, where all the binding sites are occupied. Recent work has highlighted the importance of studies at lower metal concentrations (Avery & Tobin, 1993). It appears that biomass metal affinities may exceed those of commercial ion-exchange resins at low metal concentrations and as a consequence biosorption systems have been proposed as adjuncts or polishing stages for conventional waste treatment schemes (Brierley, Goyak & Brierley, 1986; Gadd, 1993).

Adsorption isotherms are most suited to metal uptake studies with denatured biomass or live biomass which has no energy reserves or is metabolically inactive. To estimate metal uptake that is dependent on cell metabolism, continuous or frequent measurements of metal removal over time are required. In order to overcome the need for multiple discrete sampling followed by biomass separation some workers have used ion-selective electrodes to measure metal uptake continuously (de Rome & Gadd, 1987a; I. Singleton & P. Simmons, unpublished observations).

Experimental data can vary greatly depending on the method used. For example, some authors have reported that by increasing the metal: biomass ratio, that is, by decreasing the amount of biomass relative to a fixed volume and concentration of metal solution, the specific metal uptake capacity of the biomass increases (de Rome & Gadd, 1987b; Junghans & Straube, 1991; Luef et al., 1991). As discussed in the following sections, a variety of environmental factors including metal and co-ion concentrations (Luef et al., 1991; Huang et al., 1990) and solution pH (Galun et al., 1987; Kuyucak & Volesky, 1988; Mullen et al., 1992) also affect metal binding. Consequently, comparison of results from different studies is difficult since experimental conditions vary considerably. However, there is benefit to be derived from tables which give uptake comparisons provided the above factors are noted (see Table 17.1). Significant metal binding capacities are exhibited by a range of different fungal species, and where the experimental methods are comparable the results from different studies show a measure of consistency. Systems exhibiting higher than average uptake levels merit further investigation. For example, the high level of copper uptake by *Cunninghamella blakesleeana* was obtained by pregrowth of the fungus in medium containing high cobalt levels (Venkateswerlu & Stotzky, 1989). The increased uptake was attributed to higher levels of phosphate and chitosan in the cell walls of the fungus which resulted from growth in the cobalt-rich medium. This point underscores the importance of elucidating the mechanisms of metal binding and their dependency on environmental and culture conditions.

Biosorption mechanisms

Biosorption results from the physical/chemical interactions of charged metal ions in solution with the constituents and reactive groups of the cell wall. These interactions may include ion exchange, adsorption, complexation, crystallization and precipitation (Kuyucak & Volesky, 1988; McEldowney, 1990; Tobin, Cooper & Neufeld, 1984, 1990). Because the

Table 17.1. *Metal uptake capacities of various fungal species*

Metal	Organism	Metal uptake capacity (mmol metal g^{-1} biomass)	Reference
Copper	S. cerevisiae	0.68	Junghans & Straube (1991)
	Candida utilis	0.54	Junghans & Straube (1991)
	Cryptococcus terreus	1.13	Junghans & Straube (1991)
	Rhizopus arrhizus	0.25	Tobin, Cooper & Neufeld (1984)
	Cunninghamella blakesleeana	1.9	Venkateswerlu & Stotzky (1989)
Silver	Aspergillus niger	0.21	Mullen et al. (1992)
	Mucor rouxii	0.16	Mullen et al. (1992)
	Yeast species	0.004	Pumpel & Schinner (1986)
	Phoma (PT 35)	0.19	Pumpel & Schinner (1986)
	Penicillium chrysogenum	0.25	Pighi, Pumpel & Schinner (1989)
	Botrytis cinerea	0.46	Pighi, Pumpel & Schinner (1989)
	Rhizopus arrhizus	0.42	Tobin et al. (1984)
Zinc	S. cerevisiae	0.47	Kuyucak & Volesky (1988)
	P. chrysogenum	7.65	Luef et al. (1991)
	Claviceps paspali	15.30	Luef et al. (1991)
	Saccharomyces	0.12	Singleton & Simmons*
Uranium	R. arrhizus	1.29	Tsezos, McReady & Bell (1989)
	R. arrhizus	0.82	Tobin et al. (1984)
	A. niger	0.12	Tobin et al. (1984)
	A. niger	0.90	Yakubu & Dudeney (1986)
	S. cerevisiae	0.66	Kuyucak & Volesky (1988)
Cobalt	C. blakesleeana	0.52	Venkateswerlu & Stotzky (1989)
	Saccharomyces	0.15	Singleton & Simmons*
Thorium	S. cerevisiae	0.27	Gadd & White (1989)
	Penicillium italicum	0.49	Gadd & White (1989)
	P. chrysogenum	0.84	Gadd & White (1989)
	A. niger	0.28	Gadd & White (1989)
	R. arrhizus	0.50	Gadd & White (1989)

Note: *Unpublished results from authors laboratory obtained using commercially available *Saccharomyces* species.

composition and morphology of the cell wall is not only species dependent but also varies with, for example, culture conditions and age, the nature and extent of these interactions may differ greatly (Remacle, 1990a; Avery & Tobin, 1992). However, a number of consistent features of the biosorption process may be highlighted.

Biosorption is a rapid and generally temperature-independent process. Typically, near equilibrium metal uptake levels are attained within minutes and are invariant with temperature in normal ranges (5–30 °C) (Tsezos, Baird & Shemilt, 1986; Huang *et al.*, 1990; Tsezos, 1990). It is also reversible and a range of acids and complexing agents have been demonstrated to elute metal from uptake sites in biomass (McEldowney, 1990; de Rome & Gadd, 1991; Junghans & Straube, 1991). This process is frequently non-destructive so that multiple uptake/elution cycles with negligible decrease in uptake levels have been reported (Treen-Sears *et al.*, 1984; Tsezos, McCready & Bell, 1989; Zhou & Kiff, 1991).

The uptake isotherm data described above often conform to well-known adsorption models. The Langmuir adsorption model which is based on assumptions of monolayer, non-interactive adsorption has been successfully used to fit biosorption data for fungi and yeasts as well as for other microbial biomass types (Tobin *et al.*, 1984; Byerley & Scharer, 1987; Fourest & Roux, 1992). The Freundlich and BET adsorption models have also been widely applied (Tsezos & Volesky, 1981; de Rome & Gadd, 1987b) although, in cases of complex interactions such as the formation of metal precipitates (see below) or metabolic uptake, agreement with theoretical models is poor (Gadd & White, 1989a; Volesky *et al.*, 1993). Scatchard analysis has been used to further evaluate metal uptake data for yeasts and fungi. Characteristic curved plots have been reported in numerous studies (Huang *et al.*, 1990; Avery & Tobin, 1993) indicating that multiple and different binding sites contribute to biosorption. While the limitations of such model-based approaches are recognized (Gadd & White, 1989a), the analyses suggest that biosorption involves metal adsorption to a heterogeneous biomass containing a range of potential binding sites (Tobin *et al.*, 1990; Avery & Tobin, 1993). The fungal cell wall may be likened to an ion-exchange resin whose capacity depends on the presence and spatial arrangement of the complexing functional groups. Polysaccharides represent the main constituents (up to 90% by weight) of fungal walls and are usually found in association with lipid and protein fractions. In filamentous fungi, the outer layers are predominantly neutral polysaccharides (glucans, mannans) while the inner layers contain increasing levels of glucosamines (chitin and chitosan) in a microfibrillar structure (Remacle, 1990*b*). Yeast cell walls have a simpler organization and consist mainly of glucans, which confer wall rigidity, and an external layer of mannoprotein. Bud scars, left behind after asexual reproduction, are known to contain small amounts of chitin. Potential metal-binding ligands within these matrices include carboxylate, amine, phosphate, hydroxyl, sulphydryl and other functional groups (Tobin *et al.*, 1984; McEldowney, 1990). In addition, it is likely that uptake sites have several different functional groups participating to various degrees in binding the ions.

Early studies of cadmium and cobalt accumulation by yeast identified sulphydryl groups as likely binding sites and suggested that uptake specificity was related to the ionic radius of the cation (Norris & Kelly, 1977). More recently, studies of the uptake of the uranyl ion, lanthanum and a series of divalent transition metal ions by *Rhizopus arrhizus* identified phosphate or carboxyl groups or both as the principal binding ligands and demonstrated a linear relationship between uptake capacity and ionic radius (Tobin *et al.*, 1984). Studies of uranium uptake by *Aspergillus niger* and *Saccharomyces cerevisiae* have also implicated carboxyl groups in metal binding, and it has been suggested that the combined effects of these groups in dicarboxylic acids lead to enhanced binding efficiency (Shumate & Strandberg, 1985; Yakubu & Dudeney, 1986).

Numerous studies involving uranium have underscored its high affinity for biomass and have led to detailed studies of fungal radionuclide binding (Strandberg, Shumate & Parrot, 1981; Tsezos & Volesky, 1981; Tobin *et al.*, 1984; de Rome & Gadd, 1991). Tsezos and Volesky (1982*a*) have suggested a three-fold mechanism for uranyl binding to *R. arrhizus* involving coordination to chitin groups followed by additional adsorption and precipitation of uranyl hydroxide within the cell wall. In contrast, electron microscopic studies showed that thorium binding to this biomass occurred at the cell surface, and the same authors postulated a mechanism of chitin coordination followed by hydrolysis and deposition of thorium hydroxide at the cell surface (Tsezos & Volesky, 1982*b*). Deposition of uranium in the form of needle-like fibrils was observed on the cell walls of *S. cerevisiae* where individual cells accumulated up to 50% metal on a dry weight basis (Strandberg *et al.*, 1981). Similar precipitation has been reported in thorium uptake by *S. cerevisiae* (Gadd & White, 1989*b*). Significant levels of thorium binding were also observed for a range of fungal species, including yeast and notably at pH 0–1 (Gadd & White, 1989*a*) at which the chitin groups would be protonated and not expected to contribute to metal binding (Tobin *et al.*, 1984). The authors postulated the existence of alternative unidentified binding sites at lower pH ranges (Gadd & White, 1989*a*).

Strontium biosorption has been demonstrated for a range of microorganisms with *Rhizopus* and *Penicillium* strains exhibiting high uptake capacities (Watson, Scott & Falcon, 1989). More recently, strontium binding to both denatured and non-metabolising live yeast cells was investigated. For the former biomass significant although non-stoichiometric release of Mg^{2+} suggested simple ionic binding while for non-metabolizing live cells a mixed ionic/covalent mechanism was reported (Avery & Tobin, 1992).

Denatured forms of various biomass types have been shown to be effective radium adsorbents with reported uptake levels exceeding those of conventional adsorbents (Tsezos & Keller, 1983). *Penicillium chrysogenum* exhibited optimal biosorption characteristics with typical equilibrium times of up to two minutes which were not significantly affected by the presence of co-ions (Tsezos *et al.*, 1986).

Limited data on the biosorption of other radionuclides are available in the literature although they are frequently derived from differing protocols and for differing objectives (Tsezos, 1990). Attempts to formulate general biosorption mechanisms have been hampered by the non-compatibility of reported data with respect to species, strains, culture histories and experimental conditions (Siegel *et al.*, 1990). Success in correlating biosorption levels with metal characteristics such as ionic radius or position in the Irving–Williams series has been limited (Tsezos, 1983; Tobin *et al.*, 1984). More recently, the principle of hard and soft acids and bases has been considered as a basis for predicting metal/microorganism interactions (Hughes & Poole, 1991). A number of studies suggest that, although complex, these interactions can be accounted for by the hard and soft principle (Avery & Tobin 1993; Brady & Tobin, 1994).

Effects of environmental conditions

As expected from the foregoing discussion of mechanisms, environmental conditions greatly influence fungal biosorption processes. External pH conditions determine the speciation of the metal ion in solution (Baes & Mesner, 1976) and also affect the charge of the cell wall functional groups (Tobin *et al.*, 1984). For cation biosorption optimal pH values of 3–7 have been widely reported with dramatic reductions in uptake observed at pH values of 2 or lower (Strandberg *et al.*, 1981; Huang *et al.*, 1990).

Exceptions, such as the thorium uptake by various fungal species at pH 0–1 cited above, suggest alternative processes are involved (Gadd & White, 1989a). In contrast, lower pH values favour metallic anion biosorption. Molybdenum and vanadium anions were taken up by *R. arrhizus* biomass at pH 4.5 where the cell wall functional groups would be positively charged but were not adsorbed at increased pH values (Tobin *et al.*, 1984).

The presence of cationic and anionic co-ions also has a large influence on metal biosorption. Equimolar concentrations of uranyl ions were observed to cause almost complete inhibition of silver uptake by denatured *R. arrhizus* biomass, while under similar conditions silver had no effect on uranyl binding (Tobin, Cooper & Neufeld, 1988). In a separate study, uranium adsorption by *R. arrhizus* was inhibited by Fe^{2+}, Cu^{2+} and Zn^{2+} at higher pH values only, while thorium uptake was unaffected by Fe^{2+} or Zn^{2+} (Tsezos & Volesky, 1982a,b). Uranium binding by *S. cerevisiae* was affected by Ca^{2+}, and uranium binding by *Penicillium digitatum* was inhibited by Fe^{2+} (Strandberg *et al.*, 1981; Galun *et al.*, 1984). A recent study of biosorption by freeze-dried *R. arrhizus* reported that the levels of uptake competition between various divalent cations were consistent with the hard and soft principle (Brady & Tobin, 1994).

The presence of anions has been found to inhibit uptake in a range of metal/microorganism systems (Treen-Sears *et al.*, 1984; Tobin, Cooper & Neufeld, 1987; Zhou & Kiff, 1991). As would be expected, anions with high metal affinities were found to severely inhibit metal uptake at equimolar concentrations whereas anions with low metal affinities exerted little effect even when in large molar excess (Tobin *et al.*, 1987). For biosorption by *Rhizopus* strains there is general agreement in these works that inhibition effects increase in the order EDTA > SO_4^{2-} > Cl^-.

Removal of metal from biomass

The removal of adsorbed metals from the biomass for recovery and/or reuse of the biomass is an important step in any proposed fungal bioremediation application. Numerous studies have indicated that most biomass types can be fully stripped of adsorbed metals by varying treatments with little or no loss of biosorption capacity (Treen-Sears *et al.*, 1984; Tsezos *et al.*, 1989; Huang *et al.*, 1990). These findings are generally consistent with the reversible coordination/complexation mechanisms described above, although the fate of precipitated metals during regeneration treatments is unclear.

Metal elution or recovery from the biomass results when the metal/ biomass/liquid equilibrium shifts in favour of the liquid phase. The range of eluants found to be effective is in keeping with the competition mechanisms described earlier, and includes mineral acids, carbonates and bicarbonates, complexing agents such ethylenediaminetetraacetic acid (EDTA), nitriloacetic acid (NTAA) and, to lower efficiencies, organic acids.

Hydrochloric, nitric, and sulphuric acids at concentrations ranging from 1 to 0.01 M are the most widely studied eluants. Very high metal recovery efficiencies have been observed apparently resulting from the direct displacement of the metal from the biomass binding sites by the acid H^+ ions. However, loss of binding capacity in subsequent cycles has been observed for uranium biosorption by both *Rhizopus* biomass (Treen-Sears *et al.*, 1984) and *S. cerevisiae* (Strandberg *et al.*, 1981) and for radium uptake by *P. chrysogenum* (Tsezos *et al.*, 1986). These effects were observed to occur in association with losses in biomass integrity (Treen-Sears *et al.*, 1984).

The elution efficiencies of carbonates and bicarbonates are based on their high affinities for uranium and have been widely reported (Strandberg *et al.*, 1981; Tsezos, 1984). The mechanism would appear to be direct competition with the biomass sites for the uranium ions, and total uranium recovery has frequently been attained. Losses in binding capacity have been attributed to dissolution of the biomass in the eluant (Tsezos *et al.*, 1989).

The eluting potential of complexing agents such as EDTA and NTAA is apparently similarly related to their affinity for metal ions and their action is also based on direct competition with the biomass binding sites. Studies of uranium stripping from *S. cerevisiae* (Strandberg *et al.*, 1981), *Actinomyces levoris* (Horikoshi, Nakajima & Sakaguchi, 1981) and *P. digitatum* (Galun *et al.*, 1983) indicated high, although not total, removal levels, which is consistent with the level of anion competition effects reported for uranium/EDTA systems and referred to earlier (Tobin *et al.*, 1987). By contrast, EDTA completely removed radium from both *P. chrysogenum* and activated organisms (Tsezos *et al.*, 1986) and totally inhibited Cd^{2+} and Pb^{2+} uptake in direct competition studies (Tobin *et al.*, 1987). These differences may result from the variations in uptake mechanisms described earlier although the effect of pH on elution by complexing agents has also been highlighted (McEldowney, 1990).

Immobilized or pelleted biomass exhibits improved resistance to losses of biosorption efficiency during multiple uptake/elution column cycles. Tsezos *et al.* (1989) observed that, while repeated elution with 0.5 N NaHCO$_3$ was found to diminish the uranium adsorption capacity of *Rhizopus* biomass under batch conditions, the biomass remained effective when used in column studies. Similarly, Treen-Sears *et al.* (1984) showed that pelleted *Rhizopus* biomass could be effectively regenerated by either sulphuric (0.1 M) or nitric (1 M) acids and reused in columns eight times with no apparent loss in efficiency. Copper uptake was also undiminished by regeneration with 0.1 M HCl when *Rhizopus* biomass was immobilized in reticulated foam particles (Zhou & Kiff, 1991). Similar results have been observed for immobilized bacterial and algal cells (Nakajima, Horikoshi & Sakaguchi, 1982).

Current trends

Native biomass, even in a non-metabolizing state, is a fragile material which is not suited to rigorous large-scale processing applications. In contrast, pelleted or immobilized biomass, in addition to the resistance to degradation during elution cited above, has advantages of increased mechanical strength, good settling characteristics, reduced pressure drops and less likelihood of system blockage resulting in ease of scale-up. For these reasons much of the recent work in this area has focused on the production of second generation biosorbents in pelleted or immobilized form. Indeed it is generally agreed that the production of biosorbents in the form of solid granules is an essential stage in the development of successful biosorbent applications (Volesky, 1990; Tobin, L'Homme & Roux, 1993).

Certain fungal species including *Aspergillus*, *Rhizopus*, and *Penicillium* under suitable culture conditions naturally grow in spherical form (Treen-Sears *et al.*, 1984; Yakubu & Dudeney, 1986; de Rome & Gadd, 1991). Studies have shown that pelleted biomass exhibits similar biosorption characteristics to the native form (Yakubu & Dudeney, 1986). In continuous flow column operations, pellets of approximately 5 mm diameter successfully adsorbed uranium (Treen-Sears *et al.*, 1984; de Rome & Gadd, 1991), but strontium and caesium uptake was reduced as compared to batch uptake levels (de Rome & Gadd, 1991).

Biosorbent immobilization has been investigated using inert support matrices ranging from synthetic polymers and natural polysaccharide gels to coal, sand, and foam particles. For effective biosorbent performance, the biomass loading (w/w) relative to the immobilizing matrix as well as the accessibility of the metal-binding sites must be high. Natural polysaccharides, such as alginate and carageenan, which are widely used for living cell immobilization produce porous biosorbent beads resulting in negligible loss of biosorption capacity (de Rome & Gadd, 1991; Tobin *et al.*, 1993). However, the relatively low cell loadings achievable and poor mechanical strength, together with the reversibility of the crosslinking, limit their range of application.

Synthetic polymers, by contrast, can be used to produce biosorbent particles of high loading, mechanical strength and rigidity with little loss in biomass biosorption capacity. Polymers ranging from polyacrylamide to polyvinyl formal and polyethyleneimine have been used to immobilize fungal and other cells (Tsezos *et al.*, 1989; Tobin *et al.*, 1993; Brierley & Brierley, 1993) with cell loadings in excess of 80% (w/w) and minimal loss of biosorption capacities. The polymers provide a rigid supporting structure for the biomass yet at concentrations of 10–30% (w/w) do not markedly block the sites for metal binding. As evidenced in the electron micrograph of *R. arrhizus* biomass embedded in polylvinyl formal shown in Fig. 17.1, the porosity of the resulting beads minimizes the inhibiting effects of the polymer matrix.

Fig. 17.1. Scanning electron micrograph of a filamentous fungus, *Rhizopus arrhizus*, immobilized in polyvinyl formal. (Magnification × 75.)

Process applications

Biosorbents produced by immobilising cells as described above are typically beadlike or granular in form with diameters in the range 0.5 to 5 mm (Volesky, 1990). They are analogous to ion exchange resins or carbon adsorbents, and their use for metal removal from solution is essentially a conventional solid/liquid contacting and separation process.

Conventional solid/liquid processing falls into the two categories of stirred tank and packed bed systems, and the majority of laboratory and pilot plant biosorption studies employ these configurations. Well-known variants of the packed bed system such as the fluidized bed and air-lift reactors have been employed in studies with immobilized biosorbents. In these systems, the biosorbent particles are mobilized by the upflow of effluent through the column caused by pumping or a rising air stream introduced at the column base. Advantages of reduced pressure drops, and increased removal efficiencies are reported to be offset by greater space requirements and increased capital and running costs (Macaskie & Dean, 1989; White & Gadd, 1990).

Actual commercial biosorption applications based on immobilized microbial cells have employed both fixed bed canister units and fluidized bed systems for large effluent volumes (Brierley *et al.*, 1986). In the latter system, continuous operation by periodic removal of slugs of exhausted biosorbent granules and replacement with fresh biosorbent was reported. However, considerable market difficulties have been encountered in competing with ion-exchange resins (J.A. Brierley, personal communication).

Future directions

Few studies have investigated the use of actual biomass from industry as a source of biosorbent material, despite the fact that many species that exhibit biosorption potential, such as *S. cerevisiae* and *R. arrhizus*, are produced industrially in bulk. The use of waste biomass would alleviate the expense of biomass culture and merits further investigation. Recent results indicate that the biosorptive capacities of industrial strains of *S. cerevisiae* are different from those of laboratory strains and that the differences are related to cell surface properties (Avery & Tobin, 1992). Large variations in metal uptake by various commercially available *Saccharomyces* species have also been observed (I. Singleton & P. Simmons, unpublished observations, see Table 17.1) and are apparently due to differences in growth media, the age of cells, the *Saccharomyces* strain used, and the treatment of the cells (for example, autoclaving) after production.

A particular problem associated with the treatment of metal-containing effluents is the presence of metals in the form of particulates (Michaelis, 1985). One way to overcome this difficulty may be to use the ability of fungi to adsorb particulates to their surface (Wainwright, Grayston & de Jong, 1986). A variety of particulates, including sulphur, zinc dust, clays, activated carbon and metal sulphides, have been shown to be adsorbed. Larger particles are entrapped by growing hyphae whereas smaller particles are adsorbed rapidly to the mycelial surface. It has been suggested that mycelia could also be made susceptible to magnetic fields by allowing magnetite to be adsorbed by hyphae (Wainwright, Singleton & Edyvean, 1990). This would allow biomass to be separated rapidly and efficiently from solution. The potential of particle adsorption is clearly another area deserving of further study. Chemical manipulation of cells to increase metal uptake is an area that has received limited attention. A variety of protocols, including alkali and detergent treatments, has been shown to enhance biosorption capacities (Gadd & White, 1989a; Wales & Sugar, 1990). Likely mechanisms include exposure of metal binding sites by removal of cell wall components and increased cell wall permeability. Clearly, an increased knowledge of the roles of different wall constituents responsible for metal binding may allow for more directed chemical manipulations to be made. Similarly, a more direct approach to enhancement of metal uptake may be made by genetic manipulation of cell walls. Recent advances made in the understanding of the genetic basis of cell wall synthesis may allow biosorption capacities to be increased dramatically (Roemer, Delaney & Bussey,1993).

Conclusions

This research highlights both the potential of fungal cells for use in metal/radionuclide bioremediation applications and the need for standardization of methods and protocols. Although biosorption has not, to date, developed to the stage necessary for direct market competition with existing metal/recovery technologies on a wide scale, the prospects are promising. While current commercial ventures focus on niche markets, as outlined above future developments will be likely to include integration with, or polishing stages for, existing waste treatment schemes.

In the longer term, developments in genetic and strain manipulation techniques are likely to result in biomass and biopolymers of greatly increased metal specificities. This new generation of biosorbents could conceivably have the high specificities common in other biological systems. Further research and the support of committed industries will be required for the realization of this potential.

Acknowledgements

The authors would like to thank Jean-Claude Roux of the Centre d'Etudes Nucleaires, Grenoble for help with electron microscopy. I.S. gratefully acknowledges the provision of a President's Research Award from University College Dublin.

References

Avery, S.V. & Tobin, J.M. (1992). Mechanisms of strontium uptake by laboratory and brewing strains of *Saccharomyces cerevisiae*. *Applied and Environmental Microbiology*, **58**, 3883-9.

Avery, S.V. & Tobin, J.M. (1993). Mechanisms of adsorption of hard and soft metal ions to *Saccharomyces cerevisiae* and influence of hard and soft anions. *Applied and Environmental Microbiology*, **59**, 2851-6.

Baes, C.F. & Mesner, R.E. (1976). *The Hydrolysis of Cations*. New York: John Wiley & Sons.

Brady, J.M. & Tobin, J.M. (1994). Adsorption of metal ions by *Rhizopus arrhizus* biomass: characterisation studies. *Enzyme and Microbial Technology*, **16**, 671-5.

Brierley, C.L. & Brierley J.A. (1993). Immobilisation of biomass for industrial application of biosorption. In *Biohydrometallurgical Technologies*, Vol. 2, ed. A.E. Torma, M.E. Apel & C.L. Brierley, pp. 35-44. Proceedings of International Biohydrometallurgy Symposium. Wyoming, USA.

Brierley, J.A., Goyak, G.M. & Brierley C.L. (1986). Considerations for commercial use of natural products for metal recovery. In *Immobilisation of Ions by Biosorption*, ed. H. Eccles & S. Hunt, pp. 105-117. Chichester: Ellis Horwood.

Byerley, J.J. & Scharer, J.M. (1987). Uranium(VI) biosorption from process solutions. *The Chemical Engineering Journal*, **36**, B49-59.

de Rome, L. & Gadd, G.M. (1987a). Measurement of copper uptake in *Saccharomyces cerevisiae* using a Cu^{2+}-selective electrode. *FEMS Microbiology Letters*, **43**, 283-7.

de Rome, L. & Gadd, G.M. (1987b). Copper adsorption by *Rhizopus arrhizus*, *Cladosporium resinae* and *Penicillium italicum*. *Applied Microbiology and Biotechnology*, **26**, 84-90.

de Rome, L. & Gadd, G.M. (1991). Use of pelleted and immobilised yeast and fungal biomass for heavy metal recovery. *Journal of Industrial Microbiology*, **7**, 97-104.

Fourest, E. & Roux, J.C. (1992). Heavy metal biosorption by fungal mycelial by-products: mechanisms and influence of pH. *Applied Microbiology and Biotechnology*, **37**, 399-403.

Gadd, G.M. (1993). Interactions of fungi with toxic metals. *New Phytologist*, **124**, 25-60.

Gadd, G.M. & White, C. (1989a). Removal of thorium from simulated acid process streams by fungal biomass. *Biotechnology and Bioengineering*, **33**, 592-7.

Gadd, G.M. & White, C. (1989b). Uptake and intracellular distribution of thorium in *Saccharomyces cerevisiae*. *Environmental Pollution*, **61**, 187-97.

Galun, M., Galun, E., Siegel, B.Z., Keller, P., Lehr, H.& Siegel, S.M. (1987). Removal of metal ions from aqueous solutions by *Penicillium* biomass: kinetic and uptake parameters. *Water, Air and Soil Pollution*, **33**, 359-71.

Galun, M., Keller, P., Feldstein, H., Galun, E., Siegel, S. & Siegel, B. (1983). Recovery of uranium(VI) from solution using fungi. Release from uranium-loaded *Penicillium* biomass. *Water, Air and Soil Pollution*, **20**, 277-85.

Galun, E., Keller, P. Malki, D. Feldstein, H. Galun, E. Siegel, S. & Siegel, B. (1984). Removal of uranium(VI) from solution by fungal biomass: inhibition by iron. *Water, Air and Soil Pollution*, **21**, 411-14.

Horikoshi, T., Nakajima, A. & Sakaguchi, T. (1981). Studies on the accumulation of heavy metals in biological systems XIX. Accumulation of uranium by microorganisms. *European Journal of Applied Microbiology and Biotechnology*, **12**, 90-6.

Huang, C.P., Huang, C., & Morehart, A.L. (1990). The removal of Cu(II) from dilute aqueous solutions by *Saccharomyces cerevisiae*. *Water Research*, **24**, 433-9.

Hughes, M.N. & Poole, R.K. (1991). Metal speciation and microbial growth – the hard (and soft) facts. *Journal of General Microbiology*, **137**, 725-34.

Junghans, K. & Straube, G. (1991). Biosorption of copper by yeasts. *Biology of Metals*, **4**, 233-7.

Kuyucak, N. (1990). Feasibility of biosorbents application. In *Biosorption of Heavy Metals*, ed. B. Volesky, pp. 371-378. Boca Raton, Florida: CRC Press Inc.

Kuyucak, N. & Volesky, B. (1988). Biosorbents for recovery of metals from industrial solutions. *Biotechnology Letters*, **10**, 137-42.

Luef, E., Prey, T. & Kubicek, C.P. (1991). Biosorption of zinc by fungal mycelial wastes. *Applied Microbiology and Biotechnology*, **34**, 688-92.

Macaskie, L.E. & Dean, A.C.R. (1989). Microbial metabolism, desolubilisation, and deposition of heavy metals: metal uptake by immobilised cells and application to the treatment of liquid wastes. In *Biological Waste Treatment*, ed. A. Mizrahi, pp. 159-201. New York: Alan R. Liss Inc.

McEldowney, S. (1990). Microbial biosorption of radionuclides in liquid effluent treatment. *Applied Biochemistry and Biotechnology*, **26**, 159-79.

Michaelis, H.V. (1985). Integrated biological systems for effluent treatment from mine and mill tailings. In *Fundamental and Applied Biohydrometallurgy, Proceedings of Sixth International Symposium on Biohydrometallurgy*. Vancouver, ed. R.W. Lawrence, R.M.R. Branion & H.G. Ebner, pp. 311-326. Amsterdam: Elsevier.

Mullen, M.D., Wolf, D.C., Beveridge, T.J. & Bailey, G.W. (1992). Sorption of heavy metals by the soil fungi *Aspergillus niger* and *Mucor rouxii*. *Soil Biology and Biochemistry*, **24**, 129-35.

Nakajima, A., Horikoshi, T. & Sakaguchi, T. (1982). Recovery of uranium by immobilised microorganisms. *European Journal of Applied Microbiology and Biotechnology*, **16**, 88-91.

Norris, P.R. & Kelly, D.P. (1977). Accumulation of cadmium and copper by *Saccharomyces cerevisiae*. *Journal of General Microbiology*, **99**, 317-24.

Pighi, P.L., Pumpel, T. & Schinner, F. (1989). Selective accumulation of silver by fungi. *Biotechnology Letters*, **11**, 275-80.

Pumpel, T. & Schinner, F. (1986). Silver tolerance and silver accumulation of microorganisms from soil materials of a silver mine. *Applied Microbiology and Biotechnology*, **24**, 244-7.

Remacle, J. (1990*a*). Culture conditions and biomass metal-binding properties. In *Biosorption of Heavy Metals*, ed. B. Volesky, pp. 293-301. Boca Raton, Florida: CRC Press.

Remacle, J. (1990*b*). The cell wall and metal binding. In *Biosorption of Heavy Metals*, ed. B. Volesky, pp. 83-92. Boca Raton, Florida: CRC Press.

Roemer, T., Delaney, S. & Bussey, H. (1993). SKN1 and KRE6 define a pair of functional homologs encoding putative membrane proteins involved in B-glucan synthesis. *Molecular and Cellular Biology*, **13**, 4039-48.

Shumate, S.E. & Strandberg, G.W. (1985). Accumulation of metals by microbial cells. In *Comprehensive Biotechnology*, ed. M. Moo-Young, C.N. Robinson & J.A. Howell, pp. 235-247. Oxford: Pergamon Press.

Siegel, S.M. Galun, M. & Siegel, B.Z. (1990). Filamentous fungi as metal biosorbents: a review. *Water, Air and Soil Pollution*, **53**, 335-44.

Strandberg, G.W., Shumate, S.E. & Parrot, J.R. (1981). Microbial cells as biosorbents for heavy metals: accumulation of uranium by *Saccharomyces cerevisiae* and *Pseudomonas aeruginosa*. *Applied and Environmental Microbiology*, **41**, 237-45.

Tobin, J.M., Cooper, D.G. & Neufeld, R.J. (1984). Uptake of metal ions by *Rhizopus arrhizus* biomass. *Applied and Environmental Microbiology*, **47**, 821-4.

Tobin, J.M., Cooper, D.G. & Neufeld, R.J. (1987). Influence of anions on metal adsorption by *Rhizopus arrhizus* biomass. *Biotechnology and Bioengineering*, **30**, 882-6.

Tobin, J.M., Cooper, D.G. & Neufeld, R.J. (1988). The effects of cation competition on metal adsorption by *Rhizopus arrhizus* biomass. *Biotechnology and Bioengineering*, **31**, 282-6.

Tobin, J.M., Cooper, D.G. & Neufeld, R.J. (1990). Investigation of the mechanism of metal uptake by denatured *Rhizopus arrhizus* biomass. *Enzyme and Microbial Technology*, **12**, 591-5.

Tobin, J.M., L'Homme, B. & Roux, J.C. (1993). Immobilisation protocols and effects on cadmium uptake by *Rhizopus arrhizus* biosorbents. *Biotechnology Techniques*, **7**, 739-44.

Treen-Sears, M.E., Martin, S.M. & Volesky, B. (1984). Propagation of *Rhizopus javanicus* biosorbent. *Applied and Environmental Microbiology*, **48**, 137-41.

Tsezos, M. (1983). The role of chitin in uranium adsorption by *Rhizopus arrhizus*. *Biotechnology and Bioengineering*, **25**, 2025-40.

Tsezos, M. (1984). Recovery of uranium from biological adsorbents–desorption equilibrium. *Biotechnology and Bioengineering*, **26**, 973-81.

Tsezos, M. (1990). Biosorption of radioactive species. In *Biosorption of Heavy Metals*, ed. B. Volesky, pp. 45-50. Boca Raton, Florida: CRC Press.

Tsezos, M., Baird, M.H.I. & Shemilt, L.W. (1986). The kinetics of radium biosorption. *The Chemical Engineering Journal*, 33, B35-41.

Tsezos, M. & Keller, D. (1983). Adsorption of radium-226 by biological origin adsorbents. *Biotechnology and Bioengineering*, 25, 201-15.

Tsezos, M., McCready, R.G.L. & Bell, J.P. (1989). The continuous recovery of uranium from biologically leached solutions using immobilised biomass. *Biotechnology and Bioengineering*, 34, 10-17.

Tsezos, M. & Volesky, B. (1981). Biosorption of uranium and thorium. *Biotechnology and Bioengineering*, 23, 583-604.

Tsezos, M. & Volesky, B. (1982a). The mechanism of uranium biosorption by *Rhizopus arrhizus*. *Biotechnology and Bioengineering*, 24, 385-401.

Tsezos, M. & Volesky, B. (1982b). The mechanism of thorium biosorption by *Rhizopus arrhizus*. *Biotechnology and Bioengineering*, 24, 955-69.

Venkateswerlu, G. & Stotzky, G. (1989). Binding of metals by cell walls of *Cunninghamella blakesleeana* grown in the presence of copper or cobalt. *Applied Microbiology and Biotechnology*, 31, 619-25.

Volesky, B. (1990). Removal and recovery of heavy metals by biosorption. In *Biosorption of Heavy Metals*, ed. B. Volesky, pp. 7-43. Boca Raton, Florida: CRC Press.

Volesky, B., May, H. & Holan, Z. R. (1993). Cadmium biosorption by *Saccharomyces cerevisiae*. *Biotechnology and Bioengineering*, 41, 826-9.

Wainwright, M., Grayston, S. J. & de Jong, P. (1986). Adsorption of insoluble compounds by mycelium of the fungus *Mucor flavus*. *Enzyme and Microbial Technology*, 8, 597-606.

Wainwright, M., Singleton, I. & Edyvean, R.G.J. (1990). Magnetite adsorption as a means of making fungi susceptible to a magnetic field. *Biorecovery*, 2, 37-53.

Wales, D. S. & Sugar, B.F. (1990). Recovery of metal ions by microfungal filters. *Journal of Chemical Technology and Biotechnology*, 49, 345-55.

Watson, J.S. Scott, C.D. & Falcon, B.D. (1989). Adsorption of Sr by immobilised microorganisms. *Applied Biochemistry and Biotechnology*, 20/21, 699-709.

White, C. & Gadd, G. M. (1990). Biosorption of radionuclides by fungal biomass. *Journal of Chemical Technology and Biotechnology*, 49, 331-43.

Yakubu, N.A. & Dudeney, A. W. L. (1986). Biosorption of uranium with *Aspergillus niger*. In *Immobilisation of Ions by Biosorption*, ed. H. Eccles & S. Hunt, pp. 183-200, Chichester: Ellis Horwood.

Zhou, J.L. & Kiff, R.J. (1991). The uptake of copper from aqueous solution by immobilised fungal biomass. *Journal of Chemical Technology and Biotechnology*, 52, 317-30.

18

Impact of genetically modified microorganisms on the terrestrial microbiota including fungi

J. M. WHIPPS, F. A. A. M. DE LEIJ,
J. M. LYNCH AND M. J. BAILEY

Introduction

Recently there has been a considerable increase in interest in developing genetically modified microorganisms (GMMs) for a range of different purposes in the environment such as bioremediation, mineral leaching, improvement of soil nutrient status and biological control. Numerous experiments, virtually all with bacteria, have been carried out using genetic manipulation to increase or decrease expression of genes associated with these processes or to transfer or delete them (see Crawford *et al.*, 1993; Stotzky *et al.*, 1993; Lindow, Panopoulos & MacFarland, 1989). However, before commercial development of these functional GMMs can occur, the environmental risks associated with the release into the environment of GMMs lacking deliberate functional manipulation must be addressed. Consequently, for the purpose of estimating the risks of such GMM releases, a series of step-wise assessment procedures has been developed. This approach was recommended by the Organisation for Economic Co-operation and Development (OECD, 1992) and has been adopted in regulatory frameworks in most countries. These start with laboratory and glasshouse studies under contained conditions and, depending on the results, are followed by field releases and post-release monitoring procedures.

Terminology associated with risk assessment is not strictly defined (see Teng & Yang, 1993), but essentially the key initial process in the risk assessment procedure is to identify any potential hazards associated with the use of a GMM. These include pathogenicity, phenotypic and genetic stability of the parental strain, potential for survival, establishment and dissemination, potential for gene transfer and, finally, potential to affect or cause an impact on other organisms or ecological processes.

Subsequently, a risk or impact assessment can be made considering the information obtained following the hazard identification. One possible approach to this is to identify questions related to hazards (Anon, 1991). This could then lead to word models which can be the basis of generating predictive mathematical models (Bazin & Lynch, 1994). Thus, risk or biological impact can be assessed as the product of the *consequence* (or impact) of the GMM on other organisms or ecological processes and the *likelihood* (probability or risk) of an undesirable event occurring.

As far as GMMs are concerned, parental isolates are rarely pathogens and are generally chosen to be phenotypically and genetically stable. Therefore, potential for survival, establishment and dissemination, potential for gene transfer and potential to cause an impact on other organisms or ecological processes are the major hazards to be examined and form the focus of this chapter. Since the concern is initially with strains that are modified by inserting marker or reporter genes with a neutral function in the environment, results obtained from such studies will give the baseline risks associated with the release of the non-modified parental strain which will then provide a bench-mark against which future releases of functionally modified GMMs can be compared. Information relating to impact of GMMs on fungal populations will be discussed where available, but the overwhelming body of data on this subject concerns the use of genetically modified bacteria and this is reflected in the text. The authors' recent work with a GMM on wheat, which has paid more attention to fungi than most previous studies, will be discussed briefly at the end.

Experiments under contained terrestrial environments

Survival, establishment and dissemination

Initially, virtually all experiments with GMMs were concerned with monitoring survival, establishment and dissemination. This was made possible by the use of specific reporter or marker genes, antibiotic and heavy metal resistance genes or a combination of the two. Frequently, strains naturally resistant to antibiotics or heavy metals were chosen for subsequent genetic modification with marker genes or genes coding for functional traits to allow subsequent identification and selection from environmental samples. Such genetic markers enabled low numbers of cells to be detected against a large background of natural microorganisms. For example, the *lacZY* genes from *Escherichia coli* (coding for β-galactosidase and lactose permease) were introduced into the chromosome of a naturally rifampicin-resistant *Pseudomonas fluorescens* (Drahos, Hemming & McPherson, 1986). The *lacZY* genes confer the ability to utilize lactose and also to cleave the chromogenic substrate X-gal (5-chloro-4-bromo-3-indolyl-β-D-galactopyranoside) resulting in easily identifiable blue colonies on agar. This combination of *lacZY* genes and rifampicin resistance enabled detection of fewer than ten recombinant colony-forming units (cfu) g^{-1} soil and has continued to be used extensively for monitoring spread and survival of *Pseudomonas* spp. (which lack the ability to utilize lactose) in soil, plants and insects (Kluepfel & Tonkyn, 1991; Seong, Höfte & Verstraete, 1992).

Other marker systems now used regularly for identification and monitoring of GMMs in the environment include the *xylE* gene, which confers the ability to cleave catechol to yield a yellow compound (2-hydroxymuconic semi-aldehyde) (MacNaughton, Rose & O'Donnell, 1992) and the *lux* genes, which enable cells to produce light when exposed to appropriate substrates such as *n*-decyl-aldehyde (Beauchamp, Kloepper & Lemke, 1993; Boelens *et al.*, 1993). Only recently have the first marker systems suitable for monitoring spread of fungi become available. *Fusarium oxysporum* f. sp. *melonis* and *F. oxysporum* f. sp. *lini* have been transformed by introducing the *gusA* gene from *E. coli* coding for the enzyme β-D-glucoronidase in combination with the gene for nitrate reductase (*niaD*) as a selectable marker into the chromosome (Couteaudier *et al.*, 1993). Release of a blue colour from the chromogenic substrate X-gluc (5-bromo-4-chloro-3-indolyl-glucuronide) through the activity of β-D-glucuronidase has enabled these genetically modified fungi to be located on plant roots *in situ*. Similarly, *F. oxysporum* f. sp. *raphani* has been transformed for resistance to the antibiotic hygromycin A (Toyota, Tsuge & Kimura, 1992). This trait has been used to monitor populations of the recombinant in soil using dilution plating procedures onto media containing hygromycin A.

A range of other procedures based upon detecting specific genetic changes such as DNA restriction profiling and hybridization patterns as well as polymerase chain reaction (PCR) have been used to detect recombinant viruses and bacteria in environmental samples (Pickup *et al.*, 1991; Carter & Lynch, 1993; Kluepfel, 1993), but, in general, they are too expensive and time consuming for use on a large scale for long-term, quantitative monitoring.

Gene exchange

Mobile genetic elements such as plasmids and transposons have been widely used to introduce genes or affect gene expression. They are relatively simple to use and generally carry easily selectable markers such as antibiotic resistance that can be used for identification and recovery purposes. However, it is the mobility of these genetic elements which is of concern when used for genetic modification purposes. Bacteria are known to exchange plasmids and other mobile genetic elements both intra and interspecifically in a range of environments (Veal, Stokes & Daggard, 1992) and, if GMMs containing such genetic elements are released, both the frequency of transfer and the significance to the recipient of acquiring the new genes need to be assessed (van Elsas, 1992). Much of the early work with GMMs was aimed at measuring rates of gene exchange in various environments, but it was soon realized that stable genetic inserts were required for long-term use and monitoring of GMMs. Consequently, novel genes were transferred using stable transposon insertions or homologous recombination procedures into nonfunctional regions of the chromosome. For example, there is a single preferred site of integration for the transposon Tn7, used to insert the *lacZY* genes into *Pseudomonas aureofaciens*, which appears to be a noncoding, non-essential genetic region in *Pseudomonas* spp. (Barry, 1986; Drahos, Hemming & McPherson, 1986).

Alternatively, mobile genetic elements can also be used for other types of genetic modification such as disrupting functional genes on the chromosome. In this case, any marker genes carried can help in monitoring insertion into the chromosome, but their insertion *per se* is not the primary aim. For instance, antibiotic production in a biocontrol strain of *Pseudomonas fluorescens* (Voisard *et al.*, 1989) or production of plant cell wall-degrading enzymes by the pathogen *Pseudomonas solanacearum* (Hartel, Vaughn & Williamson, 1993) was achieved by transposon mutagenesis. Nevertheless, every time a GMM is created, subsequent rates of gene transfer must always be assessed when used in environmental studies to be sure of the stability of transformation.

Impact on other organisms and ecological processes

A critical feature of a risk assessment is determining the impact that a GMM has on other organisms or ecological processes. Unfortunately, this is rather a difficult procedure to carry out in comparison with monitoring survival, establishment and dissemination or gene exchange and this is reflected in the literature. For example, from a literature survey of experiments involving GMMs used for developing monitoring procedures, only studies with seven bacteria out of 30 (23%) considered impact of any kind. Of these, five estimated impact on total bacteria and only one considered impact on processes. Impact on fungi was studied in an extremely limited way with only four GMMs (Table 18.1).

The situation is somewhat different with experiments involving GMMs that have been modified functionally. These include GMMs with altered ability to degrade aromatic compounds or plant cell walls, those with altered biocontrol activity, those with added genes for nitrogen fixation and various GMMs with genes deleted. In total, 28 GMMs were found in the survey and of these, 21 (75%) considered impact of some kind. Five monitored impact on processes, mostly those with altered degradative ability, and 18 (64%) estimated impact on other organisms. Impact on fungi was considered in 11 (39%) of these studies, with target pathogens specifically examined in all those experiments concerned with altered biocontrol activity (Table 18.2). Thus, taken overall, with the exception of studies involving bacteria with altered biocontrol traits, the impact of GMMs on fungi in contained environments has been severely neglected.

Field release experiments

The number of GMMs released into the field is gradually increasing and some examples are given in Table 18.3. All studies have monitored survival, establishment and dissemination using a variety of reporter or marker systems depending on the aim of the release. Most have checked for gene exchange or stability, but only very few have monitored impact on organisms or ecological processes. Only the work of Cook *et al.* (1991) and the authors' unpublished studies with *Pseudomonas aureofaciens* have considered impact on fungi at all. Although strictly a monitoring exercise, Cook *et al.* (1991) examined effects of *P. fluorescens* on take-all of wheat caused by *Gaeumannomyces graminis* as the GMM used was a biocontrol agent of this pathogen. Interestingly, no difference in biocontrol between the parental strain and the GMM was found. The authors have examined the effect of a genetically modified *P. aureofaciens* on total bacteria, pseudomonads, actinomycetes, filamentous fungi and yeasts on the phylloplane and in the rhizosphere of wheat (*Triticum aestivum*) and sugar beet (*Beta vulgaris*), and preliminary evidence suggests that this GMM behaves similarly to its parental strain and has

Table 18.1. *Examples of impact of GMMs on organisms and ecological processes (from studies primarily concerned with assessing monitoring procedures)*

GMM	Markers	Environment	Organisms monitored	Processes monitored	Reference
Azospirillum lipoferum	Antibiotic resistance	Soil, wheat, maize	Total bacteria *Rhizobium leguminosarum*	–[a]	Bentjen *et al.*, 1989
Enterobacter cloacae	Antibiotic resistance	Soil	Range of bacterial groups	Soil respiration	Stotzky *et al.*, 1993
Escherichia coli	Antibiotic resistance + Hg resistance	Soil	Fungal propagules Protozoa	Activity of a range of soil enzymes	
Pseudomonas sp.	Antibiotic resistance	Soil, wheat	Total bacteria	–	Fredrickson *et al.*, 1989
P. aeruginosa	*lacZY* + antibiotic resistance	Soil, maize	Total bacteria Fungal propagules	–	Seong, Höfte & Verstraete, 1992
P. aureofaciens	*lacZY* + antibiotic resistance	Soil, wheat, soybean	Total bacteria *Bradyrhizobium japonicum* Mycorrhizas	–	Kluepfel & Tonkyn, 1991
Rhizobium leguminosarum bv. *trifolii*	Antibiotic resistance	Soil	Protozoa	–	Heynen *et al.*, 1988

Note: –[a] = not tested

Table 18.2. *Examples of impact of GMMs on fungi (from studies primarily concerned with studying organisms with modified functional traits)*

GMM	Markers/trait manipulated	Environment	Fungi monitored	Reference
Altered ability to degrade aromatic compounds				
Pseudomonas putida	Antibiotic and Hg resistance/ Ability to degrade 2,4-D	Soil, radish	Fungal propagules	Doyle et al., 1991; Doyle & Stotzky, 1993; Stotzky et al., 1993
Streptomyces lividans	Antibiotic resistance/ Ability to degrade lignin (+ lignin peroxidase gene)	Soil	Fungal propagules	Wang et al., 1991; Crawford et al., 1993
Altered biocontrol activity				
Escherichia coli	Antibiotic resistance/ + chitinase gene	Soil, bean	*Sclerotium rolfsii*	Shapira et al., 1989
Pseudomonas aeruginosa	Antibiotic resistance/ Decreased antibiotic or siderophore production	Wheat	*Septoria tritici*	Flaishman et al., 1990
P. aureofaciens	Antibiotic resistance/ Decreased antibiotic production	Soil, wheat	*Gaeumannomyces graminis*	Vincent et al., 1991
P. fluorescens	Antibiotic resistance/ + chitinase gene	Sand, soil, radish, wheat	*Fusarium oxysporum* f.sp. *redolens* *G. graminis* *Pythium ultimum*	Sundheim, Poplawsky & Ellingboe, 1988
	lux genes/ Increased antibiotic production	Soil, cotton	*G. graminis*	Gutterson, Howie & Suslow, 1990
	Antibiotic resistance/ Altered antibiotic production	Soil, wheat	*G. graminis*	Poplawsky & Ellingboe, 1989
	Antibiotic resistance/ Altered antibiotic production	Soil, cucumber	*P. ultimum*	Kraus & Loper, 1992
	Antibiotic resistance/ Increased antibiotic production	Soil, maize, cress, cucumber	*P. ultimum*	Maurhofer et al., 1992
	Antibiotic resistance/ Altered HCN production	Artificial soil, tobacco	*Thielaviopsis basicola*	Voisard et al., 1989

Table 18.3. *Examples of field releases of genetically-modified microorganisms*

Reporter[a] system/ selective marker	Trait(s) manipulated	Genetic[b] change(s)	Organism	Environment	Features monitored			Reference
					Survival and spread	Gene exchange	Impact	
Experiments primarily concerned with assessing monitoring procedures								
	+ novel sequence	(Chr)	*Autographa californica* nuclear polyhedrosis virus	Soil, cabbage	✓	✓	✓	Possee *et al.*, 1992
		Plasmid	*Bacillus subtilis*	Soil, wheat	✓	✓	–	van Elsas *et al.*, 1986
LacZY (+km^r+$xylE$)	Lactose uptake and utilization + catechol utilisation	(Chr)	*Pseudomonas aureofaciens*	Soil, wheat, sugar beet	✓	✓	✓	Whipps *et al.*, unpublished observations
LacZY (+rif^r)	Lactose uptake and utilization	Tn7 (Chr)	*P. aureofaciens*	Soil, wheat and other plants	✓	✓	–	Drahos *et al.*, 1988; Kluepfel *et al.*, 1991*a, b*
LacZY (+rif^r)	Lactose uptake and utilization	(Chr)	*P. corrugata*	Soil, wheat	✓	✓	–	Ryder *et al.*, unpublished observations
LacZY (+rif^r +nal^r)	Lactose uptake and utilization	Tn7 (Chr)	*P. fluorescens*	Soil, wheat	✓	✓	✓	Cook *et al.*, 1991
km^r + strep^r		Tn5 (Chr)	*P. fluorescens*	Soil, wheat, potato	✓	✓	–	van Elsas *et al.*, 1986; Bakker *et al.*, 1991
neo^r + rif^r + strep^r		Tn5 (plasmid)	*Rhizobium leguminosarum*	Soil, pea, cereals	✓	✓	–	Amarger & Delgutte, 1991; Nicholson, Jones & Hirsch, 1992
LacZ	Lactose utilization	Tn5	*R. fredii*	Soil	✓	✓	–	Rolfe *et al.*, 1989
Lux genes (+tet^r+rif^r)	Bioluminescence	Tn4431 (Chr)	*Xanthomonas campestris* pv. *campestris*	Soil, cabbage	✓	–	–	Shaw *et al.*, 1991; 1992

Table 18.3 *Continued*

Reporter[a] system/ selective marker	Trait(s) manipulated	Genetic[b] change(s)	Organism	Environment	Features monitored			Reference
					Survival and spread	Gene exchange	Impact	
Experiments primarily concerned with studying organisms with manipulated functional traits								
	tra⁻ (loss of plasmid transfer ability)	Deletion	*Agrobacterium radiobacter*	Soil, several plants	✓	✓	✓	Kerr, 1989; Jones & Kerr, 1989
	+ δ endotoxin gene from *Bacillus thuringiensis* var. *kurstaki*	(Chr)	*Clavibacter xyli* sub sp. *cynodontis*	Maize	✓	✓	–	Kostka, 1991
rif[r]	ice⁻ (loss of ice nucleating activity)	Deletion	*P. syringae*	Soil, potato	✓	✓	–	Lindow & Panopoulos, 1988

[a] km^r = kanamycin resistant; nal^r = nalidixic acid resistant; neo^r = neomycin resistant; rif^r = rifampicin resistant; strep^r = streptomycin resistant; tet^r = tetracycline resistant

[b] Tn = transposon; Chr – chromosome.

little long-term effect on any of the microbial populations studied. Monitoring of the recombinant in the field is still continuing.

The experiments carried out under contained conditions leading up to the release were highly significant in terms of the subsequent impact assessment carried out in the field, and the studies concerning wheat are described in the next section. Overall conclusions of impact studies on sugar beet were broadly similar to those of wheat.

Impact studies with a GMM on wheat

Background

The initial part of the risk assessment consisted of a survey of the microbial populations on the phylloplane and in the rhizosphere of sugar beet and wheat grown in Oxford and Littlehampton, respectively, in both the glasshouse and field. This part of the study also aimed at identifying a bacterium common to both plants which was a genetically stable, non-pathogen, and which could be marked at two well-separated sites on the chromosome with two marker gene cassettes and reintroduced into the phylloplane and rhizosphere of both crops. The organism chosen was a *P. aureofaciens* strain that was isolated from the phylloplane of sugar beet (Thompson *et al.*, 1993; Legard *et al.*, 1994).

In addition, the specific agar-based media systems used for monitoring microbial populations in soil and the phylloplane and on roots were evaluated for suitability for use in risk assessment studies. For example, on wheat, major microbial populations on the phylloplane (pink and white yeasts, filamentous fungi and bacteria) increased with time (Legard *et al.*, 1994) in a similar way to that reported in other studies on wheat, barley and ryegrass (Dickinson, Austin & Goodfellow, 1975; Fokkema *et al.*, 1979; Magan & Lacey, 1986; Magan & McLeod, 1991). Even though the dilution plate system used may have biased population estimates of fungi towards those which sporulate heavily, these studies indicated that the phylloplane of grasses acted as a fairly selective and reproducible habitat for all the major microbial groups in general. However, the variation in populations at each sampling time of even the most stable, major groups frequently varied by more than one order of magnitude (Table 18.4). Such a high level of natural variability means that, to detect an impact of any GMM on these populations, the effect would have to be fairly catastrophic. This highlights the problems of detecting impact in natural, varying populations of microbes.

Table 18.4. *Variation within microbial populations on the phylloplane of the flag leaf of wheat in the field 107 days after planting*

	Colony-forming units (g tissue)$^{-1}$
Cladosporium spp.	1.3 (2.8 – 0.6) × 10^5
Other filamentous fungi	2.6 (4.4 – 1.6) × 10^5
Pink yeasts	2.6 (4.7 – 1.5) × 10^7
White yeasts	1.4 (2.4 – 0.8) × 10^6
Bacteria	3.3 (4.6 – 2.4) × 10^6

Note: Values are means of 5 replicates with the range of 95% confidence limits in parentheses.

Consequently, subsequent impact studies with the GMM were restricted to specific groups of microorganisms where impact and gene exchange were thought likely to occur (e.g. other *Pseudomonas* spp.), and to total populations of filamentous fungi, yeasts, actinomycetes and bacteria to ensure that any unforeseen, catastrophic effects would be detected.

The basic procedure for monitoring populations of *Pseudomonas* and total culturable bacteria was also enhanced by applying ecological strategy theory to the agar plate counting process (Andrews & Harris, 1986; de Leij, Whipps & Lynch, 1993*b*). Essentially, bacterial colonies were scored as they appeared over a 7- or 10-day period into 6–7 growth classes. Those that appeared rapidly within 1–2 days were considered r-strategists (characterized as organisms that do well in uncrowded, nutrient-rich environments, are sensitive to toxins and lack ability to degrade recalcitrant substrates) and those that appeared later as K-strategists (typified by organisms that do relatively well in crowded environments that have reached their carrying capacity, are less affected by toxins and can utilize recalcitrant substrates such as lignin and cellulose). This allowed bacterial community structures to be described quantitatively in time and space and proved more sensitive than total counts alone (de Leij, Whipps & Lynch, 1993*b*).

Genetic marking and contained-environment studies

The *Pseudomonas* strain chosen for genetic modification was identified by fatty acid methyl ester analysis as *P. aureofaciens* (LOPAT group IV). This strain was coded as SBW25. By site-directed homologous recombination, the gene cassettes *lacZY* and km^r-*xylE* were introduced into two well-separated, non-essential sites of the chromosome of *P. aureofaciens* SBW25. Using a most probable number enrichment technique, a sensitivity of recovery of less than 1 cell $10 \, g^{-1}$ soil could be obtained (de Leij *et al.*, 1993*a*). The particular advantage of the dual marking system, which has not been used by other investigators, is that it facilitates tracking with greater precision and also allows any gene exchange from two distinct sites to be monitored.

Glasshouse experiments were set up with the GMM, using a range of microcosms to monitor survival, establishment and dissemination, gene exchange and impact on microbial populations as described above. Whenever possible, comparisons were made with the parental strain, *P. aureofaciens* SBW25. The work on wheat is described in detail elsewhere (de Leij *et al.*, 1994*a, b*) and the main results only are discussed briefly below.

When seeds were inoculated with 10^8 GMM colony-forming units (cfu)/seed, the recombinant survived at levels of up to 10^5 cfu g^{-1} root until harvest (4 months) after sowing, but it established on the first emerging leaves in low numbers only (approx. 10^3 cfu g^{-1} tissue). The GMM could be recovered from roots extracted from a depth of 60 cm in intact soil cores, but most GMMs (>95%) stayed associated with roots in the top 15 cm of the soil profile. However, it survived poorly in soil in the absence of roots. When the GMM was sprayed onto plants at tillering (3.7×10^7 cfu g^{-1} tissue), it could be recovered one month later (flowering) from all leaves at levels greater than 10^5 cfu g^{-1} tissue but not from the ear. At ripening, 70 d after spray application, most of the recombinants had disappeared from the phylloplane. Throughout all these experiments transfer of the novel genetic elements to other microorganisms was never detected.

Following seed application, indigenous *Pseudomonas* populations were reduced by 98% on the seed and 56% on the roots 6 d after sowing. No effect on any microbial populations on the leaves was detected at this time. As inoculated plants matured (tillering, flowering and ripening), the recombinant was also found to have no effect on total microbial populations on the roots. However, at tillering, *Pseudomonas* population structures on roots in the top 15 cm of the soil profile (assessed by the technique of de Leij *et al.*, 1993*b*) were affected significantly in both recombinant and parental treatments, but this was not maintained to ripening. Spray application of the recombinant onto the leaf surface at tillering caused no significant, long-term perturbations in any of the microbial populations present on the phylloplane.

These results formed part of the basis for a successful application to ACRE (Advisory Committee on Releases to the Environment) for permission to release this GMM into the field. The release took place in 1993 onto wheat at Littlehampton and onto sugar beet at Oxford. These were the first releases of a genetically modified non-symbiotic, saprophytic bacterium into the environment in the UK. At the time of writing, monitoring survival of the recombinant at the release sites is still taking place.

Conclusions

It is clear that fungi are a neglected group of organisms as far as studies of the impact of GMMs are concerned. Firstly, no studies are known at the time of writing of the impact of genetically modified fungi on other organisms or ecological processes. This is due to the relative difficulty in carrying out genetic-manipulations with fungi in general. However, such procedures are becoming more commonplace and it should be only a matter of time before genetically manipulated fungi are ready for risk assessment studies in a similar way to genetically modified bacteria described above. This is particularly important in view of the likely role that genetically manipulated fungi may play in agriculture in the future.

Secondly, the impact of GMMs on fungi has been largely ignored except where genetic manipulations have been carried out to alter bio-control activity. This stems, to some extent, from the background of microbial ecologists working with GMMs. Such researchers gained genetic manipulation experience using bacteria and found that it was easier and more familiar to deal with bacteria than fungi when working with environmental samples. We believe it is time to redress the balance.

Acknowledgements

We wish to thank the Agricultural and Food Research Council (AFRC), the National Environment Research Council (NERC) and the Department of the Environment (DoE) through contracts PECD 7/8/143 and PECD 7/8/161 for funding this work.

References

Amarger, N. & Delgutte, D. (1991). Monitoring genetically manipulated *Rhizobium leguminosarum* bv *viciae* released in the field. In *Biological Monitoring of Genetically Engineered Plants and Microbes*, ed. D.R.

312   *J. M. Whipps* et al.

MacKenzie & S.C. Henry, pp. 221-228. Bethesda, Maryland: Agricultural Research Institute.

Andrews, J.H. & Harris, R.F. (1986). r- and K-selection and microbial ecology. *Advances in Microbial Ecology*, **9**, 99-147.

Anon (1991) . Royal Commission on Environmental Pollution. *Fourteenth Report: GENHAZ. A System for the Critical Appraisal of Proposals to Release Genetically Modified Organisms into the Environment.* London: Her Majesty's Stationery Office.

Bakker, P.A.H.M., Schippers, B., Hoekstra, W.P.M. & Salentijn, E. (1991). Survival and stability of a Tn5 transposon derivative of *Pseudomonas fluorescens* WCS374 in the field. In *Biological Monitoring of Genetically Engineered Plants and Microbes*, ed. D.R. MacKenzie & S.C. Henry, pp. 201-204. Bethesda, Maryland: Agricultural Research Institute.

Barry, G.F. (1986). Permanent insertion of foreign genes into the chromosomes of soil bacteria. *Bio/Technology* **4**, 446-9.

Bazin, M.J. & Lynch, J.M. (eds) (1994). *Environmental Gene Release. Models, Experiments and Risk Assessment.* London: Chapman & Hall.

Beauchamp, C.J., Kloepper, J.W. & Lemke, P.A. (1993). Luminometric analyses of plant root colonization by bioluminescent pseudomonads. *Canadian Journal of Microbiology*, **39**, 434-41.

Bentjen, S.A., Fredrickson, J.K., van Voris, P. & Li, S.W. (1989). Intact soil-core microcosms for evaluating the fate and ecological impact of the release of genetically engineered microorganisms. *Applied and Environmental Microbiology* **55**, 198-202.

Boelens, J., Zoutman, D., Campbell, J., Verstraete, W. & Paranchych, W. (1993). The use of bioluminescence as a reporter to study the adherence of the plant growth promoting rhizopseudomonads 7NSK2 and ANP15 to canola roots. *Canadian Journal of Microbiology*, **39**, 329-34.

Carter, J.P. & Lynch, J.M. (1993). Introduction of new immunological and molecular techniques for microbial population and community dynamic studies in soil. In *Soil Biochemistry*, Vol. 8, ed. J-M. Bollag & G. Stotzky, pp. 249-272. New York: Marcel Dekker.

Cook, R.J., Weller, D.M., Kovacevich, P., Drahos, D., Hemming, B., Barnes, G. & Pierson, E.L. (1991). Establishment, monitoring, and termination of field tests with genetically altered bacteria applied to wheat for biological control of take-all. In *Biological Monitoring of Genetically Engineered Plants and Microbes*, ed. D.R. MacKenzie & S.C. Henry, pp. 177-187. Bethesda, Maryland: Agricultural Research Institute.

Couteaudier, Y., Daboussi, M-J., Eparvier, A., Langin, T. & Orcival, J. (1993). The GUS gene fusion system (*Escherichia coli* β-D-glucoronidase gene), a useful tool in studies of root colonization by *Fusarium oxysporum. Applied and Environmental Microbiology*, **59**, 1767-73.

Crawford, D.L., Doyle, J.D., Wang, Z., Hendricks, C.W., Bentjen, S.A., Bolton, H. Jr., Fredrickson, J.K. & Bleakley, B.H. (1993). Effects of a lignin peroxidase-expressing recombinant, *Streptomyces lividans* TK23.1, on biogeochemical cycling and the numbers and activities of microorganisms in soil. *Applied and Environmental Microbiology*, **59**, 508-18.

de Leij, F.A.A.M., Bailey, M.J., Lynch, J.M. & Whipps, J.M. (1993a). A simple most probable number technique for the sensitive recovery of a genetically engineered *Pseudomonas aureofaciens* from soil. *Letters in Applied Microbiology*, **16**, 307-310.

de Leij, F.A.A.M., Sutton, E.J., Whipps, J.M. & Lynch, J.M. (1994*a*). Effect of a genetically modified *Pseudomonas aureofaciens* on indigenous microbial populations of wheat. *FEMS Microbiology Ecology*, **13**, 249-58.

de Leij, F.A.A.M., Sutton, E.J., Whipps, J.M. & Lynch, J.M. (1994*b*). Spread and survival of a genetically modified *Pseudomonas aureofaciens* on the phytosphere of wheat and in soil. *Applied Soil Ecology*, **1**, 207-18.

de Leij, F.A.A.M., Whipps, J.M. & Lynch, J.M. (1993*b*). The use of colony development for the characterisation of bacterial communities in soil and on roots. *Microbial Ecology*, **27**, 81-97.

Dickinson, C.H., Austin, B. & Goodfellow, M. (1975). Quantitative and qualitative studies of phylloplane bacteria from *Lolium perenne*. *Journal of General Microbiology*, **91**, 157-66.

Doyle, J.D., Short, K.A., Stotzky, G., King, R.J., Seidler, R.J. & Olsen, R.H. (1991). Ecologically significant effects of *Pseudomonas putida* PPO301(pRO103), genetically engineered to degrade 2,4-dichlorophenoxyacetate, on microbial populations and processes in soil. *Canadian Journal of Microbiology*, **37**, 682-91.

Doyle, J.D. & Stotzky, G. (1993). Methods for the detection of changes in the microbial ecology of soil caused by the introduction of microorganisms. *Microbial Releases*, **62**, 63-72.

Drahos, D.J., Barry, G.F., Hemming, B.C., Brandt, E.J., Skipper, H.D., Kline, E.L., Kluepfel, D.A., Hughes, T.A. & Gooden, D.T. (1988). Pre-release testing procedures: US field test of a *lac*ZY-engineered soil bacterium. In *Release of Genetically-Engineered Micro-organisms*, ed. M. Sussman, C. H. Collins, F.A. Skinner, & D.E. Stewart-Tull, pp. 181-191. San Diego: Academic Press Ltd.

Drahos, D.J., Hemming, B.C. & McPherson, S. (1986). Tracking recombinant organisms in the environment: Ā-galactosidase as a selectable non-antibiotic marker for fluorescent pseudomonads. *Bio/Technology*, **4**, 439-44.

Flaishman, M., Eyal, Z., Voisard, C. & Haas, D. (1990). Suppression of *Septoria tritici* by phenazine- or siderophore-deficient mutants of *Pseudomonas*. *Current Microbiology*, **20**, 121-4.

Fokkema, N.J., den Houter, J.G., Kosterman, Y.J.C. & Nelis, A.L. (1979). Manipulation of yeasts on field-grown wheat leaves and their antagonistic effect on *Cochliobolus sativus* and *Septoria nodorum*. *Transactions of the British Mycological Society*, **72**, 19-29.

Fredrickson, J.K., Bentjen, S.A., Bolton, H. Jr., Li, S.W. & van Voris, P. (1989). Fate of Tn*5* mutants of root growth-inhibiting *Pseudomonas* sp. in intact soil-core microcosms. *Canadian Journal of Microbiology*, **35**, 867-73.

Gutterson, N., Howie, W. & Suslow, T. (1990). Enhancing efficiencies of biocontrol agents by use of biotechnology. In *New Directions in Biological Control: Alternatives for Suppressing Agricultural Pests and Diseases*, ed. R. R. Baker, pp. 749-765. New York: Alan R. Liss, Inc.

Hartel, P.G., Vaughn, T.M. & Williamson, J.W. (1993). Rhizosphere competitiveness of genetically altered *Pseudomonas solanacearum* in a novel gnotobiotic plant assembly. *Soil Biology and Biochemistry*, **25**, 1575-81.

Heynen, C.E., van Elsas, J.D., Kuikman, P.J. & van Veen, J.A. (1988). Dynamics of *Rhizobium leguminosarum* biovar *trifolii* introduced into soil; the effect of bentonite clay on predation by protozoa. *Soil Biology and Biochemistry*, **20**, 483-8.

314   J. M. Whipps et al.

Jones, D.A. & Kerr, A. (1989). *Agrobacterium radiobacter* strain K1026, a genetically engineered derivative of strain K84, for biological control of crown gall. *Plant Disease*, **73**, 15-18.

Kerr, A. (1989). Commercial release of a genetically engineered bacterium for the control of crown gall. *Agricultural Science*, **89**, 41-4.

Kluepfel, D.A. (1993). The behavior and tracking of bacteria in the rhizosphere. *Annual Review of Phytopathology*, **31**, 441-72.

Kluepfel, D.A., Kline, E.L., Hughes, T., Skipper, H., Gooden, D., Drahos, D.J., Barry, G.F., Hemming, B.C. & Brandt, E.J. (1991*a*). Field testing of a genetically engineered rhizosphere inhabiting pseudomonad: development of a model system. In *Biological Monitoring of Genetically Engineered Plants and Microbes*, ed. D.R. MacKenzie & S.C. Henry, pp. 189-199. Bethesda, Maryland: Agricultural Research Institute.

Kluepfel, D.A., Kline, E.L., Skipper, H.D., Hughes, T.A., Gooden, D.T., Drahos, D.J., Barry, G.F., Hemming, B.C. & Brandt, E.J. (1991*b*). The release and tracking of genetically engineered bacteria in the environment. *Phytopathology*, **81**, 348-352.

Kluepfel, D.A. & Tonkyn, D.W. (1991). Release of soil-borne genetically modified bacteria: biosafety implications from contained experiments. In *Biological Monitoring of Genetically Engineered Plants and Microbes*, ed. D.R. MacKenzie & S.C. Henry, pp. 55-65. Bethesda, Maryland: Agricultural Research Institute.

Kostka, S.J. (1991). The design and execution of successive field releases of genetically-engineered microorganisms. In *Biological Monitoring of Genetically Engineered Plants and Microbes*, ed. D.R. MacKenzie & S.C. Henry, pp. 167-176. Bethesda, Maryland: Agricultural Research Institute.

Kraus, J. & Loper, J.E. (1992). Lack of evidence for a role of antifungal metabolite production by *Pseudomonas fluorescens* Pf-5 in biological control of *Pythium* damping-off of cucumber. *Phytopathology*, **82**, 264-71.

Legard, D.E., McQuilken, M.P., Whipps, J.M., Fenlon, J.S., Fermor, T.R., Thompson, I.P., Bailey, M.J. & Lynch, J.M. (1994). Studies of seasonal changes of the microbial populations on the phylloplane of spring wheat as a prelude to the release of a genetically engineered microorganism. *Agriculture, Ecosystems and Environment*, **50**, 87-101.

Lindow, S.E. & Panopoulos, N.J. (1988). Field tests of recombinant *Pseudomonas syringae* for biological frost control in potato. In *Release of Genetically Engineered Micro-organisms*, ed. M. Sussman, C.H. Collins, F.A. Skinner, & D.E. Stewart-Tull, pp. 121-138. New York: Academic Press.

Lindow, S.E., Panopoulos, N.J. & MacFarland, B.L. (1989). Genetic engineering of bacteria from managed and natural habitats. *Science*, **244**, 1300-6.

MacNaughton, S.J., Rose, D.A. & O'Donnell, A.G. (1992). Persistence of a *xylE* marker gene in *Pseudomonas putida* introduced into soils of differing texture. *Journal of General Microbiology*, **138**, 667-73.

Magan, N. & Lacey, J. (1986). The phylloplane microflora of ripening wheat and effect of late fungicide applications. *Annals of Applied Biology*, **109**, 117-28.

Magan, N. & McLeod, A.D. (1991). Effect of open air fumigation with sulphur dioxide on the occurrence of phylloplane fungi on winter wheat. *Agriculture Ecosystems and Environment*, **33**, 245-61.

Maurhofer, M., Keel, C., Schnider, U., Voisard, C., Haas, D. & Défago, G. (1992). Influence of enhanced antibiotic production in *Pseudomonas fluorescens* strain CHA0 on its disease suppressiveness capacity. *Phytopathology*, **82**, 190-5.

Nicholson, P.S., Jones, M.J. & Hirsch, P.R. (1992). Monitoring survival of genetically-modified *Rhizobium* in the field. In *The Release of Genetically Modified Microorganisms*, ed. D.E.S. Stewart-Tull & M. Sussman, pp. 217-219. New York: Plenum Press.

OECD (1992). *Safety Considerations for Biotechnology*. Paris: OECD.

Pickup, R.W., Morgan, J.A.W., Winstanley, C. & Saunders, J.R. (1991). Implications for the release of genetically engineered organisms. *Journal of Applied Bacteriology Symposium Supplement*, **70**, 19S-30S.

Poplawsky, A.R. & Ellingboe, A.H. (1989). Take-all suppressiveness properties of bacterial mutants affected in antibiosis. *Phytopathology*, **79**, 143-6.

Possee, R.D., King, L.A., Weitzman, M.D., Mann, S.G., Hughes, D.S., Cameron, I.R., Hirst, M.L. & Bishop, D.H.L. (1992). Progress in the genetic modification and field-release of baculovirus insecticides. In *The Release of Genetically Modified Microorganisms*, ed. D.E.S. Stewart-Tull & M. Sussman, pp. 47-58. New York: Plenum Press.

Rolfe, B.G., Brockwell, J., Bolton-Gibbs, J., Clark, K., Brown, T. & Weinman, J.J. (1989). Controlled field release of genetically manipulated *Rhizobium* strains. *Australian Microbiologist*, **10**, 364.

Seong, K-Y., Höfte, M. & Verstraete, W. (1992). Acclimatization of plant growth promoting *Pseudomonas* strain 7NSK2 in soil: Effect on population dynamics and plant growth. *Soil Biology and Biochemistry*, **24**, 751-9.

Shapira, R., Ordentlich, A., Chet, I. & Oppenheim, A.B. (1989). Control of plant diseases by chitinase expressed from cloned DNA in *Escherichia coli*. *Phytopathology*, **79**, 1246-9.

Shaw, J.J., Dane, F., Geiger, D. & Kloepper, J.W. (1992). Use of bioluminescence for detection of genetically engineered microorganisms released into the environment. *Applied and Environmental Microbiology* **58**, 267-73.

Shaw, J.J., Geiger, D., Dane, F. & Kloepper, J.W. (1991). Bioluminescence for the detection of genetically-engineered microbes released into the environment. In *Biological Monitoring of Genetically Engineered Plants and Microbes*, ed. D.R. MacKenzie & S.C. Henry, pp. 229-237. Bethesda, Maryland: Agricultural Research Institute.

Stotzky, G., Broder, M.W., Doyle, J.D. & Jones, R.A. (1993). Selected methods for the detection and assessment of ecological effects resulting from the release of genetically engineered microorganisms to the terrestrial environment. *Advances in Applied Microbiology* **38**, 1-98. New York: Academic Press.

Sundheim, L., Poplawsky, A.R. & Ellingboe, A.H. (1988). Molecular cloning of two chitinase genes from *Serratia marcescens* and their expression in *Pseudomonas* species. *Physiological and Molecular Plant Pathology*, **33**, 483-91.

Teng, P.S. & Yang, X.B. (1993). Biological impact and risk assessment in plant pathology. *Annual Review of Phytopathology*, **31**, 495-521.

Thompson, I.P., Bailey, M.J., Fenlon, J.S., Fermor, T.R., Lilley, A.K., Lynch, J.M., MacCormack, P.J., McQuilken, M.P., Purdy, K.P., Rainey, P.B. & Whipps, J.M. (1993). Quantitative and qualitative seasonal changes in the

316   *J. M. Whipps* et al.

microbial community from the phyllosphere of sugar beet (*Beta vulgaris*). *Plant and Soil*, **150**, 177-91.

Toyota, K., Tsuge, T. & Kimura, M. (1992). Potential application of genetic transformants of *Fusarium oxysporum* f. sp. *raphani* for assessing fungal autecology. *Soil Biology and Biochemistry*, **24**, 489-94.

van Elsas, J.D. (1992). Environmental pressure imposed on GEMMOs in soil. In *The Release of Genetically Modified Microorganisms*, ed. D.E.S. Stewart-Tull & M. Sussman, pp. 1-14. New York: Plenum Press.

van Elsas, J.D., Dijkstra, A.F., Govaert, J.M. & van Veen, J.A. (1986). Survival of *Pseudomonas fluorescens* and *Bacillus subtilis* introduced into two soils of different texture in field microplots. *FEMS Microbiology Ecology*, **38**, 151-60.

Veal, D.A., Stokes, H.W. & Daggard, G. (1992). Genetic exchange in natural microbial communities. *Advances in Microbial Ecology*, vol. 12, ed. K.C. Marshall, pp. 383-430. New York: Plenum Press.

Vincent, M.N., Harrison, L.A., Brackin, J.M., Kovacevich, P.A., Mukerji, P., Weller, D.M. & Pierson, E.A. (1991). Genetic analysis of the antifungal activity of a soilborne *Pseudomonas aureofaciens* strain. *Applied and Environmental Microbiology*, **57**, 2928-34.

Voisard, C., Keel, C., Haas, D. & Défago, G. (1989). Cyanide production by *Pseudomonas fluorescens* helps suppress black root rot of tobacco under gnotobiotic conditions. *The EMBO Journal*, **8**, 351-8.

Wang, Z., Crawford, D.L., Magnuson, T.S., Bleakley, B.H. & Hertel, G. (1991). Effects of bacterial lignin peroxidase on organic carbon mineralization in soil, using recombinant *Streptomyces* strains. *Canadian Journal of Microbiology*, **37**, 287-94.

19

Has chaos theory a place in environmental mycology?

A. D. M. RAYNER

Introduction

Chaos theory, which strictly is only a subset of non-linear systems theory, deals with systems whose long-term behaviour or output is prone to be complex, irregular, sensitive to small changes in initial conditions and unpredictable at specific localities. Amongst several recent texts that have aimed to popularize the theory and describe its remarkable history, those by Gleick (1988) and Coveney & Highfield (1991) are perhaps the most accessible. More detailed sources of information concerning specifically biological applications of nonlinear theory are provided by Degn, Holden & Olsen (1987) and by Sleeman (1989).

The aim of this chapter is to promote appreciation of the ways in which non-linear dynamics can be expected to apply to fungal individuals, populations and communities. It will be argued that non-linear theory has more than just a *place* in environmental mycology; it provides a *basis* for understanding the complexity, interconnectedness and limits to the predictability of natural patterns of distribution and activity of fungi.

Throughout the discussion, the focus will be on understanding of the sources and ecological importance of non-linearity, rather than on rigorous mathematical treatment. At the outset, an attempt will be made to generalize about the kinds of process which underlie non-linear dynamics and the organizational properties that these processes give rise to. The mechanisms by which these processes operate at individual, population, community and sub-cellular levels of organization will then be considered and related to the ways that fungal systems respond to sources of environmental heterogeneity.

General origins and attributes of non-linearity

Basically, non-linearity is due to counteraction between expansive and constraining processes at and within a system's operational boundary. When expansive processes are driven by energy inputs greater than can be accommodated by equilibrium (linear) rates of boundary deformation, the systems become unstable, prone to partition at successively finer scales into heterogeneous subdomains. Such partitioning leads to oscillations and sub-division, and is evident in physical space in such phenomena as branching and turbulence.

The route to instability can be traced mathematically using equations in which the relationship between input to and expansion of a system changes from direct proportionality as the content increases and becomes constrained by boundary limits or resistances. In systems such as mycelial fungi, which are bounded by a restrictive physical envelope, the equations might therefore help to explore the relationship between energy gain and displacement of contents to deformable sites on the boundary.

One of the best known non-linear equations is the logistic difference equation (May, 1976). This equation has found varied biological applications, most notably in the modelling of population dynamics (May, 1987). Usually the populations in question have comprised discrete, and therefore easily countable individual organisms. However, any bounded system containing increasing 'numbers' of mutually exclusive energetic entities can be described following the same principles.

The equation relates the size (in terms of numbers) of a current generation of organisms (x) to that of the next generation (x_{next}) in terms of the net per capita rate of increase (r). In the simplest applications, each generation is treated as discrete from the previous one, i.e. there is no overlap or carry over from one generation to the next. Here,

$$x_{\text{next}} = rx - rx^2 \tag{1}$$

with x being normalized, allowed to vary only between limits of 0 and 1. The potential for increase or 'expansion' of x, due to the proliferative drive (r) resulting from resource acquisition is counteracted by the negative feedback or boundary constraint (e.g. due to resource limitation), rx^2.

If the equation is *iterated* from some low initial value of x, i.e. if each output value of x_{next} that the equation produces is used as input, x, to calculate a subsequent value of x_{next}, then the increasing value of rx^2 will progressively constrain growth. In fact, if $1 < r < 3$, then a value of

$x = 1 - 1/r$ is eventually attained, often (i.e. if $r > 2$) after a series of progressively smaller fluctuations (damped oscillations). At this stage, the population reaches its equilibrium size or 'carrying capacity' (K) at which it neither increases nor decreases.

However, for values of $r > 3$, i.e. when the population is being driven increasingly hard by its own proliferative power, then instabilities set in which prevent x from ever attaining an equilibrium value. Instead, as r is increased, x oscillates around first two, then four, then eight, then sixteen . . . 2^n values in a so-called period-doubling cascade before becoming (at $r = 3.57$) fully chaotic (unrepetitive) and eventually ranging unpredictably between all values between 0 and 1 (at $r = 4$). The sequence of values of x in this chaotic region appears to be random, even though being deterministically derived, and is extremely sensitive to small changes in the initial values of x prior to iteration.

The production of such complex patterns from so simple an equation contains some important object lessons (May, 1976, 1987). Firstly, there is the impossibility of predicting the long-term behaviour of chaotic systems; due to counteractive feedback even the minutest change in initial conditions may result in radical alterations in the itinerary of the system over time. Whilst the general limits of chaotic behaviour can be outlined readily, what will happen at a particular location becomes increasingly uncertain as the future in question becomes more remote.

A second point of interest is that systems which proliferate most rapidly, i.e. with high r values, are the ones most prone to instability. Many theorists equate proliferative power with 'fitness', so that any mechanism which reduces proliferation becomes regarded as having an evolutionary 'cost'. However, in nonlinear systems a 'cost' may in fact be a 'stabilizer'. This would apply particularly to long-lived or 'K-selected' populations in persistent niches as opposed to 'r-selected' populations in temporary niches (note that the terms r- and K-selection are actually derived from the logistic equation; see Pianka, 1970).

A third point is that, in non-linear systems, strongly interdependent variables may appear to be uncorrelated. However, an enormous body of environmental science uses correlation to demonstrate the connection between variables and a lack of correlation is often taken to imply that the variables are independent (cf. Sugihara & May, 1990).

An understanding of how the complexities generated by iterating nonlinear equations arise can be obtained by plotting each successive solution to these equations as points on a map on which the co-ordinates define the state of the system in 'phase space'. By joining the points, a 'trajectory' is produced which defines the changing state of the system. A simple way of showing the trajectories of solutions to the logistic equation can be achieved by drawing graphs showing the relationship between x_{next} and x for particular values of r (Fig. 19.1). The graphs take the form of parabolas which climb from zero to a maximum value, $\frac{1}{4}r$, before declining symmetrically back to zero. If a line is drawn at 45° to the x axis, then, as shown by Feigenbaum (1978), this can be used to find values of x and x_{next} in successive rounds of iteration. It can then readily be seen that for r values between 1 and 3 (below 1 the population becomes extinct), a trajectory can be traced which homes in on the equilibrium point where the 45° line and the parabola intersect. With r values > 3, however, the trajectory generally cannot home in in this way, and will either cycle around the equilibrium point forming one or more closed loops or it will circumnavigate this point indefinitely without exact repetition.

The three kinds of trajectories just described correspond with three kinds of 'attractors', an attractor being a state towards which the trajectory of a dynamic system evolves as if drawn by a magnet. The simplest attractors are fixed point attractors, equilibrium states which once arrived at cease to change. K values obtained when $1 < r < 3$ represent such attractors. When a trajectory ends up oscillating repetitively between two or more states, the attractor is a 'limit cycle'. Where the trajectory is non-repetitive but none the less never exceeds certain bounds, it is known as a 'strange', 'chaotic' or 'fractal' attractor.

The term fractal, used here, relates to a kind of geometry of irregular structures which cannot be described in Euclidean terms of smooth curves, surfaces and volumes arranged in integral dimensions of 1, 2 and 3 (Mandelbrot, 1982). An important property of these structures is that when they are examined at closer and closer range, more and more irregularities come into view. Their lengths or areas are therefore infinite when viewed at infinitesimal scales, even though they can be circumscribed within finite planes or volumes. 'Ideal' fractals exhibit exact 'self-similarity', such that however small a portion of the structure may be, it will, when magnified, appear identical to larger portions. Many naturally irregular structures, whilst not conforming absolutely to this ideal, are better approximated to fractals than to smooth figures. A classic example is the coastline of an island. Many branching structures are also most effectively approximated to fractals, at least down to a minimum scale corresponding with the width of their axes.

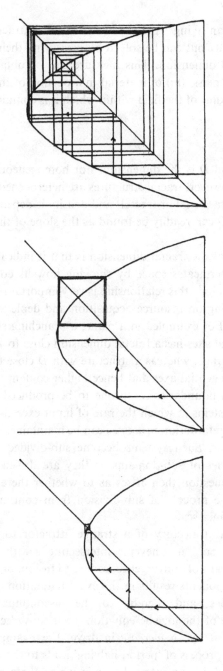

Fig. 19.1. Trajectories produced by iterating solutions of the logistic difference equation, $x_{next} = rx - rx^2$ for various values of r. The horizontal axis represents values of x that serve as input, the vertical axis represents values of x_{next}. Arrows represent the process of iteration from initial values of x. Trajectories leading to equilibrium (fixed point, with $r = 2.5$), repetitive oscillation (limit cycle, with $r = 3.1$) and non-repetitive dynamics (chaos, with $r = 3.8$) are shown from left to right. See text for further explanation.

The problem of quantifying fractal structures is intractable if approached conventionally but can be solved by measuring their degree of irregularity or fractal dimension. This measure can be obtained by relating the 'content' or 'mass' (m) of material in a portion of the structure to its 'extent', the radius of the field within which it is contained (R), according to the formula:

$$m = kR^D \tag{2}$$

where k is a constant and D is the dimension. For homogeneous structures, D is an integer. However, fractal structures are heterogeneous, D is not an integer (i.e. it is fractional or fractal) and density is dependent on R. For such structures, D can readily be found as the slope of the graph of $\ln m$ against $\ln R$.

A very useful feature of the fractal dimension is that it indicates how completely a structure permeates space by showing how its content is related to its extent. In biology this relationship is very important because it has considerable bearing on resource acquisition and depletion rates and competitive ability. For example, on a plane, a branching structure with predominantly radial axes has a fractal dimension close to 1, corresponding with wide coverage, whereas a structure with D close to 2 has proportionately more tangential axes and hence higher content.

To summarize, fractal boundaries are prone to be produced around and within dynamical systems in which the rate of input exceeds that at which advancement and retardation processes can be maintained in equilibrium with one another. Such systems become sub-divided into an increasingly complex series of sub-domains as they are driven further from equilibrium. The question then arises as to whether there is anything which can stop the process of sub-division from continuing indefinitely, on finer and finer scales.

The propensity for the trajectory of a strange attractor to wander indefinitely depends on it never intersecting with itself. Correspondingly, any means of convergence leading to fusion, or anastomosis, of trajectory components results in the exact reiteration of pathways, as in limit cycles, and thereby to the discontinuation of proliferation. In the case of the logistic equation, such convergence can actually be brought about by an *increase* in drive. For example, at r values just below 3.83, a process of 'period-halving' leads to the opening of a 'window of stability', i.e. an interval of r in which three repeated population values or states occur.

In physically bounded systems, fusion between subdomains results in the formation of entire or partial true networks. The resultant connection of boundary resistances in parallel rather than in series has profound repercussions for input–throughput relationships and subsequent patterns of expansion or regression of the systems. It stabilizes the systems by increasing their throughput capacity whilst allowing multiple options for redistribution and potentiating large amplifications in organizational scale. The latter result from the increase in power which can be delivered through a network to a local site of expansion.

Individual heterogeneity and interconnectedness

Generally, individual fungi proliferate either as colonies or as discrete, particulate entities that separate readily from one another, such as unicells. In liquids, particulate proliferation allows rapid dispersion and maximum exposure of absorptive surface to the environment. Here, the boundary of the system is defined either at liquid–air or liquid–solid interfaces. Proliferation of the discrete entities within this boundary would be expected to follow logistic dynamics providing that nutrient input is maintained (as in continuous culture) and the entities remain dispersed. Initially exponential proliferation may therefore be superseded by maintenance of an equilibrium population size, or, if r values are allowed to become high enough, regular or chaotic oscillations could result. The systems may be homogeneous in space, but they are liable to become heterogeneous in time.

Colonies form when individual entities do not dissociate. This may happen when the entities remain attached to one another – as in a branching system, or because they are attracted to one another, or because they are constrained by an external boundary such as a solid or semi-solid interface. Here, the colony boundary defines the system, and the colony itself begins to acquire the characteristics of an individual in its own right, albeit an indeterminate one potentially capable of indefinite self-proliferation. Colonies are therefore liable to generate heterogeneity both in themselves and in their environment and both in space and in time. They are non-linear systems whose outward form is dictated by counteraction between self-proliferation and limits to rates of uptake and deformation imposed at their boundaries.

Fungal colonies are either assemblages of unicells such as yeasts, or mycelia consisting of branching, tubular, apically extending hyphae.

Mycelia are often described as being multicellular, but this can be confusing, even in septate forms, in view of their fundamentally coenocytic organization.

In the absence of motility and some means, such as chemical attraction, of achieving social coherence, assemblages of unicells are limited in the range of dynamic patterns that they can assume. However, the possession of motility and a pheromonal guidance system in cellular slime moulds, for example, enables them to generate diverse patterns on a par with those in societies of multicellular organisms such as army ant raids. Such patterns result from what can be thought of as social indeterminacy and are founded on *behavioural* mechanisms that control the responses of individual entities both to one another and to their environmental circumstances. These behavioural mechanisms cause the components of the system to interact dynamically as if they were bounded by a deformable physical envelope, even though these components may appear to an external observer to remain distinct from one another.

Hyphae, on the other hand, have walls and their contents are not resolvable into individual cells but form a partitionable continuum containing populations of discrete organelles. Mycelia can generate patterns of equivalent complexity and versatility to those apparent in social organizations, possibly for the same fundamental reasons (Rayner & Franks, 1987). However, since the boundary of the system is identifiable as the boundary of the organism, mycelia may be described as developmentally indeterminate.

From a logistic viewpoint, the difference between *behavioural* and *developmental* mechanisms therefore merely reflects identification of different reference boundaries. It is more important to understand how these mechanisms affect the fundamental pattern-generating processes (energy assimilation, conservation, distribution and redistribution) that operate at, and within, these boundaries and are common to collective organizations of all kinds.

Some suggestions as to how these processes may contribute to the non-linear dynamics of mycelial systems will be made below. However, it is important first to appreciate that the predominant approach over the last two decades has been to treat mycelia, at least 'ideally', as purely additive assemblages of regularly duplicating, functionally independent hyphal growth units (e.g. Trinci, 1978; Prosser, 1993). The hypothesis that mycelia can be so treated, despite the interconnectedness of their components, is attractive because it eases predictive calculation of their growth kinetics.

For such a treatment to be valid, it is essential that the distribution of mycelial structure is homogeneous, at least on average, so that a change in the scale of observation results in a proportional increase or decrease in the amount of biomass present (Equation 2). In fact, actual measurements of the fractal dimensions of mycelial structures have indicated that this does not necessarily apply, and indeed that these structures are both heterogeneous and coherent (Ritz & Crawford, 1990; Obert, Pfeifer & Sernetz, 1990; Obert, Neuschulz & Sernetz, 1991; Ainsworth & Rayner, 1991; Bolton & Boddy, 1993; Matsuura & Miyazima, 1993; Mihail *et al.*, 1994; Crawford & Ritz, 1994). Even so, such heterogeneity may still be regarded as the result of imperfection or 'noise'. Either the mycelium itself is deemed to be wayward, or it becomes so as a result of the complicating effects of a changing environment on what would otherwise be homogeneous growth patterns. To improve predictability, efforts have therefore been made either to maintain as constant conditions as possible, or to restrict observation of growth patterns to situations where environmental gradients have not yet been established – as in germlings and mature colony margins (Trinci, 1978). Observations of exponential and linear growth made in these circumstances have reinforced duplication cycle and peripheral growth zone models that relate mycelial growth patterns directly to a specific rate of increase in biomass for any particular set of microenvironmental conditions.

These additive models cannot, however, account for many properties, including what Prosser (1993) describes as 'atypical' features such as annulations, spirals and sectors, that appear even when mycelia are grown in near homogeneous conditions. They do not explain, except tautologically, the onset of branching or polarity. Nor can they account for the maintenance of an evenly growing margin at a constant radial increment in more than one spatial dimension. This is because of the need to generate space-filling biomass that keeps up with leader hyphae as the colony radius increases. Moreover, additive models cannot explain such radical shifts in organizational pattern as occur between septate and coenocytic, slow–dense and fast–effuse, submerged and emerged, diffuse and aggregated developmental modes (e.g. Rayner & Coates, 1987; Stenlid & Rayner, 1989; Rayner, Griffith & Wildman, 1994*a*). Much less can they account for the high degree of coordination and attunement to environmental circumstances which are often evident when mycelia grow in biotically and abiotically heterogeneous environments (see below).

Non-linear systems theory, on the other hand, has the potential to explain all these properties, as well as encompassing exponential growth of systems with low initial input rates and linear growth of systems operating at carrying capacity. It may therefore be better to regard heterogeneity not as the complicated by-product of imperfection but rather as the complex consequence of the built-in nonlinearity of mycelial systems – and the means by which these systems respond with such sensitivity and versatility to changes in their environmental circumstances.

To illustrate this possibility, and without discounting other hypotheses (such as the reaction-diffusion mechanism regulating cytosolic calcium gradients envisaged by Crawford & Ritz, 1994), an approach will now be described which treats mycelia as hydrodynamic non-linear systems. Proliferative drive in these systems is supplied by uptake of water and nutrients from the environment. This drive results in varied patterns of displacement of hyphal contents and boundaries depending on the way that it is regulated by three basic parameters.

The first basic parameter is the *resistance to deformation* of hyphal envelopes, and depends on the interplay between counteractive plasticization and rigidification processes. This interplay has attracted much attention with respect to understanding the apical extension of hyphae (e.g. Bartnicki-Garcia, 1973; Wessels, 1986), but its relation to nonlinearity does not seem to have been widely recognized.

The second parameter is the *resistance to passage of solutes and water* across hyphal walls, which can be varied by depositing and releasing, polymerizing and depolymerizing hydrophobic compounds. Such variation would be logical because any effective energy-transducing system would be expected to possess mechanisms that enhance or maintain assimilation when and where there is plenty, but reduce leakage when and where there is shortage. In concert with variations in boundary strength, such mechanisms would fundamentally affect patterns of uptake of resources, their distribution to sites of expansion or discharge, and therefore the outward form adopted by mycelia as hydrodynamic systems.

However, these possibilities seem largely to have been neglected in considerations of mycelial dynamics. Indeed, in discussing peripheral growth zone and duplication cycle concepts, Prosser (1991, 1993) relates hyphal extension purely to the *synthesis*, delivery and incorporation of wall and membrane material carried in vesicles. These processes are said, tautologically, to be determined by specific growth rate, by the *volume* of hypha supplying material and by the size of the extension zone. The ultimate dependence of such synthesis and delivery (by whatever mechanism) upon rates of uptake of resources across hyphal boundaries is not acknowledged, implying that the properties of these boundaries have either negligible or uniform effects on assimilation rates. Neither of these implications seems likely.

There is therefore a critical need for more information about possible variations in hyphal permeability, the materials involved, and how the production, sequestration and release of these materials is regulated so as to generate appropriate distributive patterns. According to one current theory (Rayner, Griffith & Wildman, 1994*a*,*b*), hydrophobic materials, in the form of polypeptides known as hydrophobins (Wessels, 1991, 1992) and/or polyphenolics and terpenoid compounds, are produced when transductive (secondary) metabolism is initiated by a drop in internal energy charge. Correspondingly, hyphal boundaries (walls and membranes) are maintained in permeable form whilst there are sufficient external supplies to drive active uptake, involving inductive, ATP-generating (primary) metabolism, but become sealed in the absence of these supplies. The action of enzymes, such as phenoloxidases, which can lead to both polymerization and depolymerization of hydrophobic compounds via the formation of free radicals, may play a key role here. When expressed within the protoplasm, it may also be expected to lead to localized cell death (Rayner *et al.*, 1994*a*).

The two resistances, just described, to the deformation and passage of water and solutes across hyphal walls and membranes, comprise what may be thought of as the 'insulation' of hyphal boundaries (Rayner *et al.*, 1994*a*). The third important parameter regulating displacement (throughput) patterns is *the degree of protoplasmic continuity or partitioning* of hyphae. This is determined by the presence of septa and anastomoses, by septal sealing and localised protoplasmic death. The ways in which variations in insulation and hyphal partitioning could regulate four fundamental processes involved in energy capture and transduction by mycelial systems are illustrated in Fig. 19.2. The possible involvement of these processes in mycelial development will now be discussed.

Spore germination and hyphal branching

Spores germinate by taking up water and nutrients from their environment. The resultant displacement of their boundaries usually leads to initially isotropic expansion. In some circumstances, notably at elevated temperatures, this expansion may continue, so producing giant cells capable of sporulating directly. In other cases, budding yeast-like or mon-

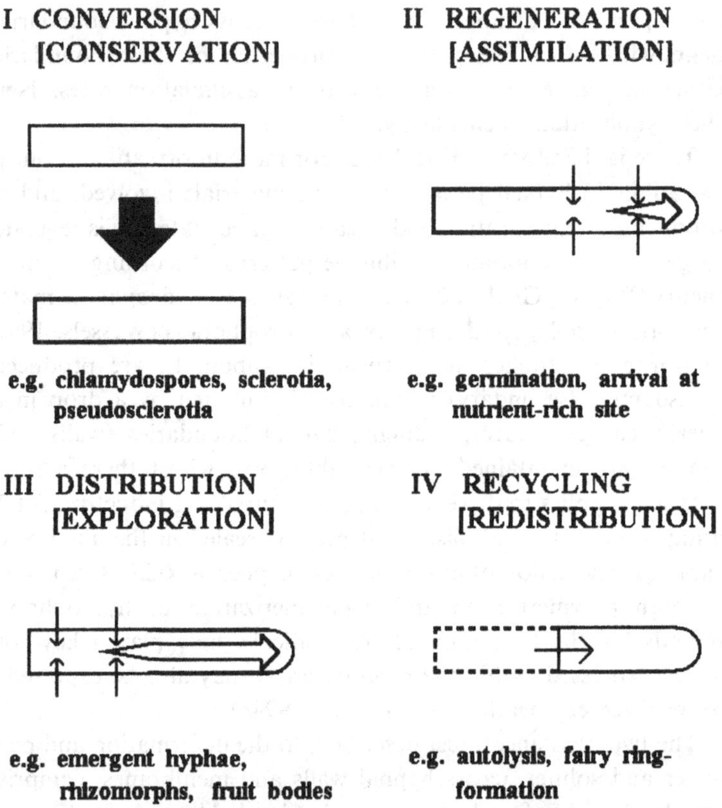

328   *A. D. M. Rayner*

I CONVERSION
   [CONSERVATION]

II REGENERATION
    [ASSIMILATION]

e.g. chlamydospores, sclerotia,
pseudosclerotia

e.g. germination, arrival at
nutrient-rich site

III DISTRIBUTION
     [EXPLORATION]

IV RECYCLING
    [REDISTRIBUTION]

e.g. emergent hyphae,
rhizomorphs, fruit bodies

e.g. autolysis, fairy ring-
formation

Fig. 19.2. Four fundamental processes in elongated hydrodynamic systems, as determined by boundary deformability, permeability and internal partitioning. Rigid boundaries are shown as straight lines, deformable boundaries as curves, impermeable boundaries by thicker lines, degenerating boundaries as broken lines and protoplasmic disjunction by an internal dividing line. Simple arrows indicate input across permeable boundaries into metabolically-active protoplasm, tapering arrows represent throughput due to displacement of contents towards sites of deformation. Devised during discussions with Philip Drazin and David Griffel.

ilioid chains of inflated cells can be produced. Indeterminate (mycelial) development is initiated by the emergence of one or more polarized hyphae either from the germinating spore itself, or from cells that it has given rise to, as in yeast–mycelial transitions. In the latter case it appears that there may be a critical diameter of the parent cells above which the ability to produce a *hypha* is not possible, and the same applies to series of monilioid cells developing from regenerating protoplasts of the basidiomycete *Stereum hirsutum* (M. Ramsdale pers. comm.; cf. Peberdy, 1979).

The emergence of hyphae, due to the cessation of isotropic expansion and onset of polarity, represents a transition from growth in integral dimensions to growth in fractal dimensions and corresponds with a process known in non-linear systems theory as symmetry breaking. It may result from some kind of threshold phenomenon where the relation between input and radius of curvature of the structure causes local stress points in the boundary to become amplified as instabilities.

The initiation of branching leads to a progressive increase in the fractal dimension of the emerging hyphal system and clearly also involves a threshold phenomenon. Most efforts to explain hyphal branching have been based on the adaptive requirement for a chemotropic mechanism in order to exploit resources maximally by directing growth along gradients and away from sites of depletion (e.g. other hyphae). However, branching can also be explained as the consequence of uptake into the system exceeding throughput (carrying) capacity to existing sites of expansion of hyphal boundaries (Rayner *et al.*, 1994*a*; Rayner, Griffith & Ainsworth, 1994).

In a fully assimilative hypha, the propulsion of the deformable apex that follows uptake would also have the effect of increasing the absorptive surface so that the rate of uptake would increase without an accompanying increase in the number of distribution points. Initially, the increase both in uptake and absorptive surface would be exponential. However, the in-series resistances posed by the diameter of the tube and the deforming apex would progressively impede throughput until a limit where no additional net increase in extension rate could be sustained. The hypha would then extend linearly, whilst any additional input would be prone to induce branching. The statement that branching is due to the 'inability of a hypha extending at a linear rate to accommodate biomass synthesized at an exponential rate' (Prosser, 1993) expresses a superficially similar argument to the above but does not identify the causal counteraction.

In the kind of distributive system just described, irregularities in the lateral wall could serve as stress accumulation sites where processes leading to deformation are initiated. Moreover, vesicles containing enzymes and components involved in wall formation could accumulate in those regions where additional input cannot be displaced axially because of their remoteness from the apex. These regions would therefore be those at which septa and branches would be expected to form. The same process of extension to maximum carrying capacity prior to branching would then be repeated in the branches themselves. The result would be a racemose pattern. Such patterns are common in mycelia, and would be expected particularly in fully assimilative systems produced following arrival of spores or mycelium in nutrient-rich sites, corresponding with the regenerative state in Fig. 19.2. However, a mycelium proliferating only in this way would become very irregular and have restricted abilities to explore its environment. In being unable to prevent losses in nutrient-poor or depleted sites, it would also be extremely dissipative (wasteful).

Insulation, conservation and exploration

The problem of conserving resources in energy-poor environments can be solved by sealing hyphal boundaries against losses and/or by converting solutes into insoluble (and therefore temporarily unusable) storage compounds such as glycogen. Such processes convert assimilative biomass into non-assimilative biomass. Wholesale conversion necessitates subsequent inactivity but enables temporal survival of shortage, in the form of exogenously or endogenously dormant spores, sclerotia, etc. Partial conversion, providing that protoplasmic continuity is maintained, allows distribution of resources from assimilative phases to non-assimilative phases. Were the contents of the latter to contain fewer solutes than at assimilative sites, they will act as translocation sinks so long as some part of their boundary remains permeable. However, such gradients cannot, in themselves, drive extension (cf. Jennings, 1984, 1987). If, on the other hand, the boundary of non-assimilative hyphae is relatively impermeable, then, providing that their apices remain deformable, they can be driven by throughput from assimilative sites. Such hyphae provide means for exploration of territory. When and if they branch (due to input at source exceeding distributive capacity at sink), they will do so like the distributaries in a river delta.

By combining explorative and assimilative phases, mycelia can increase in circumference at a constant overall rate whilst maintaining an even distribution of biomass. Were these phases to remain in full protoplasmic continuity, then high throughput rates could be maintained. This would allow the rapid extension rates which are characteristic of many coenocytic fungi, and of the coenocytic phases of certain basidiomycetes, such as *Phlebia* and *Phanerochaete* species (Boddy & Rayner, 1983; Ainsworth

& Rayner, 1991). However, there is little scope for differentiation in coenocytes, and the maintenance of continuity with redundant components could drain resources and so inhibit growth.

Cycling and recycling

The partitioning of hyphae by septa facilitates differentiation of separate compartments. It also provides a means of limiting distribution away from assimilative sites and, if associated with membrane disjunction, of allowing redistribution of resources from degenerating compartments. Septa would thereby fulfil an important role as valves and isolators in mycelial systems. However, they will also reduce throughput capacity, particularly in a radial or in-series resistance system.

It therefore may not be coincidental that septate mycelial systems, including the septate phases of *Phlebia* and *Phanerochaete* species mentioned above, should, unlike coenocytic systems, characteristically form abundant anastomoses. By replacing an in-series system with an in-parallel system, anastomoses greatly increase throughput capacity, hence limiting branching whilst allowing enhanced supply to local sites of expansion. The latter enhanced delivery may, in turn, allow the system to amplify its operational scale, for example through the emergence of large diameter hyphae (e.g. Ainsworth & Rayner, 1990), fan-like sectors (cf. Coggins *et al.*, 1980) and well insulated, cable-like mycelial cords and rhizomorphs (e.g. Rayner *et al.*, 1985). The ability of the latter structures to extend at rates of an order of magnitude greater than that of individual hyphal elements, inexplicable by present theories of mycelial dynamics, could thereby be understood. Likewise, the rapid expansion rates of large fruit bodies may well necessitate high rates of delivery through a networked system.

However, once a network becomes sufficiently large and complete, it is liable to act as a massive and persistent sink, cycling resources within itself. A fully networked mycelium may therefore regain the symmetry broken at germination. The system may then only be able to continue to expand through the onset of degenerative processes that allow redistribution to occur to emergent sites on the network's boundary. Just such a process may explain the formation of fairy rings such as those of *Clitocybe nebularis* where explorative mycelial cords at the margin of the annulus are superseded by dense and degenerative mycelial zones (Dowson, Rayner & Boddy, 1989).

Combine harvesting: mycelial distribution patterns in
heterogeneous environments

Fairy ring formation illustrates a particular kind of *foraging* pattern which results from the integration of explorative, exploitative, conservative and degenerative processes in natural environments. As noted by Dowson *et al.* (1989) this pattern is apposite to situations where resources are plentiful, widespread and in close proximity to one another. In other cases, resources are more patchily distributed and processes that enable mycelia to channel resources between assimilative sites become increasingly important. Studies of these processes in wood-decomposing fungi have highlighted the importance of explorative, redistributional and consolidative processes during the formation of mycelial cord networks between wood blocks in soil (e.g. Boddy, 1993).

Another way of demonstrating and analysing the feedback relationships between environment, metabolism and gene expression that define pattern-generation is to grow mycelia in heterogeneous matrices. The system illustrated in Fig. 19.3 was designed and tested at Bath by Erica Bower and Louise Owen. It consists of a set of chambers that are isolated from one another with respect to diffusion through the growth medium, but interconnected by passageways that allow particular portions of the mycelium to grow between and across the chambers. The design therefore combines the discreteness that enables key stages of development to be analysed in a particular locale with the continuity which is fundamental to the operation of the mycelium as an integrated system.

To illustrate the kinds of pattern which can emerge, and how these are related to ecological niches, stages in the formation of mycelial systems by four ecologically distinctive wood-inhabiting basidiomycetes grown in matrices containing high and low nutrient media will be outlined. In each case, the observed behaviour is very different from that of a purely assimilative system, which would be expected simply to vary its density in proportion to nutrient supply.

Coprinus picaceus and *Coprinus radians* are both relatively rapidly growing fungi that cause brown rot decay (due to cellulose degradation) of angiospermous wood. *C. picaceus* produces fruit bodies on soil some distance away from colonized wood, to which they are attached by white mycelial cords. *C. radians* produces both fruit bodies and a covering of orange-brown mycelial cords, the *Ozonium* state, directly on the surface of colonized wood.

In both *C. picaceus* (Fig. 19.3(*a*)–(*d*)) and *C. radians* (Fig. 19.3(*e*)–(*h*)), production of diffuse, appressed mycelium in high nutrient chambers was followed by proliferation of aerial mycelium across previously uncolonized low nutrient chambers. This aerial mycelium was hydrophobic, initially being covered with water droplets, and gave rise to white (*C. picaceus*) or orange-brown (*C. radians*) mycelial cords. The latter formed along primary and, in *C. picaceus*, also along secondary communication paths between high nutrient sites, resulting in the formation of almost symmetrical networks. Degeneration of non-connective mycelium accompanied formation of the cords, and was particularly pronounced in *C. radians*, where fruit body initials emerged, and sometimes subsequently matured, in low nutrient chambers.

Mycelial patterns generated by *Phlebiella vaga* and *Phallus impudicus* (not shown), both relatively slow-growing species that cause white rot (degradation of lignin and cellulose), were less symmetrical and less closely attuned to local nutrient supply than the *Coprinus* species. *P. impudicus* is like *C. picaceus* in producing fruit bodies attached to mycelial cords, some distance away from colonized wood, whereas *P. vaga* produces resupinate fruit bodies directly on the wood surface, where its mycelium is organized into diffuse mats, fan-like sheets or cords.

Whilst the appearance of mycelium of *P. vaga* within any one chamber was relatively uniform, there were many changes in pattern across the matrix, particularly with respect to whether it developed as hydrophobic yellow-white mycelial fans or brownish mycelial mats. Brown mycelial cords developed in some chambers, accompanied by the degeneration of non-connective mycelium, and eventually produced persistent networks interconnecting sequences of five or more chambers. There were strong tendencies for growth in a particular direction to be terminated whilst other parts of the system continued to proliferate, and 'focusing' effects, such that mycelium emerging from a partition was much denser than that entering it. In *P. impudicus*, asymmetry was related to the readiness with which mycelial cords formed prior to, rather than after, establishment of connections across chambers.

Population and community structure and dynamics: organismal coevolution

Like the individuals from which they are composed, fungal populations and communities are bounded but incompletely sealed systems. As else-

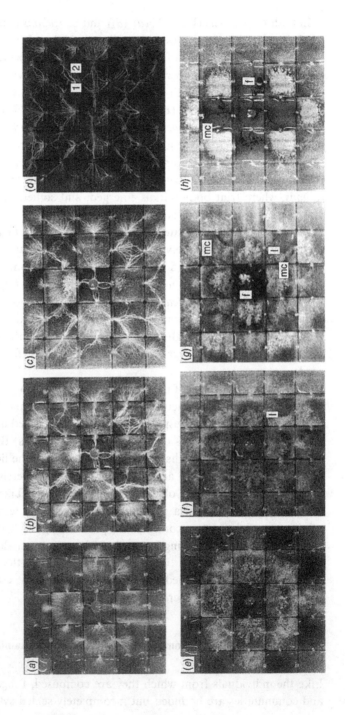

Fig. 19.3. Mycelial patterns of *Coprinus picaceus* (*a*)–(*d*) and *Coprinus radians* (*e*)–(*h*) grown in heterogeneous matrices within 100 cm² plastic Repli dishes containing 25 chambers. A hole, 2 mm wide, was cut through the centre of each partition, just above the level of the agar medium within each well, the fungi were inoculated into the central chamber and the dishes incubated in darkness at 20 °C. (*a*)–(*c*) Patterns produced after 20, 30 and 57 d in a matrix containing a 'chessboard' of alternating low (distilled water agar) and high (2% malt agar) nutrient media, starting with low nutrients in the central chamber. After sparse growth across the central chamber, *C. picaceus* produced dense, diffuse growth, closely appressed to the nutrient medium in the adjacent high nutrient chambers. Growth in the high nutrient chambers was accompanied by formation of mycelial cords along connective pathways across the central chamber, and followed by dense, diffuse, aerial growth bearing water droplets across the succeeding low nutrient chambers. Upon contacting the next set of high nutrient chambers, diffuse, appressed growth was again initiated and accompanied by consolidation of connective mycelial cords and degeneration of all other mycelium across the preceding low nutrient chambers. In this way assimilative mycelial phases in high nutrient chambers became interlinked by persistent distributive networks across intervening low nutrient chambers. (*d*) Pattern in a chessboard matrix after 60 d, starting from a high nutrient chamber. Successive rounds of assimilative and distributive growth have again resulted in formation of a persistent network of cords, with both secondary (2) and primary (1) connective routes across low nutrient chambers. (*e*)–(*g*) Patterns of *C. radians* after 15, 20 and 30 d in a chessboard matrix starting with low nutrients. After sparse initial growth, alternating phases of dense, diffuse, appressed mycelium and dense, diffuse, hydrophobic aerial mycelium were produced in succeeding high and low nutrient chambers. Connection between high nutrient chambers is followed by lysis (1) of non-connective mycelium, formation of orange–brown mycelial cords (mc) along communication channels and emergence of fruit body initials (f) in low nutrient chambers. (*h*) Pattern after 30 d in a matrix containing a central quincunx of five low nutrient chambers. The pattern is similar to that on the chessboard, but emphasizes that dense mycelium is only produced in low nutrient chambers that are reached *after* growth across high nutrient chambers. (Courtesy of Erica Bower and Louise Owen.)

where in ecology, the tendency to treat these systems as strictly hierarchical, with any particular level capable of being studied in isolation from the others has been unfortunate. It has led to simplistic considerations of such important concepts as individual fitness (without regard to population constraints) and succession (as a predictable temporal sequence insensitive to small changes in initial conditions). In reality population and community dynamics are as sensitive to the properties of the component individuals as are the individuals to the constraints and opportunities afforded by their neighbours within the system limits. Competitive, parasitic and mutualistic relationships affect one another and are affected by environmental circumstances. A change in conditions may bring about a change in relationships, for example from mutualistic to pathogenic. Pathogens, and pathogens of pathogens, can have markedly destabilizing effects on host population dynamics (May, 1985, 1990; Shaw, 1994).

There is therefore both 'top–down' and 'bottom–up' management; no two fungal populations or communities can be exactly alike, and given very similar initial inputs they can indeed become very dissimilar, both structurally and functionally. This has provided the perfect excuse for repetitive investigations of particular kinds of populations or communities, each revealing different details and slight or major contradictions of previous work.

Like a maturing fungal mycelium, fungal communities can be seen to progress through stages of openness, closure and degeneration, with each stage characterized by a different ecological strategy (Cooke & Rayner, 1984). In open communities, there is either continual renewal of resources or unoccupied domain still available for colonization. Proliferation of individual community members is therefore relatively unconstrained, either by the system boundary or by mutual effects. In closed communities, boundary constraints around and between members are very strong, and the interaction between any two neighbours may have repercussions throughout the networked system. Cycles of relative stability and rapid change are therefore not unexpected in these systems (Rayner & Todd, 1979) as the emphasis changes from r- to K-selection in correspondence with the proliferative and carrying capacity terms of the logistic equation.

Not only can the interaction between two neighbouring mycelia have repercussions for the whole community, but this interaction itself is also liable to be extraordinarily sensitive to initial conditions. After all, what is involved here is an engagement between two complex, indeterminate partial networks – a meeting of fields, not particles! For example, in a recent study, three fundamentally different outcomes were identified amongst twenty replicate pairings made under as closely similar conditions as possible between the same two strains of *Coriolus versicolor* and *Peniophora lycii* on malt agar. Equally, there can be varying outcomes along the length of the interaction interface between two fungi (Rayner, Griffith & Wildman, 1994*b*). Such results do not imply absolute unpredictability and unrepeatability, because similar patterns of interaction can be identified. However, they do illustrate that the interaction outcome between complex non-linear systems can itself be complex and therefore have complex repercussions.

Just as interactions between different species can be complex, so too can those within populations, i.e. between neighbouring individuals of the same species. At the population level, varied interaction outcomes have important implications for patterns of genetic heterogeneity within and between gene pools. At the individual level they reflect the fact that fungal mycelia are organized as bounded populations of genomic organelles.

Subcellular population dynamics: genomic coevolution

In higher fungi, hyphal fusions between genetically different mycelia bring subcellular populations of nuclei and mitochondria into protoplasmic communion. Subsequent patterns of coexistence of these populations depend on whether and how they interact stably or unstably with one another, and can have a profound influence on phenotypic expression. That certain combinations are able to coexist stably is evident in sexually outcrossing higher basidiomycetes, where vigorous, independently growing heterokaryons containing nuclei with complementary mating type genes are formed. However, how these heterokaryons maintain stability is not known, and they do not in any case seem able to contain combinations of more than two genetically disparate kinds of nuclei and more than one mitochondrial type (e.g. Rayner, 1991). This may reflect the fact that systems containing more than two competing attractors are chaotic.

On the other hand, the existence of somatic rejection responses resulting in protoplasmic degeneration, and probably involving the generation of free radicals (see above), indicates the potential for genomic disparity to result in instability. A dynamic, counteractive, relationship exists between these rejection responses and somatic acceptance responses that allow genetic nonself access, both in ascomycetes and basidiomycetes (e.g. Rayner, 1991). This relationship is critical to understanding how

338   *A. D. M. Rayner*

boundaries between populations are configured. Amongst basidiomycetes, the failure of somatic acceptance to override rejection results in reproductive isolation, and hence in speciation and the evolution of non-outcrossing breeding strategies (Rayner *et al.*, 1984; Rayner, 1991). Conversely, allowing access between strains which for one reason or another are not able to maintain stable genomic interactions can lead to extensive degeneracy, genomic replacement and complex patterns of phenotypic expression (Ainsworth *et al.*, 1990*a,b*; 1992). The potential for chaos, implicit in nonlinear or counteractive relationships, can therefore be found at subcellular as well as at individual, population and community levels of organization, both defining and being defined by the boundary limits of fungal systems.

Conclusions: chaos, fungi and environmental change

The indeterminate, energy-distributing systems of fungi can in many ways be regarded as the mainstream of ecosystem function, interconnecting and influencing the lives and deaths of other organisms in innumerable and sometimes surprising ways. To leave fungi out of the equations when trying to predict the causes and consequences of environmental change is unrealistic, to put it mildly. On the other hand, if fungi are included in the equations, it becomes impossible to ignore the subtleties, sensitivities and long-term unpredictability of non-linear, delicately counterpoised systems of all kinds. Reality is uncertain, but not incomprehensible; hindsight reveals the limitations of foresight.

Acknowledgements

I would like to thank Martyn Ainsworth, John Beeching, Erica Bower, John Broxholme, John Crawford, John Crowe, Neil Cryer, Philip Drazin, Fran Fox, Nigel Franks, David Griffel, Gwyn Griffith, Louise Owen, Mike Mogie, Neil Porter, Mark Ramsdale, Stuart Reynolds, Karl Ritz, Brian Sleeman, Zac Watkins and Howard Wildman, amongst others, for many helpful discussions.

References

Ainsworth, A. M., Beeching, J. R., Broxholme, S. J., Hunt, B. A., Rayner, A. D. M. & Scard, P. T. (1992). Complex outcome of reciprocal exchange of nuclear DNA between two members of the basidiomycete genus *Stereum*. *Journal of General Microbiology*, **138**, 1147–57.

Ainsworth, A. M. & Rayner, A. D. M. (1990). Mycelial interactions and outcrossing in the *Coniophora puteana* complex. *Mycological Research*, **94**, 627–34.

Ainsworth, A. M. & Rayner, A. D. M. (1991). Ontogenetic stages from coenocyte to basidiome and their relation to phenoloxidase activity and colonization processes in *Phanerochaete magnoliae*. *Mycological Research*, **95**, 1414–22.

Ainsworth, A. M., Rayner, A. D. M., Broxholme, S. J. & Beeching, J. R. (1990*a*). Occurrence of unilateral genetic transfer and genomic replacement between strains of *Stereum hirsutum* from non-outcrossing and outcrossing populations. *New Phytologist*, **115**, 119–28.

Ainsworth, A. M., Rayner, A. D. M., Broxholme, S. J., Beeching, J. R., Pryke, J. A., Scard, P. T., Berriman, J., Powell, K. A., Floyd, A. J. & Branch, S. K. (1990*b*). Production and properties of the sesquiterpene, (+)-torreyol, in degenerative mycelial interactions between strains of *Stereum*. *Mycological Research*, **94**, 799–809.

Bartnicki-Garcia, S. (1973). Fundamental aspects of hyphal morphogenesis. *Symposia of the Society for General Microbiology*, **23**, 245–67.

Boddy, L. (1993). Saprotrophic cord-forming fungi: warfare strategies and other ecological aspects. *Mycological Research*, **94**, 641–55.

Boddy, L. & Rayner, A. D. M. (1983). Mycelial interactions, morphogenesis and ecology of *Phlebia radiata* and *Phlebia rufa* in oak. *Transactions of the British Mycological Society*, **80**, 437–48.

Bolton, R. G. & Boddy, L. (1993). Characterization of the spatial aspects of foraging mycelial cord systems using fractal geometry. *Mycological Research*, **97**, 762–8.

Coggins, C. R., Hornung, U., Jennings, D. H. & Veltkamp, C. J. (1980). The phenomenon of 'point growth', and its relation to flushing and strand formation in the mycelium of *Serpula lacrimans*. *Transactions of the British Mycological Society*, **75**, 69–76.

Cooke, R. C. & Rayner, A. D. M. (1984). *Ecology of Saprotrophic Fungi*. London: Longman.

Coveney, P. & Highfield, R. (1991). *The Arrow of Time*. London: Flamingo.

Crawford, J. W. & Ritz, K. (1994). Origin and consequences of colony form in fungi: a reaction-diffusion mechanism for morphogenesis. In *Shape and Form in Plants and Fungi*, ed. D. S. Ingram, pp. 311–327. London: Academic Press.

Degn, H., Holden, A. V. & Olsen, L. F. (1987). *Chaos in Biological Systems*. New York and London: Plenum Press.

Dowson, C. G., Rayner, A. D. M. & Boddy, L. (1989). Spatial dynamics and interactions of the woodland fairy ring fungus, *Clitocybe nebularis*. *New Phytologist*, **111**, 501–9.

Feigenbaum, M. J. (1978). Quantitative universality for a class of nonlinear transformations. *Journal of Statistical Physics*, **19**, 25–52.

Gleick, J. (1988). *Chaos*. London: Heinemann.

Jennings, D. H. (1984). Water flow through mycelia. In *The Ecology and Physiology of the Fungal Mycelium*, ed. D. H. Jennings & A. D. M. Rayner, pp. 143–164. Jennings: Cambridge University Press.

Jennings, D. H. (1987). Translocation of solutes in fungi. *Biological Reviews*, **62**, 215–43.

Mandelbrot, B. B. (1982). *The Fractal Geometry of Nature*. San Francisco: W. H. Freeman.

Matsuura, S. & Miyazima, S. (1993). Colony of the fungus *Aspergillus oryzae* and self-affine fractal geometry of growth fronts. *Fractals*, 1, 11–19.

May, R. M. (1976). Simple mathematical models with very complicated dynamics. *Nature*, 261, 459–67.

May, R. M. (1985). Regulation of populations with non-overlapping generations by microparasites: a purely chaotic system. *American Naturalist*, 125, 573–84.

May, R. M. (1987). Chaos and the dynamics of biological populations. *Proceedings of the Royal Society of London*, A413, 27–44.

May, R. M. (1990). Population biology and population genetics of plant-pathogen associations. In *Pests, Pathogens and Plant Communities*, ed. J. J. Burdon & S. R. Leather. Oxford: Blackwell.

Mihail, J. D., Obert, M., Taylor, S. J. & Bruhn, J. N. (1994). The fractal dimension of young colonies of *Macrophomina phaseolina* produced from microsclerotia. *Mycologia*, 86, 350–6.

Obert, M., Neuschulz, U. & Sernetz, M. (1991). Comparison of different microbial growth patterns described by fractal geometry. In *Fractals in the Fundamental and Applied Sciences*, ed. H. -O. Peitgen, J. M. Henriques & C. F. Penedo, pp. 293–306. B. V. North-Holland: Elsevier Science.

Obert, M., Pfeifer, P. & Sernetz, M. (1990). Microbial growth patterns described by fractal geometry. *Journal of Bacteriology*, 172, 1180–5.

Peberdy, J. F. (1979). Fungal protoplasts, isolation, reversion and fusion. *Annual Review of Microbiology*, 33, 21–39.

Pianka, E. R. (1970). On r- and K-selection. *American Naturalist*, 104, 592–7.

Prosser, J. I. (1991). Mathematical modelling of vegetative growth of filamentous fungi. In *Handbook of Applied Biology, vol. 1*, ed. D. H. Arora, B. Rai, K. G. Mukerji & G. R. Knudsen, pp. 591–623. New York: Marcel Dekker.

Prosser, J. I. (1993). Growth kinetics of mycelial colonies and aggregates of ascomycetes. *Mycological Research*, 97, 513–28.

Rayner, A. D. M. (1991). The challenge of the individualistic mycelium. *Mycologia*, 83, 48–71.

Rayner, A. D. M. & Coates, D. (1987). Regulation of mycelial organisation and responses. In *Evolutionary Biology of the Fungi*, ed. A. D. M. Rayner, C. M. Brasier & D. Moore, pp. 115–136. Cambridge: Cambridge University Press.

Rayner, A. D. M., Coates, D., Ainsworth, A. M., Adams, T. J. H., Williams, E. N. D. & Todd, N. K. (1984). The biological consequences of the individualistic mycelium. In *The Ecology and Physiology of the Fungal Mycelium*, ed. D. H. Jennings & A. D. M. Rayner, pp. 509–40. Cambridge: Cambridge University Press.

Rayner, A. D. M. & Franks, N. R. (1987). Evolutionary and ecological parallels between ants and fungi. *Trends in Ecology and Evolution*, 2, 127–33.

Rayner, A. D. M., Griffith, G. S. & Ainsworth, A. M. (1994). Mycelial interconnectedness and the dynamic life styles of filamentous fungi. In *The Growing Fungus*, ed. N. A. R. Gow & G. M. Gadd, pp. 21–40. London: Chapman & Hall.

Chaos theory and environmental mycology 341

Rayner, A. D. M., Griffith, G. S. & Wildman, H. G. (1994a). Differential insulation and the generation of mycelial patterns. In *Shape and Form in Plants and Fungi*, ed. D. S. Ingram, pp. 293–312. London: Academic Press.

Rayner, A. D. M., Griffith, G. S. & Wildman, H. G. (1994b). Induction of metabolic and morphogenetic changes during mycelial interactions among species of higher fungi. *Biochemical Society Transactions*, **22**, 391–6.

Rayner, A. D. M., Powell, K. A., Thompson, W., & Jennings, D. H. (1985). Morphogenesis of vegetative organs. In *Developmental Biology of Higher Fungi*, ed. D. Moore, L. A. Casselton, D. A. Wood & J. C. Frankland, pp. 249–279. Cambridge: Cambridge University Press.

Rayner, A. D. M. & Todd, N. K. (1979). Population and community structure and dynamics of fungi in decaying wood. *Advances in Botanical Research*, **7**, 333–420.

Ritz, K. & Crawford, J. (1990). Quantification of the fractal nature of colonies of *Trichoderma viride*. *Mycological Research*, **94**, 1138–41.

Shaw, M. W. (1994). Seasonally induced chaotic dynamics and their implications in models of plant disease. *Plant Pathology*, **43**, 790–801.

Sleeman, B. D. (1989). Complexity in biological systems and hamiltonian dynamics. *Proceedings of the Royal Society of London*, **A425**, 17–47.

Stenlid, J. & Rayner, A. D. M. (1989). Environmental and endogenous controls of developmental pathways: variation and its significance in the forest pathogen, *Heterobasidion annosum*. *New Phytologist*, **113**, 245–58.

Sugihara, G. & May, R. M. (1990). Nonlinear forecasting as a way of distinguishing chaos from measurement error in time series. *Nature*, **344**, 734–41.

Trinci, A. P. J. (1978). The duplication cycle and vegetative development in moulds. In *The Filamentous Fungi, vol. 3*, ed. J. E. Smith & D. R. Berry, pp. 132–163. London: Arnold.

Wessels, J. G. H. (1986). Cell wall synthesis in apical hyphal growth. *International Review of Cytology*, **104**, 37–79.

Wessels, J. G. H. (1991). Fungal growth and development: a molecular perspective. In *Frontiers in Mycology*, ed. D. L. Hawksworth, pp. 27–48. Kew, Surrey, UK: CAB International.

Wessels, J. G. H. (1992). Gene expression during fruiting of *Schizophyllum commune*. *Mycological Research*, **96**, 609–20.

Index of generic and specific names

344   *Index of generic and specific names*

346   *Index of generic and specific names*

Subject index